MATHEMATICAL MODELING OF MELTING AND FREEZING PROCESSES

MATHEMATICAL MODELING OF MELTING AND FREEZING PROCESSES

Vasilios Alexiades

The University of Tennessee and Oak Ridge National Laboratory

Alan D. Solomon

Consultant
Formerly at Oak Ridge National Laboratory

HEMISPHERE PUBLISHING CORPORATION

A member of the Taylor & Francis Group

Washington Philadelphia London

USA	Publishing Office:	Taylor & Francis 1101 Vermont Avenue,N.W., Suite 200 Washington, DC 20005-2531 Tel: (202)289-2174 Fax: (202)289-3665
	Distribution Center:	Taylor & Francis 1900 Frost Road, Suite 101 Bristol, PA 19007-1598 Tel: (215) 785-5800 Fax: (215)785-5515
UK		Taylor & Francis Ltd. 4 John Street London WC1N 2ET, UK Tel:071 405 2237 Fax: 071 831 2035

MATHEMATICAL MODELING OF MELTING AND FREEZING PROCESSES

1 2 3 4 5 6 7 8 9 0 BRBR 9 8 7 6 5 4 3 2

Cover design by Michelle Fleitz.
A CIP catalog record for this book is available from the British Library.
∞ The paper in this publication meets the requirements of the ANSI Standard Z39.48-1984(Permanence of Paper)

Library of Congress Cataloging-in-Publication Data

Alexiades, Vasilios
Mathematical modeling of melting and freezing processes / Vasilios Alexiades, Alan D. Solomon.
p. cm.
Includes bibliographical references and index.
1. Solidification—Mathematical model. 2. Fusion—Mathematical models. 3. Phase transformations (Statistical physics)—Mathematical models. I. Solomon, Alan D. II. Title.
QC303.A38 1993
536'.42'01518—dc20 92-28180
ISBN 1-56032-125-3 CIP

CONTENTS

PREFACE

In the late 19th century, J. Stefan formulated the problem of finding the temperature distribution and freezing front history of a solidifying slab of water. Over the last century, particularly in the last 30 years, the problem bearing his name has been extended to include such complex phenomena as the solidification of alloy systems, supercooling, melting due to Joule heating and laser irradiation. As the range of application grows, so too does the mathematical interest, the numerical challenge, and the engineering utility in problems involving melting and solidification. To the mathematician the Stefan problem is one which takes us slightly into the world of nonlinearity. To the computational scientist, modeling phase change processes leads beyond intuition to the need for advanced graphics and computing tools. In the same vein, the engineer must grapple with ever more complex processes arising from dynamically growing technological challenges.

For many years the authors have been increasingly fascinated by the wealth of challenges found under the heading of "phase change processes." They have been fortunate to have been involved in work on processes arising from such diverse areas as cryosurgery and alloy solidification in the microgravity of a space station. In the course of work on mathematical, computational and engineering questions in this area we decided several years ago to undertake the preparation of a book reflecting the breadth of activities in modeling melting and freezing processes. The book was to serve as a vehicle for presenting the key physical, mathematical, computational and modeling concepts to a wide audience of readers which would include mathematicians interested in learning about the physical phenomena involved as well as engineers seeking solutions to problems. It is intended as an interdisciplinary introduction to this, by nature, interdisciplinary subject. We have strived to minimize the prerequisite knowledge by presenting both the physical and the mathematical-computational background in sufficient detail for someone with generic training in science or mathematics or engineering to understand the material. We hope that we have attained our goal.

The book is comprised of five chapters. We begin in the first chapter with a precise formulation of the Stefan problem for a melting/freezing process, based on the underlying physical processes taking place. We turn, in CHAPTER 2, to the study of the classical Stefan Problem and its explicit (similarity) solution, and then progressively introduce additional physical phenomena, but in a setting where explicit solutions are still possible so that the effect on the solution can be seen clearly. The underlying physics and effects of a volume change, of supercooling, and of coupling with interdiffusion when the material is a binary alloy are examined. Also, explicit solutions in axially and spherically symmetric geometries are presented. In CHAPTER 3 we examine a number of analytical approximation techniques available for phase change processes. These approaches are commonly the methods of choice for working mechanical and chemical engineers and we believe that mathematicians interested in contributing to our abilities to deal with Stefan-type problems should be aware of these approaches. The fact that such basic questions as accuracy of these methods are still open, attest to the basic challenges implicit in them. CHAPTER 4 is devoted to the numerical simulation of phase change processes. The standard "control volume" discretization of conservation laws is presented first on plain heat conduction, leading to discrete heat balances expressed as finite difference equations. Then we undertake a detailed description of the most general, versatile, and easily programmable numerical method available for phase change problems, the so called "enthalpy" method. Its theoretical foundations in global weak formulations are presented and the convergence of the enthalpy scheme to the weak solution is established in the simplest and most easily understood setting. We close in CHAPTER 5 with three engineering applications arising in the area of latent heat thermal energy storage.

A notable omission in the list of topics treated is that of "convection in the melt". It arises whenever the density of the liquid is not constant, and it is most notable when melting is induced by heating from below a liquid of relatively low viscosity. While it has been tempting to discuss natural convection in this book, we felt that a proper treatment of the physical, mathematical and computational aspects of the Navier-Stokes equations that it entails would almost double the size of the book and would thus occupy a disproportionate place to its importance in the subject. Indeed, only a small proportion of the literature on Stefan-type problems deals with convection. This is of course partly due to the serious mathematical and computational difficulties one encounters when dealing with convective flows, but also due to the fact that most phase-change processes of interest are not "convection limited". Often its primary effect is to enhance heat transfer (possibly by several orders of magnitude, in which case the liquid is effectively isothermal), and in many cases it may be accounted for approximately by simply enhancing the thermal diffusivity in the melt. In any case, ignoring convection in a melting process will yield a conservative estimate (lower bound) of the speed of melting, and certainly one must know how to deal with phase changes without convection before attempting to incorporate convection.

A two-semester course, using evolving drafts of the book as text, has been taught three times to mixed groups of engineering and mathematics graduate students at the University of Tennessee, Knoxville. Similarly it has been used as the basis for a graduate course for mathematicians at the Weizmann Institute of Science. Many of the methods described in it have been used by the authors in applied engineering settings, and examples based on questions arising in such applications are presented.

The authors owe many thanks to a number of people. A special gratitude is felt towards D. George Wilson with whom we were fortunate to actively collaborate on much of this work over many years. A.D. Solomon is grateful to Eugene Isaacson and Milton Rose who introduced him to this fascinating subject over 25 years ago. We are extremely grateful to a long list of colleagues and former colleagues at the Oak Ridge National Laboratory. A complete list of these wonderful scientists and engineers would be overly long. Of these friends, special thanks are directed to Bob Ward, John Tomlinson, Bob Kedl, Mitch Olszewski and Jim Martin. Our collaborative work with Ralph Deal of Kalamazoo College was of particular importance and pleasure to us, as was later work with Bob Wichner of Oak Ridge National Laboratory. Special appreciation for his help in numerous ways over the years goes to John Drake of ORNL. The miracle of unix, troff, Xwindows, xfig, xprism and GNU software made possible (but not easy) the preparation of the manuscript in this ready-for-copying form, with thanks to Michael Stevens for making it all work together.

Last and probably most significantly, we express our appreciation to our families for the long hours away from home and for their support of our efforts, to our wives, Linda and Faiza, and to our children Nikolas, Alexander, Anthy, Julianna, and Anneli and Joey.

Vasilios Alexiades
Oak Ridge, Tennessee

Alan D. Solomon
Omer, Israel

CHAPTER 1

PROBLEM FORMULATION

The purpose of simulation is to gain understanding of the process being simulated. Building understanding is complex, involving iterative use of experiment and observation on the "physical plane" and model building and analysis on the "conceptual plane." The act of formulating a mathematical model tests our understanding of the physical process. Do we know what is important, and what may be ignored? Do we believe that we know the underlying relationships between the entities making up our process well enough to formulate them in mathematical terms?

This book is concerned with simulating the processes of melting and freezing on a macroscopic scale. Very few processes are more familiar to us than these. Yet our interest in them, motivated by processes of increasing complexity, has grown steadily in recent years, while our intuition is increasingly tested. On the one hand we know that it will take half a day to defrost a piece of meat. On the other hand, we have not the slightest intuition about the performance of a material used to store solar energy as the latent heat of melting as it cycles through repeated freeze/thaw cycles. The growing use of Silicon brings us into intimate contact with processes involving freezing of supercooled liquid; alloy processing in space leads us to surface-tension driven convection; laser induced melting brings us to the edge (and perhaps beyond it) of credibility of the heat equation, our fundamental tool for heat transfer modeling on a macroscopic scale. Time scales have broadened. When Stefan formulated the classical phase change model, the credible time scale for processes of interest could be measured in days to several years. In recent years the authors have dealt with problems whose time scales are measured in pico-seconds, and with problems whose time scales are literally 20,000,000 years. Thus our need to know has expanded, and with it, our intuition about even simple processes has waned. To build our intuition we need experiment and observation on the one hand, and better simulation tools on the other.

In this chapter we formulate the classical Stefan-type model of melting and freezing. Historically this problem has been regarded as lying on the border between tractable and intractable problems. It is nonlinear, and its principal difficulty lies in the fact that one of its unknowns is the region in which it is to be solved. For this reason it is called a "moving boundary problem." Our intuition tells us that for most reasonable phase change processes energy conservation prevails. This is manifested in two ways: the heat equation, its differential expression, holds within liquid or solid, and a "jump" condition expressing energy conservation, prevails along the curves separating solid from liquid. Are there such curves? Sometimes we really do see "nice" curves separating solid from liquid, and sometimes instead, we see "fuzzy" or even "fat" regions in place of sharp curves. The classical formulation is based on an underlying assumption, which we will later discard, that the front is indeed of zero thickness.

The chapter is divided into three sections. The first presents an overview of the physical ideas relevant to phase-change processes. In the second we introduce the precise mathematical formulation of the basic physical facts leading to the "Stefan Problem", the prototype of all phase-change models and a central subject of this book. In the third section we discuss some of the complications and difficulties encountered in more realistic phase-change processes.

1.1. AN OVERVIEW OF THE PHENOMENA INVOLVED IN A PHASE CHANGE

Several mechanisms are at work when a solid melts or a liquid solidifies. Such a change of phase involves heat (and often also mass) transfer, possible supercooling, absorption or release of latent heat, changes in thermophysical properties, surface effects, etc. We shall briefly discuss these in *rough* qualitative terms in order to orient the reader and introduce the terminology.

It is important to note that although the focus of our discussions is **solidification** and **melting**, the *principles*, *ideas*, and *many of the results* apply as well to other first-order phase transitions, including certain solid-to-solid transitions, vapor condensation and evaporation (gas-liquid transition), sublimation (gas-solid transition), or even certain magnetization phenomena. A clear qualitative picture of first-order phase transitions is presented in [CALLEN], [PIPPARD], [WALDRAM].

Both solid and liquid phases are characterized by the presence of cohesive forces keeping atoms in close proximity. In a **solid** the molecules vibrate around fixed equilibrium positions, while in a **liquid** they may "skip" between these positions. The macroscopic manifestation of this vibrational energy is what we call **heat** or **thermal energy**, the measure of which is **temperature**. Clearly atoms in the liquid phase are more energetic (hotter) than those in the solid phase, all other things being equal. Thus before a solid can melt it must acquire a certain amount of energy to overcome the binding forces that maintain its solid structure. This energy is referred to as the **latent heat** L (heat of fusion) of the material and represents the difference in thermal energy (enthalpy) levels between liquid and solid states, all other things being equal. Of course, solidification of liquid requires the removal of this latent heat and the structuring of atoms into more stable lattice positions. In either case there is a major re-arrangement of the entropy of the material, a characteristic of first-order phase transitions.

There are three possible modes of heat transfer in a material: **conduction, convection** and **radiation**. Conduction is the transfer of kinetic energy between atoms by any of a number of ways, including collision of neighboring atoms and the movement of electrons; there is no flow or mass transfer of the material. This is how heat is transferred in an opaque solid. In a liquid heat can also be transferred by the flow of particles, i.e. by convection. Radiation is the only mode of energy transfer that can occur in a vacuum (it requires no participating medium). Thermal radiation, emitted by the surface of a heated solid, is radiation of wave-length roughly in the range 0.1 to 10 microns (μm). An excellent source on the fundamentals of heat transfer modes is [BIRD et al], see also [WHITAKER].

The transition from one phase to the other, that is, the absorption or release of the latent heat, occurs at some temperature at which the stability of one phase breaks down in favor of the other according to the available energy. This **phase-change**, or **melt temperature** T_m depends on pressure. Under fixed pressure, T_m may be a particular fixed value characteristic of the material (for example, 0°C for pure water freezing under atmospheric pressure), or a function of other thermodynamic variables (for example, of glycol concentration in an anti-freeze mixture).

Most solids are crystalline, meaning that their particles (atoms, molecules, or ions) are arranged in a repetitive lattice structure extending over significant distances in atomic terms. In this context atoms may be regarded as spheres of diameter 2 to 6 Angstroms (1 Angstrom = 10^{-10} meters). Since formation of a crystal may require the movement of atoms into the solid lattice structure, it may well happen that the temperature of the material is reduced below T_m without formation of solid. Thus **supercooled liquid**, that is, liquid at temperatures below T_m, may appear; such a state is thermodynamically *metastable* [CALLEN], [TURNBULL] (see **§2.4**). We note that melting requires no such structuring, possibly explaining why "superheating" is rarely observed [CHRISTIAN]. To see the possible effect of supercooling (also called *undercooling* sometimes), typical cooling curves for both normal freezing and supercooling are shown schematically in **Figure 1.1.1**. These curves are meant to show the temperature of a sample of material as a function of time, as heat is extracted from the sample at a constant rate. Note that for the supercooling curve of **Figure 1.1.1(b)**, the temperature rapidly rises back to the melt temperature T_m when crystallization does take place. This can occur only if the latent heat released upon freezing is sufficient to raise the temperature to T_m, i.e., the liquid was not cooled too much. Liquid cooled to a temperature so low that the latent heat is not sufficient to raise its temperature to T_m, is referred to as being **hypercooled**.

The phase-transition region where solid and liquid coexist is called the **interface**. Its thickness may vary from a few Angstroms to a few centimeters, and its microstructure may be very complex, depending on several factors (the material itself, the rate of cooling, the temperature gradient in the liquid, surface tension, etc.). For most pure materials solidifying under ordinary freezing conditions at a fixed T_m the interface appears (locally) planar and of negligible thickness. Thus it may be thought of as a "sharp front," a surface separating solid from liquid at temperature T_m. In other cases, typically resulting from supercooling, the phase-transition region may have apparent thickness and is referred to as a "mushy zone"; its microstructure may now appear to be dendritic or columnar (shown schematically in **Figure 1.1.2**). The local freezing temperature at a curved solid surface facing the liquid becomes depressed by an amount depending on the solid-liquid surface tension and the local curvature. This so-called **Gibbs-Thomson effect** is small for the overall freezing process, but crucial for the resulting microstructure of the interface [PORTER-EASTERLING], [KURZ-FISHER], [GURTIN] (see **§2.4**).

Most thermophysical properties of a material (usually varying smoothly with temperature) undergo more or less sudden changes at T_m. For example the heat capacity of aluminum changes by 11% at its melt temperature (of 659°C), but that of silicon changes by only 0.3% (at 1083°C). Such **discontinuities in thermophysical properties** complicate the mathematical problems because they

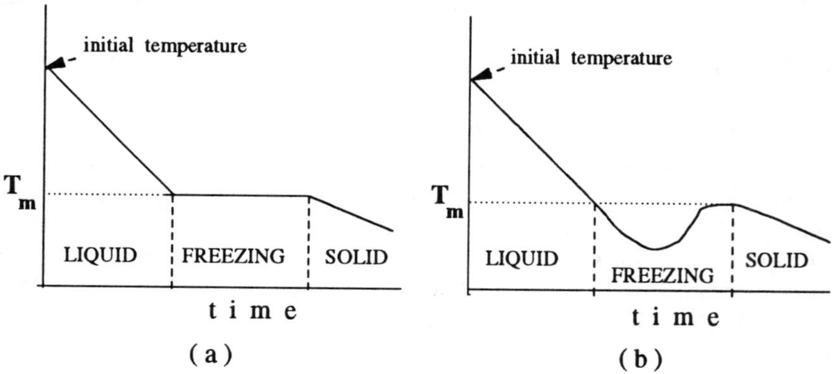

Figure 1.1.1. Cooling curves for (a) normal freezing and (b) with supercooling.

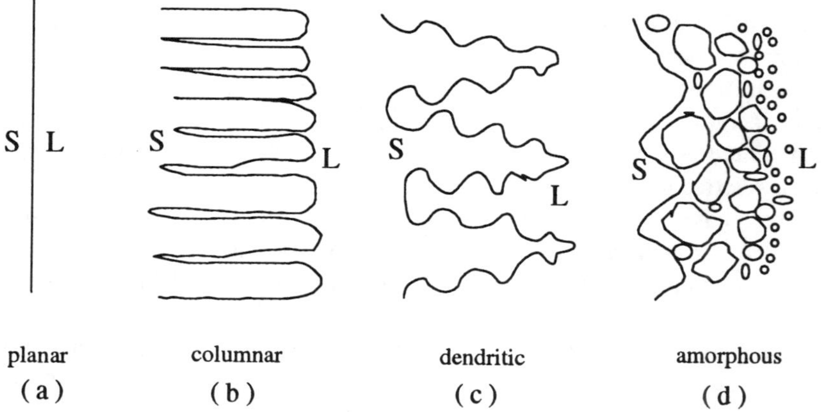

Figure 1.1.2. Common interfacial morphologies (schematic).

induce discontinuities in the coefficients of differential equations. However the most fundamental and pronounced effects are due to changes in density.

Typical **density changes** upon freezing or melting are in the range of 5% to 10% but can be as high as 30%. For most materials the solid is denser than the liquid, resulting in possible formation of voids in freezing or breaking of the container in melting (**§2.3**). On the other hand water expands on freezing, resulting in broken pipes on cold days and ice floating instead of filling the bottom of the oceans. The density variation with temperature induces flow by natural convection in the presence of gravity, rapidly equalizing the temperature in the liquid and greatly affecting heat transfer. In microgravity there is no natural convection but "Marangoni" convection [MYSHKIS et al], due to surface tension (capillary) effects, may arise instead and dominate heat transfer. All these effects may complicate a phase-change process beyond our ability to analyze effectively.

The explanation of why and how the phenomena mentioned above occur is the subject of physical theories, some still under development. Discussions may be found in the relevant literature [CALLEN], [CHALMERS], [FLEMINGS], [KURZ-FISHER], [LANDAU-LIFSHITZ], [PORTER-EASTERLING], [ROSENBERGER], [TILLER], [TURNBULL], [WALDRAM].

The subject of this book is the mathematical modeling and analysis of phase change processes at the **macroscopic** level. The purpose of mathematical modeling is to quantify the process in order to be able to predict (and ultimately control) the evolution of the temperature field in the material, the amount of energy used and stored, the interface location and thickness, and any other quantity of interest. Thus the equations and conditions that express the physics of the process must be formulated subject to certain accepted approximations; these result from the simplifying assumptions made in order to obtain a manageable problem. We will attempt to state such assumptions as clearly as possible so that the reader will be aware of the limitations of the resulting model.

When referring to a material that undergoes a phase change we will often use the abbreviation "PCM" to represent the term "Phase Change Material."

PROBLEMS

PROBLEM 1. It is asserted that if the density of ice were greater than that of water then life on earth would be impossible. Why?

PROBLEM 2. Ice-making is a major industrial activity. It is said that the person developing a method for pumping supercooled liquid without it freezing would become "rich and famous." Why?

PROBLEM 3. In your experience is it more common or less common to observe sharp freezing fronts. Give an example.

PROBLEM 4. It was very cold yesterday and yet the snow on the roof partly melted. Where could the heat have come from?

PROBLEM 5. Suppose that you are flying in a plane at a very high altitude and you can fire a water droplet into the cold exterior air. Describe what you think will happen to it.

PROBLEM 6. Describe the appearance of the curve corresponding to that of **Figure 1.1.1(b)** for the case of superheating of a solid.

1.2. FORMULATION OF THE STEFAN PROBLEM

1.2.A Introduction

We begin by formulating the mathematical model of a simple melting or freezing process incorporating only the most basic of the phenomena mentioned in §1.1. This model is known as the **Classical Stefan Problem** and constitutes the foundation on which progressively more complex models can be built by incorporating some of the effects initially left out. The characteristic of phase-change problems is that, in addition to the temperature field, the **location** of the interface is unknown. Problems of this kind arise in fields such as molecular diffusion, friction and lubrication, combustion, inviscid flow, slow viscous flow, flow in porous media, and even optimal decision theory. Overviews of the origins of such problems, referred to as "moving boundary problems" or "free boundary problems" may be found in [RUBINSTEIN] and in [ELLIOTT-OCKENDON], [CARSLAW-JAEGER] and [CRANK]; see also the conference proceedings [OCKENDON-HODGKINS], [WILSON-SOLOMON-BOGGS], [FASANO-PRIMICERIO, 1983], [BOSSAVIT-DAMLAMIAN-FREMOND], [HOFFMANN-SPREKELS], [CHADAM-RASMUSSEN] on moving boundary problems.

We start with a discussion of the physical assumptions leading to the Stefan Problem. Then we introduce the basic concepts and equations of heat conduction, boundary conditions, and interface conditions, ending up with the precise mathematical formulation of the Classical Stefan Problem.

1.2.B Assumptions

It is crucial to have a clear picture of exactly which phenomena are taken into account and which are not, because a model can at best be as good as its underlying physical assumptions. To make the assumptions as clear as possible we present, in **Table 1.1**, a summary of the physical factors involved in a phase change and the simplifying assumptions that will lead us to the classical Stefan Problem.

Our phase change process involves a PCM (phase change material) with constant density ρ, latent heat L, melt temperature T_m, phase-wise constant specific heats c_L, c_S, and thermal conductivities, k_L, k_S. Heat is transferred only isotropically (see below) by conduction, through both the solid and the liquid; the phases

Table 1.1

Physical Factors Involved in Phase change Processes	*Simplifying Assumptions for the Stefan Problem*	*Remarks on the Assumptions*
1. Heat and mass transfer by conduction, convection, radiation with possible gravitational, elastic, chemical, and electro-magnetic effects.	*Heat transfer isotropically by conduction only, all other effects assumed negligible.*	*Most common case. Very reasonable for pure materials, small container, moderate temperature gradients.*
2. Release or absorption of latent heat	*Latent heat is constant; it is released or absorbed at the phase-change temperature.*	*Very reasonable and consistent with the rest of the assumptions.*
3. Variation of phase-change temperature	*Phase-change temperature is a fixed known temperature, a property of the material.*	*Most common case, consistent with other assumptions.*
4. Nucleation difficulties, supercooling effects	*Assume not present.*	*Reasonable in many situations.*
5. Interface thickness and structure	*Assume locally planar and sharp (a surface separating the phases) at the phase-change temperature.*	*Reasonable for many pure materials (no internal heating present).*
6. Surface tension and curvature effects at the interface	*Assume insignificant.*	*Reasonable and consistent with other assumptions.*
7. Variation of thermo-physical properties	*Assume constant in each phase, for simplicity* ($c_L \neq c_S$, $k_L \neq k_S$).	*An assumption of convenience only. Reasonable for most materials under moderate temperature range variations. The significant aspect is their discontinuity across the interface, which is allowed.*
8. Density changes	*Assume constant* ($\rho_L = \rho_S$).	*Necessary assumption to avoid movement of material. Possibly the most unreasonable of the assumptions.*

are separated by a sharp interface of zero thickness, an isotherm at temperature T_m, where the latent heat is absorbed or released.

This situation arises very often in applications with the result that the Stefan Problem is by far the most frequently applied model of a phase change process.

An additional assumption of mathematical nature often goes unmentioned. In deriving the model it will be assumed that all the functions representing physical quantities are as smooth as required by the equations in which they appear. Of course this has to be assumed a-priori in order to proceed with the formulation of the mathematical model, but has to be justified a-posteriori by proving that the resulting mathematical problem does indeed admit such smooth solutions. It turns out that this may not always be the case, making "weak solution" reformulations necessary (see **§4.4**).

EXAMPLE. The specific heat (in $kJ/kg\ K$) of ice/water is well approximated by the relation

$$c = \begin{cases} 7.16\times 10^{-3}\, T + .138\,, & T \le 273\,K \text{ (ice)} \\ 4.1868\,, & T \ge 273\,K \text{ (water)}\,. \end{cases} \tag{1}$$

Similarly the thermal conductivity (in $kJ/m\,s\,K$) is given by

$$k = \begin{cases} 2.24\times 10^{-3} + 5.975\times 10^{-6}(273-T)^{1.156}\,, & T \le 273\,K \\ 1.017\times 10^{-4} + 1.695\times 10^{-6}\, T\,, & T \ge 273\,K \end{cases} \tag{2}$$

Note that at the melt temperature $T_m = 273\ K$, $c_S = 2.0927$ is only about half of $c_L = 4.1868$, while $k_S = 2.24\times 10^{-3}$ is about four times larger than $k_L = 0.5644\times 10^{-3}$.

1.2.C Heat Conduction

The fundamental quantities involved in conduction of heat through a material are: **temperature, heat** (thermal energy) and **heat flux**.

The **temperature** is a macroscopic measure of perpetual molecular movement. Its meaning and measurement are subjects of Thermodynamics [CALLEN], [REYNOLDS], [WALDRAM], and Heat Transfer [ECKERT-DRAKE], [Mc-ADAMS]. It is measured in degrees Kelvin (K), Celsius (°C) or Fahrenheit (°F).

The **heat** absorbed by a material under constant pressure is a thermodynamic quantity called the **enthalpy**. We shall denote the enthalpy (thermal energy) per unit mass by e and per unit volume by E, measuring the former in kJ/kg, and the latter in kJ/m^3. For a pure material heated under *constant pressure*, when there are *no* volume changes, the enthalpy represents the *total energy* (see **§2.3**), and the heat absorbed is related to the temperature rise by

$$de = c\ dT\ . \tag{3}$$

The quantity

$$c(T) = \frac{de}{dT},$$

is the **specific heat** (heat capacity per unit mass) ($kJ/ kg°C$) under constant pressure and represents the heat needed to raise the temperature of 1 kg of the material by 1°C. It is a property of the material, always positive, varies slowly with temperature, and is usually higher for the liquid phase than for the solid phase. Thus for example, the specific heat of water as given in (1) above can be taken as 4.1868 kJ/kg°C, while that of ice at 0°C (273 K) is 2.0927 kJ/kg°C. The heat accompanying a temperature rise is referred to as **sensible heat**, in contrast with the **latent heat of phase change**, absorbed or released at constant temperature (the melt temperature). In the context of a phase change process the most convenient reference temperature, relative to which temperature changes are measured, is the **phase-change** temperature of the material, T_m (provided it is constant). From (3) the sensible heat in a unit mass of material at temperature T is then given by

$$\textit{sensible heat} = \int_{T_m}^{T} c(\bar{T})\, d\bar{T}\,, \tag{4}$$

with $c = c_L$ if the material is liquid ($T > T_m$) and $c = c_S$ if it is solid ($T < T_m$). At $T = T_m$ the material may be either liquid or solid. What distinguishes the two is the energy (enthalpy); liquid at T_m contains the latent heat L per unit mass, whereas solid at T_m has no latent heat. Hence, if we fix the energy scale by choosing $e = 0$ *for solid at* $T = T_m$, the enthalpy is given by

$$e = \begin{cases} e^L(T) := L + \int_{T_m}^{T} c_L(\bar{T})\, d\bar{T}, & \text{for } T > T_m \quad (\textit{liquid}) \\[2ex] e^S(T) := \int_{T_m}^{T} c_S(\bar{T})\, d\bar{T}, & \text{for } T < T_m \quad (\textit{solid}) \end{cases} \tag{5}$$

Therefore the phases are characterized by *either* temperature *or* enthalpy as

$$\begin{array}{lcccc} \text{solid} & \Longleftrightarrow & T < T_m & \Longleftrightarrow & e < 0\,, \\ \text{liquid} & \Longleftrightarrow & T > T_m & \Longleftrightarrow & e > L\,, \end{array} \tag{6}$$

At $T = T_m$ the enthalpy undergoes a jump of magnitude L. If $e = 0$ the material is solid while for $e = L$ it is liquid. Any region where

$$T = T_m \qquad \text{and} \qquad 0 < e < L \tag{7}$$

is referred to as a **mushy zone**. According to our Assumption 5 (**§1.2.B**) this region has zero thickness (a surface) and therefore the enthalpy jump across it is

$$[\![e]\!]_{solid}^{liquid} := e^L(T_m) - e^S(T_m) = L . \tag{8}$$

When the specific heats c_S, c_L are constants, (5) reduces to

$$e = \begin{cases} e^L(T) := L + c_L[T - T_m], & T \geq T_m \quad (liquid), \\ e^S(T) := \quad c_S[T - T_m], & T \leq T_m \quad (solid), \end{cases} \tag{9}$$

the graph of which is shown in **Figure 1.2.1(a)**, and of its inverse in **Figure 1.2.1(b)**. For water/ice, (1) may be used to relate T directly to e; when graphed we obtain the curve of **Figure 1.2.2**.

The amount of heat crossing a unit area per unit time is called the **heat flux,** denoted by $\vec{q}$, and measured in $kJ/\, s\, m^2 = kW/\, m^2$. The heat flux is a vector pointing in the direction of heat flow, and given by

Fourier's Law:
$$\vec{q} = -\underset{\sim}{k}\, \nabla T \tag{10}$$

where the tensor $\underset{\sim}{k}$ is the **thermal conductivity** of the material, measured in kJ / m s °C or kW / m K. In general it is a tensor with positive components varying with temperature. We shall assume **isotropic conduction**, i.e., that $\underset{\sim}{k}$ is a scalar $k > 0$. Typically, its value for the solid is higher than for the liquid. For water/ice, for example, we see from (2) that k_S is substantially greater than k_L. Regarding the isotropicity assumption, we note that in reality, even for ice frozen off a flat surface, there is a direction dependence of conductivity.

In circular cylindrical coordinates $(r\, ,\, \theta\, ,\, z)$,

$$x = r\cos\theta\ , \quad y = r\sin\theta\ , \quad z = z\ , \tag{11a}$$

Fourier's law for an isotropic medium takes the form

$$\vec{q} = -\, k\, (T_r\, ,\, \frac{1}{r} T_\theta\, ,\, T_z)\ , \tag{11b}$$

while in spherical coordinates $(r\, ,\, \phi\, ,\, \theta)$

$$x = r\cos\theta\sin\phi\ , \quad y = r\sin\theta\sin\phi\ , \quad z = r\cos\phi\ , \tag{12a}$$

it takes the form

$$\vec{q} = -\, k\, (T_r\, ,\, \frac{1}{r} T_\phi\, ,\, \frac{1}{r\sin\phi} T_\theta)\ . \tag{12b}$$

In one space dimension Fourier's law is expressed by

$$q = -\, k\, T_x \tag{13a}$$

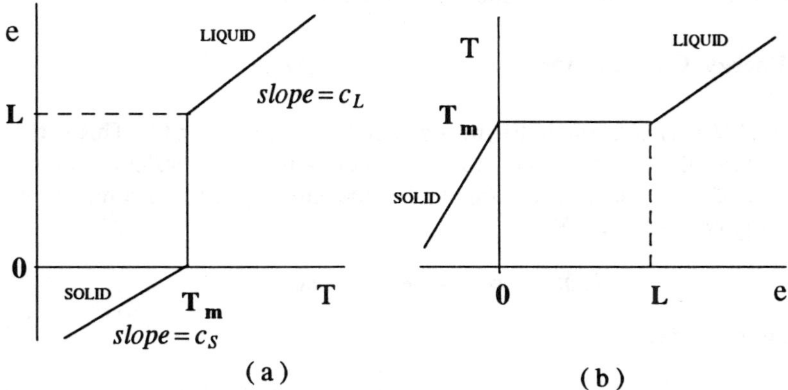

Figure 1.2.1. Typical state relation, $e = e(T)$, and its inverse, $T = T(e)$, (with constant c_L, c_S).

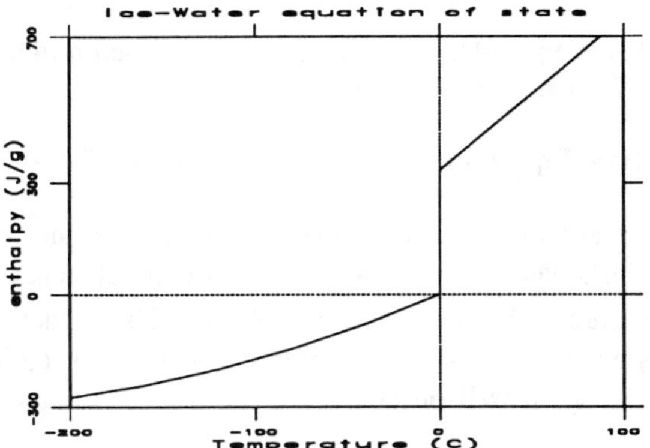

Figure 1.2.2. State relation, $e = e(T)$, for ice-water.

for the cartesian coordinate x, and by

$$q = -k\,T_r \tag{13b}$$

for the r direction in both the circular cylindrical and spherical systems.

The **heat flow rate** across a surface of area A and unit normal $\vec{n}$ is given by

$$(\vec{q}\cdot\vec{n})\,A = -\underset{\sim}{k}\,\nabla T\cdot\vec{n}\,A, \tag{14}$$

and represents the heat crossing the area A in the direction normal to the surface per unit time. Thus the radial heat flow rate through the surface of a circular cylinder of radius r and height z is $(-k\,T_r)\,2\pi rz$, and that through the surface of a sphere of radius r is $(-k\,T_r)\,4\pi r^2$.

The fundamental law of conductive heat transfer is the

Energy Conservation Law: $$(\rho\, e)_t + div\,\vec{q} = F\,, \tag{15}$$

where F $(kJ/m^3 s)$ is a volumetric heat source (> 0) or sink (< 0). This is the differential form (the subscript t denotes partial derivative with respect to time) of the **First Law of Thermodynamics** (applied to the enthalpy, since this represents the total energy here, see **§2.3.F**):

$$heat\ increase = heat\ in - heat\ out,$$

or, in terms of rates:

$$\frac{d}{dt}\iiint_V \rho\, e\, dV = \iiint_V F\, dV - \iint_S \vec{q}\cdot\vec{n}\, dS\,, \tag{16}$$

where V is a volume, S its surface and $\vec{n}$ the *outward* unit normal to S. Since, by the Divergence Theorem, the surface integral equals $\iiint_V div\,\vec{q}\, dV$ and the volume is arbitrary, (15) is equivalent to (16). Expressing e and $\vec{q}$ in terms of temperature via (3) and (10) we obtain the general

Heat Conduction Equation: $$\rho\, c\, T_t = div(k\,\nabla T) + F \tag{17}$$

expressing local conservation of heat conducted isotropically through the material according to Fourier's law. This is a partial differential equation obeyed by $T(\vec{x}, t)$, the temperature at location $\vec{x}$ at time t. It holds at each point $\vec{x}$ of the region occupied by a material of density ρ, specific heat c, and conductivity k.

We recall the concept of **well-posedness.**

DEFINITION. A mathematical problem is said to be **well-posed** if it admits a unique solution that depends continuously on the data (small changes in the data result in small changes in the solution).

The well-posedness concept mimics the corresponding requirements on the physical problem being modeled. For most "standard" physical processes we expect a unique solution that will not "jump" in response to slight changes in input data.

A well-posed problem for equation (17) in a spatial domain Ω and for positive time $t > 0$ (**Figure 1.2.3**) requires the following information:

(a) partial differential equation (17), valid for $\vec{x}$ in Ω, $t > 0$,

(b) initial condition $T(\vec{x}, 0) = T_{init}(\vec{x})$, $\vec{x}$ in Ω,

(c) boundary conditions (specify T, or its normal derivative, or a combination of both, see **§1.2.D**), at every point of the boundary, $\partial\Omega$, of Ω for $t > 0$.

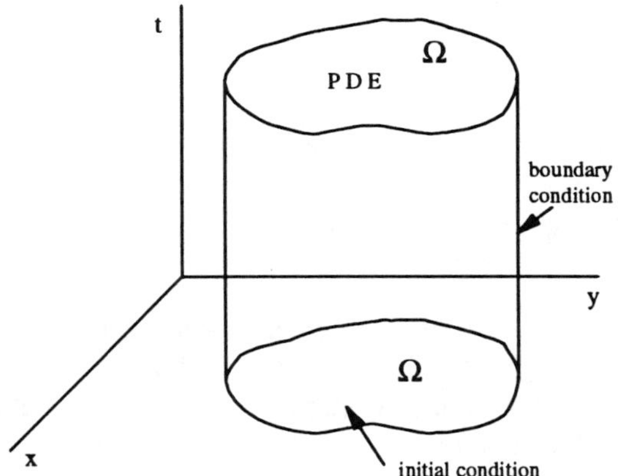

Figure 1.2.3. Space-time region for the Heat Conduction Equation.

Here T_{init} is a known initial temperature distribution.

The **boundary conditions** that most often arise in heat transfer are discussed in **§1.2.D**. Different types of boundary conditions may be imposed on different parts of the boundary, but **some** condition is needed on **each** part.

When the thermal conductivity k can be considered constant, then it is convenient to introduce the

Thermal Diffusivity: $$\alpha = \frac{k}{\rho c} \qquad (m^2/s), \tag{18}$$

and write (17) in the form known as the

Heat Equation: $$T_t = \alpha \nabla^2 T + f, \tag{19}$$

which is now a **linear** parabolic equation, (see (23)).

In one-space dimension, when no internal sources are present, the Fourier Law, general Heat Conduction Equation, and Heat Equation take the forms

$$q = -k T_x, \tag{20}$$

$$\rho c T_t = (k T_x)_x, \tag{21}$$

$$T_t = \alpha T_{xx}. \tag{22}$$

Let us briefly recall the place the heat equation occupies among the "standard" partial differential equations. The classical second order partial differential equations of mathematical physics [COURANT-HILBERT], [CHESTER] [DUCHATEAU-ZACHMANN], [ZACHMANOGLU-THOE], [STRAUSS], are epitomized by the

Laplace equation: $$0 = div(\nabla u), \tag{23a}$$

the

wave equation: $$u_{tt} = div(\nabla u)\ ,\tag{23b}$$

and the

heat equation: $$u_t = div(\nabla u)\ ,\tag{23c}$$

where, for each, the right hand side contains derivatives only with respect to the spatial variables.

The Laplace equation is said to be of **elliptic** type, a class of equations for which time as such plays no role. Elliptic equations characterize steady state behavior of a system. To be well-posed a problem involving an elliptic equation in a region Ω requires data around the entire boundary of the region, and will tell us what is happening in the interior.

EXAMPLE 1. A well-posed problem for the Laplace equation

$$\frac{1}{r}(ru_r)_r + \frac{1}{r^2}u_{\theta\theta} = 0 \tag{24a}$$

on the unit circle Ω of the x, y plane in polar coordinates, is to seek a solution u to the equation within Ω which on the boundary $r = 1$ of Ω is equal to a given function $f(\theta)$ (**Figure 1.2.4**). It is easily seen by separation of variables that if $f(\theta)$ can be expanded in a Fourier series

$$f(\theta) = \sum_{n=0}^{\infty} \alpha_n cos(n\theta) + \beta_n sin(n\theta)\ , \tag{24b}$$

then the unique solution to this problem is (PROBLEM 10)

$$u(r,\theta) = \sum_{n=0}^{\infty} r^n[\alpha_n cos(n\theta) + \beta_n sin(n\theta)]\ . \tag{24c}$$

where α_n and β_n are the Fourier coefficients of $f(\theta)$.

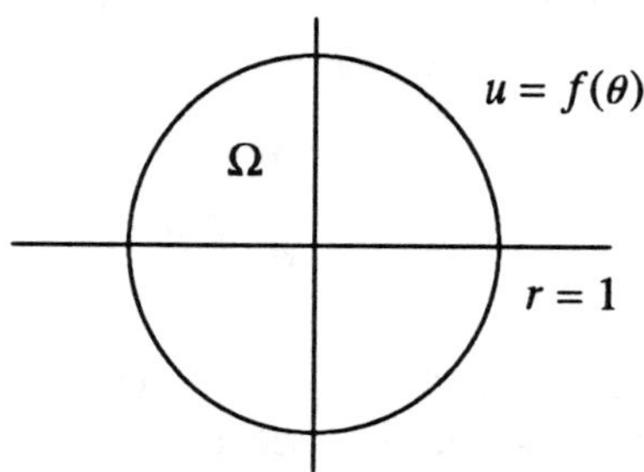

Figure 1.2.4. Region for Laplace's equation (24a).

The Wave Equation is said to be of **hyperbolic** type, and corresponds to time-dependent movement of a body due to conditions at its boundary. Solutions to the equation are functions $u(\vec{x}, t)$ defined on the set of points $(\vec{x}, t)$ where $\vec{x}$ varies in some spatial domain Ω and t varies on some interval $0 \leq t \leq t_{\max}$. Well-posedness requires the assignment of a condition everywhere on the boundary of Ω throughout the time interval, together with an appropriate initial condition. Since this equation represents the second derivative u_{tt} in terms of spatial information, comparison with the case for ordinary differential equations implies that throughout Ω an initial distribution of both u and u_t is needed for well-posedness.

EXAMPLE 2. A well posed problem for the wave equation in one space dimension for Ω the interval $0 \leq x \leq \pi$ is (**Figure 1.2.5**)

$$
\begin{aligned}
u_{tt} &= u_{xx}, & 0 < x < \pi, \ t > 0, & \quad (25a)\\
u(0, t) &= 0, & t > 0 \ \textit{(boundary condition at } x = 0), & \quad (25b)\\
u(\pi, t) &= 0, & t > 0 \ \textit{(boundary condition at } x = \pi), & \quad (25c)\\
u(x, 0) &= 0, & 0 \leq x \leq \pi \ \textit{(initial condition)}, & \quad (25d)\\
u_t(x, 0) &= \sin x, & 0 \leq x \leq \pi \ \textit{(initial condition)}. & \quad (25e)
\end{aligned}
$$

The unique solution to this problem is $u(x, t) = \sin t \sin x$, (PROBLEM 11).

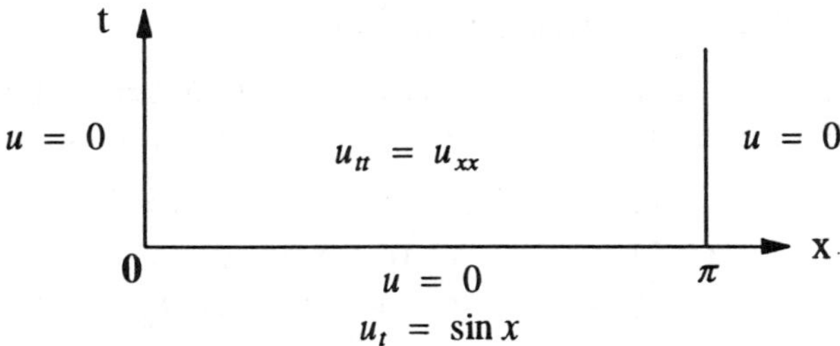

Figure 1.2.5. Region for the wave equation problem (25).

The heat equation is said to be of **parabolic** type. Now the time derivative appears only to first order. As with the wave equation a condition is still needed everywhere on the boundary of Ω, but only one initial condition, the specification of the value of $u(\vec{x}, t)$ for all $\vec{x}$ in Ω at the initial time $t = 0$ is required (see below).

EXAMPLE 3. A well-posed problem for the one-dimensional heat equation on the interval $0 \leq x \leq \pi$ is (**Figure 1.2.6**)

$$
\begin{aligned}
u_t &= u_{xx}, & 0 < x < \pi, \ t > 0 & \quad (26a)\\
u(0, t) &= 0, & t > 0 \ \textit{(boundary condition at } x = 0), & \quad (26b)\\
u(\pi, t) &= 0, & t > 0 \ \textit{(boundary condition at } x = \pi), & \quad (26c)\\
u(x, 0) &= \sin x, & 0 < x < \pi \ \textit{(initial condition)}. & \quad (26d)
\end{aligned}
$$

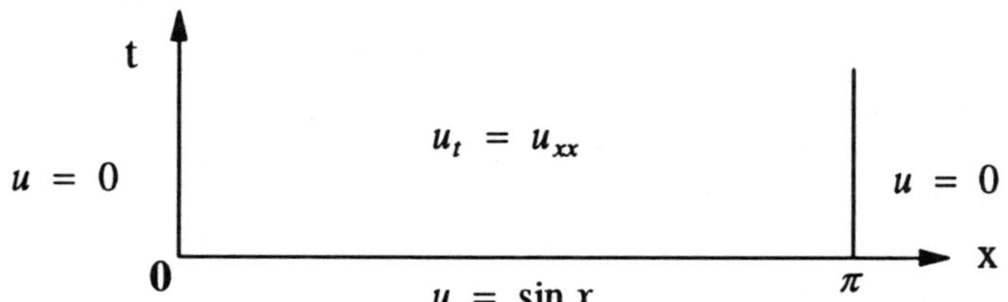

Figure 1.2.6. Region for the heat equation problem (26).

The unique solution to this problem is $u(x, t) = e^{-t} \sin x$, (PROBLEM 12).

The wave equation and the heat equation differ in two fundamental aspects. The first is that while time is "reversible" for the wave equation, meaning that exchanging t with $-t$ will not effect the behavior of the solution, it is not so for the heat equation. Indeed, "backward" problems involving the determination of an earlier thermal state from "final" data, are in general not well-posed. The second difference is that signals moving according to the wave equation have a finite speed, while for the heat equation signals are propagated at infinite speed. (see below).

EXAMPLE 4. Heat conduction in a semi-infinite slab:

Assume the slab $0 \leq x < \infty$ is initially at a uniform temperature T_0 and a temperature T_L is imposed at the face $x = 0$. Thus, we seek $T(x, t)$ such that

$$
\begin{aligned}
T_t &= \alpha T_{xx}, & 0 < x < \infty,\ t > 0 && \text{(27a)}\\
T(x, 0) &= T_0, & 0 < x < \infty && \text{(27b)}\\
T(0, t) &= T_L, & \lim_{x \to \infty} T(x, t) = T_0,\ t > 0\ . && \text{(27c)}
\end{aligned}
$$

This is also a well-posed problem with solution (PROBLEM 14)

$$T(x, t) = T_L - (T_L - T_0)\,\mathrm{erf}\left(\frac{x}{2\sqrt{\alpha t}}\right), \tag{28}$$

where the **error function** is given by [ABRAMOWITZ-STEGUN], **§2.1**,

$$\mathrm{erf}(z) = \frac{2}{\sqrt{\pi}} \int_0^z e^{-s^2} ds\ . \tag{29}$$

The solution is discontinuous at $x = 0$, $t = 0$, but infinitely smooth for any $t > 0$. We see that heat conduction is a *smoothing* process, a property of parabolic equations not shared by hyperbolic ones (and a reason why "backward" parabolic problems are ill-posed, generally).

Note that as long as the face $x = 0$ is held also at temperature T_0, the whole slab remains at T_0. The moment the face temperature is changed to a different temperature, T_L, say higher than T_0, the temperature at *every* point of the *entire* slab also rises *immediately*, according to (28). Hence, the disturbance at $x = 0$ is felt immediately everywhere! In this sense, the thermal signal travels with *infinite* speed, also a property of parabolic equations not shared by hyperbolic ones, as mentioned earlier. Clearly this contradicts our experience and suggests that the heat equation (actually Fourier's law) may be a poor model for heat conduction. Why is it then universally used as such? The reason is that this unphysical behavior is more "theoretical" than "practical": The error function increases to 1 extremely rapidly (e.g. $\mathrm{erf}(5) \approx 1 - .15 \times 10^{-11}$) and therefore, for moderate $T_L - T_0$, $T(x,t)$ is *detectably* different from T_0 *only* for points x very close to $x = 0$ (say, $x < 10\sqrt{\alpha t}$); the rest of the slab will essentially be at T_0, as far as any measuring device can detect, in accord with experience.

Notice, however, that for *very large* temperature gradients (e.g. very large $T_L - T_0$) this will not be the case and then the validity of Fourier's law does become questionable even from the "practical" point of view. Such situations arise in various applications, involving very intense local heating/cooling, e.g. laser annealing, and there have been various attempts at examining alternatives leading to "hyperbolic heat transfer" and "hyperbolic Stefan Problems", [OZICIK], [SOLOMON et al,1985], [SHOWALTER-WALKINGTON], [FRIEDMAN-HU], [LI].

1.2.D Boundary Conditions

In addition to knowing the initial temperature distribution, the thermal conditions on *every* exterior surface of the body must be known at all times. This requirement is not only physically obvious, but also mathematically necessary in order to have a well-posed mathematical problem for the parabolic equation (17).

Surface heating or cooling can be achieved by the following methods, each one inducing a boundary condition appropriate for equation (17).

I. Imposed temperature:

$$T(\vec{x},t) = T_{boundary}(\vec{x},t) \quad \text{for} \quad \vec{x} \in \partial\Omega, \quad t > 0, \tag{30}$$

i.e., the temperature at each boundary point is specified. This amounts to assuming perfect contact between the surface and a heat source of known temperature.

II. Imposed Flux:

$$-k\,\frac{\partial T}{\partial \vec{n}_{in}}(\vec{x},t) = q_{boundary}(\vec{x},t) \quad \text{for} \quad \vec{x} \in \partial\Omega, \quad t > 0, \tag{31}$$

where $\vec{n}_{in}(\vec{x})$ is the inward normal to $\partial\Omega$ at $\vec{x} \in \partial\Omega$, and $\dfrac{\partial T}{\partial \vec{n}_{in}} = \nabla T \cdot \vec{n}_{in}$ denotes the directional derivative. This means we know the amount of heat entering the surface per unit area per unit time. Of course, what we really know is the amount of heat delivered to the surface by, say, a heater lamp, but not how much actually inside unless there is perfect contact and no reflective losses. If the **absorptivity** a of the surface is known, then the right-hand side will be $a \cdot q_{boundary}(\vec{x}, t)$. In particular, (perfect) thermal insulation is expressed by

$$-k \frac{\partial T}{\partial \vec{n}_{in}} = 0 . \tag{32}$$

III. Convective Flux:

$$-k \frac{\partial T}{\partial \vec{n}_{in}} (\vec{x}, t) = h \left[T_{\infty} (\vec{x}, t) - T(\vec{x}, t) \right] \quad \text{for} \quad \vec{x} \in \partial \Omega, \quad t > 0 , \tag{33}$$

that is, the incoming flux is proportional to the temperature difference between the surface of the material and an imposed ambient temperature T_{∞} (Newton's law of cooling). This is the most realistic means of heat input, achieved by pumping a heat transfer fluid (such as water or air) of known (and regulated) temperature T_{∞} over a container wall. The wall creates in fact a thin boundary layer and is modeled effectively by an experimentally determined convective **heat transfer coefficient,** h. The units of h are kJ/m^2s°C, and it depends on the material, geometry, and roughness of the wall, as well as the velocity and properties of the heat transfer fluid. Determination of heat transfer coefficients occupies a sizable part of the science of Heat Transfer, and there exist formulas for effective heat transfer coefficients under a myriad of physical situations (see [McADAMS], [CHAPMAN]).

IV. Radiative Flux:

$$-k \frac{\partial T}{\partial \vec{n}_{in}} (\vec{x}, t) \;=\; \hat{h} \left[T_{\infty} (\vec{x}, t)^4 - T(\vec{x}, t)^4 \right] \quad \text{for} \quad \vec{x} \in \partial \Omega, \quad t > 0. \tag{34}$$

A hot body radiates heat with a flux which, according to Stefan's Fourth Power Law, is proportional to the difference of the fourth powers of the temperatures of the surfaces exchanging heat. The actual situation is very complicated, but (34) is often a satisfactory approximation [BIRD et al, Chap. 14]. The radiative heat transfer coefficient $\hat{h}$ depends on the emissivities of the participating media, the geometry of walls and gaps, etc. Frequently radiative flux exists in conjunction with convective and/or conductive fluxes in which case the total incoming flux is their sum. Note that

$$T_{\infty}^4 - T^4 \;=\; (T_{\infty}^2 + T^2)(T_{\infty} + T) \cdot [T_{\infty} - T] ,$$

so that by absorbing $(T_{\infty}^2 + T^2)(T_{\infty} + T)$ into the heat transfer coefficient the radiative boundary condition can be written as a convective boundary condition but with h depending *nonlinearly* on the ambient and surface temperatures.

We conclude that a general expression for flux input is

$$-k \frac{\partial T}{\partial \vec{n}_{in}} \;=\; h(T_{\infty}, T, t) \cdot [T_{\infty} - T] , \quad \text{on} \quad \partial \Omega , \quad t > 0 , \tag{35}$$

where the heat transfer coefficient may depend nonlinearly on the imposed ambient temperature, the surface temperature, the time, as well as the emissivities, conductivities, geometry of the wall and air layers, and velocity and properties of the heat transfer fluid. Then (35) includes both (33) and (34). In addition, we can interpret $h \to \infty$ in (33) as implying $T = T_\infty$, thus including the imposed temperature boundary condition (30). Finally $h \equiv 0$ implies the insulated boundary case (32).

We close with the reminder that *different* boundary conditions may be imposed on *different* parts of the boundary. What is important is that **some** boundary condition must be imposed on **each** part of the boundary (PROBLEM 18).

1.2.E Interface Conditions

In a melting or solidification process, conforming to our assumptions in §**1.2.B**, the physical region Ω occupied by the phase change material will be subdivided into two phases, liquid and solid, separated by a sharp interface Σ (of zero thickness, **Figure 1.2.7**).

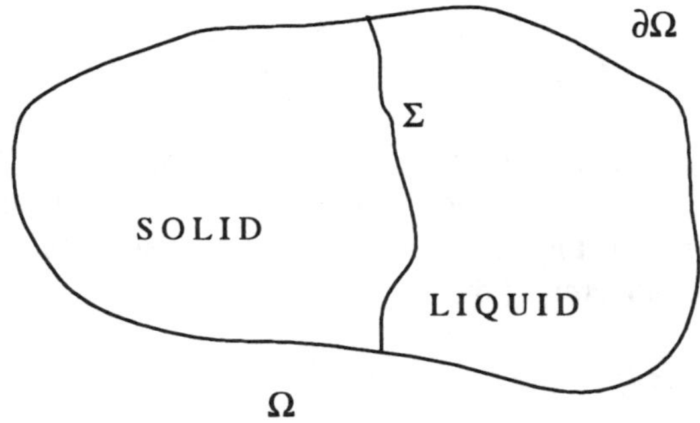

Figure 1.2.7. Solid and liquid phases separated by a sharp interface.

According to our discussion in §**1.2.C** conservation of energy in each phase demands that a heat conduction equation (17) be satisfied there. In the liquid, the coefficients will be c_L and k_L and in the solid c_S and k_S (see §**1.2.F**); for simplicity in notation we shall not use different symbols for the temperatures of liquid and of solid. The interface constitutes part of the boundary of both the liquid and solid regions; hence we need a boundary condition from each side in order to complete the initial-boundary value problem in each phase. Since the temperature must always be continuous, and by Assumption 5 the interface is an isotherm at T_m, we have the interface conditions

$$\lim_{\substack{\vec{x} \to \text{interface} \\ \vec{x} \in \text{liquid}}} T(\vec{x}, t) = T_m \quad \text{and} \quad \lim_{\substack{\vec{x} \to \text{interface} \\ \vec{x} \in \text{solid}}} T(\vec{x}, t) = T_m \,, \tag{36}$$

that is, the temperature is prescribed there. If the interface location were known we would have enough conditions to determine the temperature inside the liquid and the solid regions. Since the location is unknown, we need one more condition to determine it. Such an additional condition results from energy conservation **across** the interface, and it can be derived in several ways. In order to avoid needless complications of notation we will postpone the derivation until **§4.4** in **CHAPTER 4**. We will merely note that jump conditions for conservation laws can be derived in general as Rankine-Hugoniot "shock" conditions (Kotchine's Theorem, see [TRUESDELL-TOUPIN], [SEGEL], [ARIS], [CHORIN-MARSDEN], [DELHAYE]). For a conservation law of the general form

$$A_t + div\,\vec{B} = f\,, \tag{37}$$

the jump condition across a smooth surface (fixed in space-time),

$$\Sigma(\vec{x},t) = 0\,, \tag{38}$$

can be shown to be

$$[\![A]\!]^+_- N_t + [\![\vec{B}]\!]^+_- \cdot \vec{N}_x = 0\,.$$

Here

$$\vec{N} = (\vec{N}_x\,, N_t) = (\nabla\Sigma\,, \Sigma_t)\,\frac{1}{\sqrt{\Sigma_t^2 + |\nabla\Sigma|^2}}$$

denotes the unit normal to the surface pointing towards the "+" side, and $[\![A]\!]^+_- = A^+ - A^-$ denotes the difference between the limiting value of a quantity A on Σ from the "+" side and its value on Σ from the "–" side. More conveniently, this may be expressed in terms of the

normal velocity $$\mathrm{v} := \frac{d\vec{x}}{dt}\cdot\vec{n} = -\frac{\Sigma_t}{|\nabla\Sigma|} \tag{39}$$

of the moving surface (see next paragraph) as

$$[\![A]\!]^+_- \mathrm{v} = [\![\vec{B}\cdot\vec{n}]\!]^+_-\,, \tag{40}$$

where $\vec{n} := \dfrac{\nabla\Sigma}{|\nabla\Sigma|}$ denotes the unit normal to the moving surface at each time.

Equation (39) may be seen as follows. Firstly (38) can be thought of either as representing a **fixed** surface in space-time, with unit normal $\vec{N} = (\vec{N}_x\,, N_t)$, or, a **moving** surface in space, with unit normal $\vec{n} = \dfrac{\nabla\Sigma}{|\nabla\Sigma|}$ at each fixed time. From (38), we have $\Sigma_t + \nabla\Sigma\cdot\dfrac{d\vec{x}}{dt} = 0$, where $\dfrac{d\vec{x}}{dt}$ is the velocity of a point on the surface. Hence $\mathrm{v} := \dfrac{d\vec{x}}{dt}\cdot\vec{n}$ represents the normal component of the velocity of the moving interface and therefore

$$\mathrm{v} = \frac{d\vec{x}}{dt} \cdot \frac{\nabla\Sigma}{|\nabla\Sigma|} = -\frac{\Sigma_t}{|\nabla\Sigma|} \,.$$

Applying this to the energy conservation law, (15), we find

$$[\![\rho e]\!]_-^+ \, \mathrm{v} = [\![\vec{q} \cdot \vec{n}]\!]_-^+ \,. \tag{41}$$

Since both the energy, ρe, and the flux, $\vec{q}$, across any surface Σ entirely inside the liquid (or the solid) are continuous within each one-phase region, the jumps $[\![\rho e]\!]_-^+$ and $[\![\vec{q}]\!]_-^+$ are both zero, and (41) degenerates to $0 = 0$. But, if Σ is the interface, then the jumps are *not* zero. Indeed, let us choose the liquid as "+" side and the solid as "–" side of the interface Σ. By (8), the enthalpy jump is $[\![\rho e]\!]_{solid}^{liquid} = \rho L > 0$, and therefore (41) becomes the so-called

Stefan Condition:
$$\rho L \, \mathrm{v} = [\![\vec{q} \cdot \vec{n}]\!]_{solid}^{liquid} \tag{42}$$

on the interface. This states that the latent heat released due to the interface displacement equals the net amount of heat delivered to (or from) the interface per unit area per unit time ($\vec{q} \cdot \vec{n} =$ flux normal to the moving surface). Thus, the Stefan Condition is a statement of heat balance across the interface. For more general versions see **§2.3.E** and **§2.4.F**.

Now, we restrict our considerations to the 1-dimensional case and derive the Stefan Condition directly from *global* energy balance. Consider a slab of material, $0 \le x \le l$, of constant cross-sectional area A. Heat is input or output at the faces $x = 0$ and $x = l$ by some means, resulting in, say, liquid in $0 \le x < X(t)$ and solid in $X(t) < x \le l$, separated by a sharp interface at $x = X(t)$, at each time $t > 0$. We assume constant density and phase-wise constant properties. The total enthalpy in the slab at time $t > 0$, referred to the melt temperature T_m (see PROBLEM 22) is

$$E(t) = A \left\{ \int_0^{X(t)} \{\rho c_L[T(x,t) - T_m] + \rho L\} dx + \int_{X(t)}^{l} \rho c_S[T(x,t) - T_m] dx \right\}. \tag{43}$$

Global heat balance demands (see (16))

$$\frac{dE}{dt} = \text{net heat flow into the slab} = A\{q(0,t) - q(l,t)\} \,, \tag{44}$$

where $q(0,t)$ and $-q(l,t)$ are the heat fluxes *into* the slab through the faces $x = 0$ and $x = l$, respectively. Leibnitz's rule enables us to compute

$$\frac{1}{A}\frac{dE}{dt} = \rho c_L[T(X(t),t) - T_m] \cdot X'(t) + \int_0^{X(t)} \rho c_L T_t(x,t) dx$$

$$+ \rho LX'(t) - \rho c_S[T(X(t),t) - T_m] \cdot X'(t) + \int_{X(t)}^{l} \rho c_S T_t(x,t)dx \,. \tag{45}$$

Using $T(X(t), t) = T_m$ and substituting the heat equation (21) for each phase we obtain

$$\frac{1}{A}\frac{dE}{dt} = k_L T_x(X(t)^-, t) - k_L T_x(0,t) + \rho LX'(t) + k_S T_x(l,t) - k_S T_x(X(t)^+, t)\,,$$

where $T_x(X(t)^{\mp}, t)$ denotes the values of $T_x(x,t)$ as $x \to X(t)^{\mp}$, i.e. from the left or right. But, $-k_L T_x(0,t)$ and $+k_S T_x(l,t)$ are precisely the fluxes $q(0,t)$ and $-q(l,t)$. Therefore, (44) yields the

$$\textbf{Stefan Condition:} \quad \rho\, L\, X'(t) = -\, k_L\, T_x(\, X(t)^-, t) + k_S\, T_x(\, X(t)^+, t), \tag{46}$$

expressing energy conservation across the interface $x = X(t)$ in the 1-dimensional case. In fact, since the velocity of the interface is $v = X'(t)$, and

$$(q^{liquid} - q^{solid})|_{x=X(t)} = -\, k_L T_x(X(t)^-, t) + k_S T_x(X(t)^+, t),$$

we see that the 1-dimensional version of (42) is indeed (46).

The Stefan Condition says that the rate of change in latent heat $\rho\, L\, X'(t)$, equals the amount by which the heat flux jumps across the interface. In particular, the heat flux can be continuous across the interface if and only if either $L = 0$ or the interface does *not* move. For an axially symmetric phase-change in a cylinder, or a spherically symmetric phase-change in a sphere, the Stefan condition across the interface $r = R(t)$ has the same form, namely (PROBLEMS 24, 25)

$$\rho\, L\, R'(t) = -\, k_L\, T_r(\, R(t), t) + k_S\, T_r(\, R(t), t)\,. \tag{47}$$

1.2.F The Stefan Problem

Now we have all the ingredients we need to state the mathematical problem modeling a phase-change process that satisfies the assumptions of **§1.2.B**. We will assume that no internal heating sources are present in the material.

As a very simple model process, we consider the following

PHYSICAL PROBLEM. A slab, $0 \le x \le l$, initially solid at temperature $T_{init} < T_m$, is melted by imposing a hot temperature $T_L > T_m$ at the face $x = 0$ and keeping the back face, $x = l$, insulated. We assume constant thermophysical parameters ρ, c_L, k_L, c_S, k_S (hence constant diffusivities $\alpha_L = k_L/\rho c_L$ and $\alpha_S = k_S/\rho c_S$), and a sharp interface $x = X(t)$.

A schematic picture of the process is shown in **Figure 1.2.8**. At each time t, liquid occupies $[0, X(t))$ and solid $(X(t), l]$. The curve $x = X(t)$ represents the interface location, schematically, demarcating the liquid and solid space-time regions. The heat equation is to be satisfied in each phase-region and the initial and boundary conditions of the problem are shown.

The mathematical model for this process is the following

Two-Phase Stefan Problem *(for a slab melting from the left)*:

Find the temperature $T(x, t)$, $0 \leq x \leq l$, $t > 0$, and interface location $X(t)$, $t > 0$, such that the following are satisfied (**Figure 1.2.8**):

Partial differential equations

$$T_t = \alpha_L T_{xx} \quad \text{for} \quad 0 < x < X(t),\ t > 0 \quad (\textit{liquid region}) \tag{48a}$$

$$T_t = \alpha_S T_{xx} \quad \text{for} \quad X(t) < x < l,\ t > 0 \quad (\textit{solid region}) \tag{48b}$$

Interface Conditions

$$T(X(t), t) = T_m, \qquad t > 0 \tag{49a}$$

$$\rho L X'(t) = -k_L T_x(X(t)^-, t) + k_S T_x(X(t)^+, t), \quad t > 0 \tag{49b}$$

Initial Conditions

$$X(0) = 0 \tag{50a}$$

$$T(x, 0) = T_{init} < T_m, \quad 0 \leq x \leq l, \ (\textit{and initial state is solid}) \tag{50b}$$

Boundary Conditions

$$T(0, t) = T_L > T_m, \quad t > 0 \quad (\textit{imposed temperature}) \tag{51a}$$

$$-k_S T_x(l, t) = 0, \quad t > 0 \qquad (\textit{insulated boundary}) \tag{51b}$$

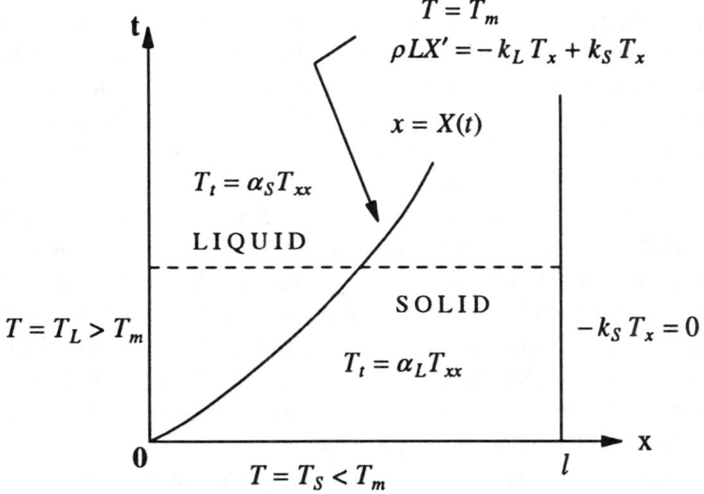

Figure 1.2.8. Space-time diagram for the Two-Phase Stefan Problem.

DEFINITION : We say that the functions $T(x, t)$, $X(t)$ constitute a **classical solution of the above Stefan Problem** up to a time t^* if the functions have continuous derivatives to all orders appearing in the problem formulation and satisfy the conditions of the problem.

Here, t^* is the time up to which the solution is desired (global solution).

Clearly, T_{init} and T_L could be non-constant: $T_{init}(x) \leq T_m$, $T_L(t) \geq T_m$. From the mathematical point of view, we could allow c_L, c_S, k_L, k_S to be (known) functions of (x, t, T) by replacing the heat equations in (48) by heat conduction equations $\rho c_i T_t = (k_i T_x)_x$, $i = L, S$. Naturally, the boundary conditions (51) can be replaced by any combination of the standard conditions described in **§1.2.D**.

All such generalizations of (48-51) are still referred to as Stefan Problems or **Stefan-type** problems. They serve as *prototypes* of the, so-called, **moving boundary problems** whose basic feature is that the regions in which the partial differential equations are to hold are unknown and must be found as part of the solution of the problem. This amounts to a **non-linearity** of **geometric nature**, apparent in (49a,b), even when the rest of the equations appear to be linear, and is the source of the mathematical difficulties that moving boundary problems present. Their non-linearity destroys the validity of the Superposition Principle, and "separation of variables" is no longer applicable ! The underlying geometric nonlinearity can be made to appear algebraically in the partial differential equations by a change of variables: replacing x by $\xi = x / X(t)$ transforms the varying region $0 < x < X(t)$ to the fixed region $0 < \xi < 1$, and the linear equation $T_t = \alpha_L T_{xx}$ to the nonlinear equation $X^2 T_t - \xi X X' T_\xi = \alpha_L T_{\xi\xi}$ (see **§3.3**).

The well-posedness (existence of a unique classical solution depending continuously on the data) of reasonably general 1-dimensional Stefan Problems without undue restrictions on the data was established only during the mid 1970's! (CANNON-HENRY-KOTLOW, see **§4.4**). We note that **local** solvability (meaning: there exists a time t^* up to which a unique classical solution exists) was already proved by Rubinstein in 1947 (see [RUBINSTEIN] for a historical survey of the mathematical development up to the mid 1960's). Of course, particular, well-behaved problems (those with a monotonic interface, like (52-55) below) were treated earlier by various methods.

Stefan-type problems can also be formulated classically in two or three dimensions (see **§4.4**), but such formulations may admit no (classical) solution, as the breakup of a piece of ice into two or more pieces indicates physically. Fortunately "weak" or "generalized" ("enthalpy") formulations which are well-posed (and computable) came to the rescue in the early 1960's, as we shall see in CHAPTER **4**. Let us also note that even 1-dimensional problems with either internal sources or variable T_m may develop mushy regions rendering the above sharp-front classical formulation inappropriate.

Certainly however, the great majority of phase-change processes lead to 1-dimensional Stefan Problems as just described. In fact one frequently deals with the so-called **One-Phase Stefan Problem**, in which only one "active" phase is present. If, in the physical problem, we assume that initially the slab is solid

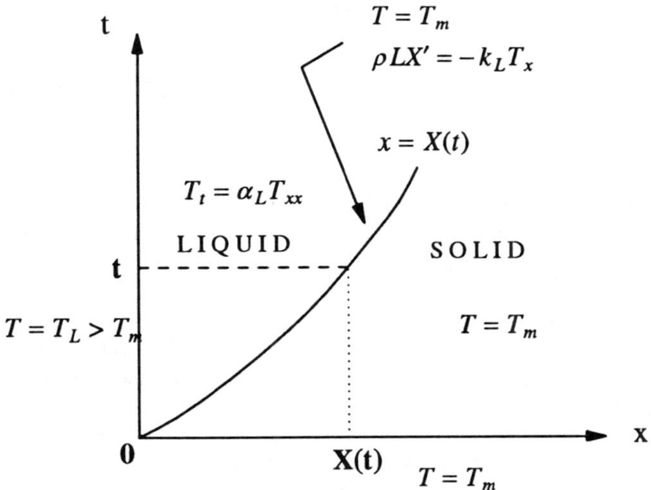

Figure 1.2.9. Space-time diagram for the One-Phase Stefan Problem.

at the melt temperature: $T_{init}(x) \equiv T_m$, then we obtain the

One-Phase Stefan Problem *(for a slab melting from the left)*:

Find $T(x,t)$ and $X(t)$ such that

$$T_t = \alpha_L T_{xx}, \quad 0 < x < X(t), \quad t > 0, \quad \text{(liquid region)} \tag{52}$$

$$T(X(t),t) = T_m, \quad t > 0, \tag{53a}$$

$$\rho L X'(t) = -k_L T_x(X(t),t), \quad t > 0, \tag{53b}$$

$$X(0) = 0, \tag{54}$$

$$T(0,t) = T_L(t) > T_m, \quad t > 0. \tag{55}$$

The temperature needs to be found only in the liquid $0 < x < X(t)$, $t > 0$, because it is identically T_m in the solid, and there is no initial condition for $T(x, t)$ because initially there is no liquid. Since the back face $x = l$ now plays *no role*, the solid can be considered effectively semi-infinite, occupying $(X(t), \infty)$. The diagram for this problem is shown in **Figure 1.2.9**. Clearly few realistic phase-change processes will actually lead to the one-phase situation, with **ablation** (instantaneous removal of melt) and induced stirring of liquid while freezing, being notable exceptions. On the other hand, molecular diffusion, filtration, and other processes commonly lead to one-phase problems [RUBINSTEIN], [ELLIOTT-OCKENDON]. The inherent simplicity of one-phase problems makes them amenable to a variety of simple approximation methods, as we shall see in CHAPTER **3**, and an entire book [HILL] is devoted to them.

PROBLEMS

PROBLEM 1. Using relation (1) for the specific heat $c(T)$ of water/ice, find the relation between the enthalpy e and the temperature T. Also express T as a function of e and plot.

PROBLEM 2. Recall (**§1.1**) that a liquid is hypercooled if the sensible heat extracted from it is more than the latent heat. Show that its temperature should be no more than $T_m - \frac{L}{c_L}$.

PROBLEM 3. Describe an experiment for the determination of the thermal conductivity of a solid; of a liquid. On what factors do you expect it to depend? [CHAPMAN], [ECKERT-DRAKE].

PROBLEM 4. The thermal conductivity of aluminum is $k = .213\ kJ/m\,s°C$. Assuming that the temperature distribution through a plate of Aluminum is a straight line, how thick should the plate be in order that a $100°C$ temperature drop across the plate will induce a heat flux of $q = 1000\ kJ/m^2\ s$?

PROBLEM 5. A heat source has been placed at the center of an aluminum ball of radius 10 cm emitting heat at a rate of 1000 kJ/s. What should the flux of heat at the surface of the ball be under the condition that the ball contains no heat sources or sinks?

PROBLEM 6. What is the heat flux along an isotherm for an isotropic medium? What if the medium is not isotropic?

PROBLEM 7. Derive (15) as the limiting case of energy conservation on a box $x_0 \le x \le x_0 + \Delta x,\ \ y_0 \le y \le y_0 + \Delta y,\ \ z_0 \le z \le z_0 + \Delta z$ over a time interval $t_0 \le t \le t_0 + \Delta t$ as $\Delta x, \Delta y, \Delta z, \Delta t \to 0$ [WHITAKER] , [BIRD et al].

PROBLEM 8. The thermal diffusivity of most materials is a non-constant function of the temperature. Describe in qualitative terms under what conditions you could "safely" pass from the nonlinear equation of (21) to the linear equation (22) without making too great an error.

PROBLEM 9. Show that the initial value problem $y' = \sqrt{y}$, $y(0) = 0$ is *not* well-posed.

PROBLEM 10. Using separation of variables, derive the solution (24c) for the problem of EXAMPLE 1, **§1.2.C**. Verify (at least formally) that it is indeed a solution.

PROBLEM 11. Derive the solution to Problem (25) of EXAMPLE 2, **§1.2.C**.

PROBLEM 12. Derive the solution to Problem (26) of EXAMPLE 3, **§1.2.C**.

PROBLEM 13. Show that the error function, (29), EXAMPLE 4, §**1.2.C**, satisfies $\text{erf}(0) = 0$, $\text{erf}(\infty) = 1$. [ABRAMOWITZ-STEGUN].

PROBLEM 14. (a) Seek the solution to Problem (27), EXAMPLE 4, §**1.2.C**, in the form $T(x,t) = F(\xi)$ with $\xi = x/2\sqrt{\alpha t}$ (similarity solution). Show that the unknown $F(\xi)$ must satisfy $F'' + 2\xi F' = 0$, $F(0) = T_L$, $F(\infty) = T_0$. Solve this ODE to obtain (28). (b) Verify that (28) solves problem (27).

PROBLEM 15. Using the value T_0 as the reference temperature for zero energy, find the total energy of the semi-infinite slab of EXAMPLE 4, §**1.2.C**, as a function of time (see PROBLEM 17, §**2.2**). [*Answer*: $E(t) = 2(T_L - T_0)(\rho c k t/\pi)^{1/2}$].

PROBLEM 16. Verify that the function

$$T(x, t) = A + B\left(1 - e^{-\frac{U}{\alpha}(x - Ut)}\right) \tag{56}$$

solves the heat equation (22), for any constants A, B, U. It represents a (thermal) front traveling with constant speed U (to the right if $U > 0$, to the left if $U < 0$).

PROBLEM 17. Leave a cup of hot coffee to cool in a room while using a thermometer to monitor its temperature $T(t)$ with passing time t. Taking Newton's Law of cooling in the form $T'(t) = h\cdot(T_{room} - T(t))$, find h for your cup of coffee, where T_{room} is the room temperature.

PROBLEM 18. A solid slab $0 \le x \le l$ is initially at a uniform temperature $T_{init} < T_m$. Heat is withdrawn from the front face, $x = 0$, and an experimenter can measure both the temperature, $T(0,t)$ (by a thermocouple) and the flux $q(0,t)$ (by a flux meter) at this face. The experimenter wants to determine the temperature distribution $T(x, t)$ in the slab. Unfortunately, the back face $x = l$, is inaccessible, so the only known data are $T(x,0) = T_{init}$, $0 \le x \le l$, $T(0,t) = T_{face}(t)$, and $-k\,T_x(0,t) = q_{face}(t)$, $t > 0$. Explain why he **cannot** determine $T(x, t)$ throughout the slab without any information about $x = l$.

PROBLEM 19. One of the more important and elusive types of information that we should be able to extract from models of physical processes is the sensitivity of these processes with respect to the system specifications. For example, one might wish to obtain information about the dependence of the temperature distribution history on the conductivity of a material. For the case of constant thermophysical properties and heat transfer in a finite slab use the heat equation to find relations that will yield the dependencies of the temperature and total heat content of the slab with respect to each of the parameters α, k, c, ρ and the slab length.

PROBLEM 20. Verify the units in equations (43), and (45).

PROBLEM 21. Derive the Stefan Condition for a slab, $0 \le x \le l$, freezing from the left. [Hint: Proceed as in (43-46)].

PROBLEM 22. Let $T_{ref} < T_m$ be a reference temperature. Show that the total energy in the slab $0 \le x < l$ at any time $t > 0$, referred to the temperature T_{ref} is

$$E(t) = A\{\int_0^{X(t)} [\rho c_S(T_m - T_{ref}) + \rho c_L(T - T_m) + \rho L]dx + \int_{X(t)}^{l} \rho c_S(T - T_{ref})dx\}.$$

Compare with (43).

PROBLEM 23. Show that the Stefan Condition remains unchanged no matter what we choose as reference temperature. [see the previous problem, proceed as in (43-46)].

PROBLEM 24. Derive the Stefan Condition, (47), across the interface $r = R(t)$ for an axially symmetric phase-change process in a cylinder. For definiteness, consider $0 < r < R(t)$ as liquid and $R(t) < r < R_0$ as solid. Note that the total enthalpy (per unit height) may be written as

$$\int_0^{R(t)} \{\rho c_L[T(r,t) - T_m] + \rho L\}2\pi r dr + \int_{R(t)}^{R_o} \rho c_S[T(r,t) - T_m]2\pi r dr \; .$$

PROBLEM 25. Derive the Stefan Condition across the interface $r = R(t)$ for a spherically symmetric phase-change process in a sphere. For definiteness, consider $0 < r < R(t)$ as liquid and $R(t) < r < R_0$ as solid. Then we have

$$E(t) = \int_0^{R(t)} \{\rho c_L[T(r,t) - T_m] + \rho L\}4\pi r^2 dr + \int_{R(t)}^{R_0} \rho c_S[T(r,t) - T_m]4\pi r^2 dr \; .$$

PROBLEM 26. Consider the 1-phase Stefan Problem for the slab (when the material is initially solid at its melt temperature $T = T_m$). Using the fact that the phase change front is an isotherm, show that along the front we have

$$T_{xx} = \frac{c_L}{L} T_x^2 \tag{57}$$

exhibiting vividly the nonlinearity of the problem. Observe that (57) is a special case of (40) for one space dimension and one phase, if we regard the interface as an isotherm for the temperature, permitting the function Σ of (40) to be defined as $\Sigma = T - T_m$.

PROBLEM 27. A slab $0 < x < l$ is filled with material having melting temperature T_m. The material has constant thermophysical parameters, different for liquid and solid phases. At $x = 0$ a temperature $T_0 < T_m$ is imposed for all time, while at $x = l$ a temperature $T_l > T_m$ is imposed for all time. Find the *steady state* temperature distribution and the location of the front separating solid and liquid phases.

PROBLEM 28. (A 1-phase Stefan Problem with straight-line interface)
Fix U=constant and assume a straight-line interface $X(t) = Ut$, separating liquid in $0 \leq x < X(t)$ from solid at the melt temperature T_m (for $x > X(t)$).

(a) Determine the constants A, B so that the traveling-front solution (56) also satisfies the interface conditions (53). Then, the resulting $T(x, t)$ will satisfy (52-53), with $X(t) = Ut$.

(b) Show that the boundary temperature, $T_L(t)$, necessary for the $T(x, t)$ (found in (a)) to satisfy the 1-phase problem (52-55) is an exponentially increasing function of time, given by

$$T_L(t) = T_m - \frac{L}{c_L}[1 - e^{U^2 t/\alpha}], \quad t > 0 \ .$$

1.3. GENERAL MELTING AND SOLIDIFICATION PROCESSES

The mathematical model of a melting process formulated in **§1.2** is disarmingly simple, since it concerns a process for which, at any time there are two distinct regions, one solid, the other liquid, separated by a single phase front $X(t)$ (**Figures 1.2.6,7**). Moreover the phase boundary is always moving in the same direction, while the temperature at any point is always rising. Such simplicity is not typical of the kinds of problems that arise in melt/freeze scenarios. Thus for example, molten metal placed in a cast will partly contract away from the cast wall because of the increase in density upon solidification; this reduces the cooling effect of the cast and can induce remelting of the solid skin of the metal adjacent to the cast. More dramatically, in latent heat thermal energy storage applications where we might wish to store solar energy as the latent heat of melting of a PCM during daylight hours for use at night, the PCM will go through repeated melt/freeze cycles possibly producing a multitude of phase change fronts separating zones of liquid and solid. Similarly, placing two cold ice cubes sufficiently close to each other in water can induce the formation of an "ice bridge" linking them, due to local freezing of the water; if the water is sufficiently warm this bridge, together with the cubes themselves, will melt.

Let us examine the possibility of multiple melt/freeze fronts for a process in which a material is subjected alternatively to temperatures above and below its melt temperature.

EXAMPLE 1. A slab of length l is intermittently heated and cooled to the extent that freezing and melting take place. The periods of heat input are referred to as *charging times* while those of heat withdrawal are *discharging*

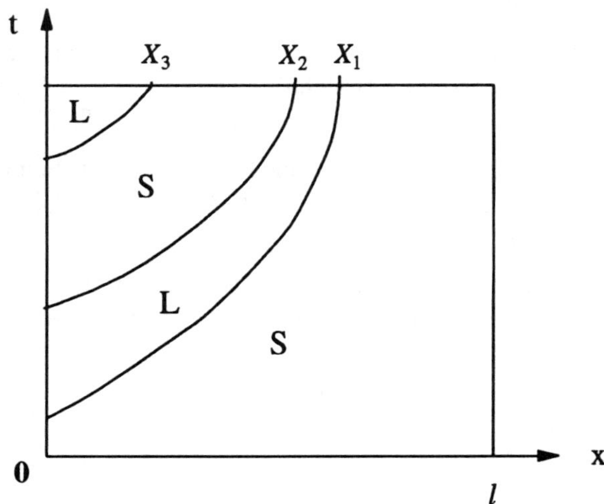

Figure 1.3.1. Repeated melting and freezing of a slab.

times. Assume that heat input and withdrawal are always carried out at the left-hand face of the slab, while the right-hand face is insulated (**Figure 1.3.1**). Let the slab be entirely solid at the beginning of our scenario.

Suppose that heat is input to the slab over a period of time t_1^C with the superscript C corresponding to the "charging" mode. If the initial temperature was not too low, and the heat flux into the slab is not too small, a melt front will appear at $x=0$ moving monotonically into the slab. For an imposed temperature the front will appear at $t=0$ while for flux, convection or radiation input, the front will only appear when the material temperature at the face $x=0$ has reached T_m. In any event, a front X_1 with liquid on its left and solid at its right may well appear before the end of the first charge interval, $t=t_1^C$.

Now suppose that the material begins to discharge heat at time $t=t_1^C$ in a process whose duration is t_1^D (the superscript D representing "discharge"). Depending on the mechanism of heat withdrawal, a new front X_2 will appear at some time $t>t_1^C$ separating solid material (on its left) from liquid material (on its right). Meanwhile the earlier appearing melt front X_1 will continue to move into the solid; the velocity of this front will be far less than it was during the charging process, since its driving force (the surface heat flux into the slab) has now been replaced by the low temperature gradient in the liquid. Eventually, it will stop advancing (if the right-hand face were cooled instead of being insulated, it could even move leftward!) Hence during the time of discharge several possible events may occur, two of which are: a) the freezing front X_2 moving to the right meets the melt front X_1; when this occurs the liquid region formed during the charge interval will disappear, as will both fronts; b) the freezing front X_2 will move to the right, but *not* overtake X_1 before the discharge period ends. Now when the new charge period begins we may find ourselves in the situation shown in **Figure 1.3.1**, wherein three fronts separating four zones exist: the first melt front X_1, the second freeze front X_2, and a third melt front X_3. As the number of cycles increases a myriad of possible phase configurations may occur because of variations in heat input and extraction rates and durations.

EXAMPLE 1 was concerned with a one-dimensional problem. For two or three dimensions the complexity of the geometry of the phase regions and their boundaries goes beyond the stage where simple physical intuition can be of much use. A hint of what may occur in two space dimensions is given in EXAMPLE 2, concerned with a charge-discharge cycle with a reversal of flow of the heat transfer fluid.

EXAMPLE 2. *Two Dimensional Phase Change*: Consider a two-dimensional material slab as seen in **Figure 1.3.2**. Heat exchange with a transfer fluid channel will take place at face A while faces B,C,D are assumed insulated. We also assume that the material is initially solid. During a charge period a hot transfer fluid is to flow upward- in the direction of increasing y ; as the transfer fluid flows upward it will cool down; thus a melt front will be formed, as seen in the Figure, moving further into the slab for smaller y than for larger y . If the discharge process is carried out with a downward flow of a cold transfer fluid, then at the end of the discharge cycle, we may well find a "V" shaped front extending only partly in the y direction. Clearly further cycles, if they correspond to intermittent heat sources and sinks, may well result in isolated regions of one phase, within material of the second phase.

Note that the qualitative description of EXAMPLE 2 is not surprising. If one were to immerse an ice-sculpture of, say, a unicorn, in a hot- water bath, we might well find after some moments that the single ice statue has melted in such a way that several distinct pieces of ice result.

Our examples imply that in processes involving alternate melt-freeze cycles we must seek modeling techniques in which we do not have to know apriori the qualitative behavior of the process. In general, we will encounter multiple fronts, disappearing phases, and extremely complex geometries. Whether the actual physical process would yield such multiple solid/liquid regions depends on the physical makeup of the material and the conditions of the process. Thus for example in a microgravity space environment one could expect a much reduced tendency for solid particles to float or settle in the liquid.

Of course the complexity of the phase change process is only increased when other, sometimes extremely realistic physical phenomena are taken into account. Thus for example liquid paraffin wax easily dissolves air; on solidification the air is entrapped in the solid. On melting, the solid particles are alternately buoyant- when they contain bubbles of air, and sink in the liquid when the bubbles are released from the melting solid. In engineering applications one may have to accommodate such behavior, if possible, through suitable assumptions in a model not explicitly incorporating the phenomenon, or in the development of a more complete model.

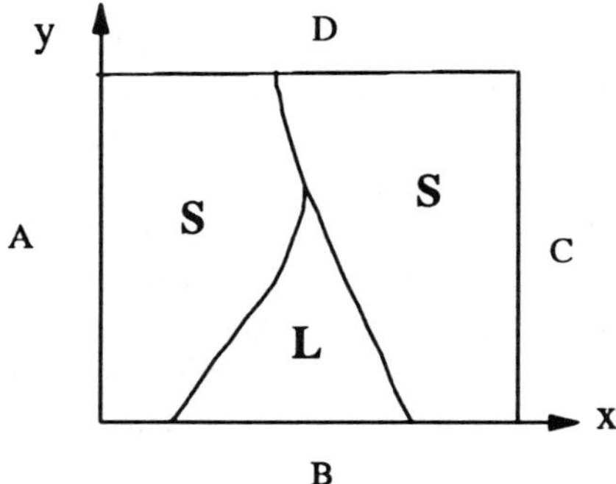

Figure 1.3.2. A two-dimensional charge/discharge process.

PROBLEMS

PROBLEM 1. Discuss the phase change history of a sample of radioactive material placed in an insulated container. Assume that the sample is cubically shaped and that it is initially at a temperature T_S below its melt temperature T_m. Will any phase change front appear in the material?

PROBLEM 2. In addition to the radioactive decay of PROBLEM 1 internal heating may arise from radiative transfer through semi-transparent materials. Describe a charge-discharge process similar to that in Example 1, but with radiation taken into account.

PROBLEM 3. Two samples of a material, one of them liquid at a high temperature, the second solid at a low temperature, are placed in contact. Discuss what factors will determine whether the liquid freezes or the solid melts.

PROBLEM 4. For several days the temperature in the town was on the average $0\ ^\circ F$. Then suddenly the weather changed, the temperature rose to the low 40's and an ice layer formed on all the streets, paralyzing movement of cars and people for a while. What happened?

CHAPTER 2

PROBLEMS WITH EXPLICIT SOLUTIONS

The formulation of Stefan Problems as models of basic phase-change processes was presented in §**1.2**. Under certain restrictions on the parameters and data such problems admit **explicit solutions in closed form**. These simplest possible, explicitly solvable Stefan problems form the backbone of our understanding of all phase-change models and serve as the only means of validating approximate and numerical solutions of more complicated problems.

Unfortunately, closed-form explicit solutions (all of which are of similarity type) may be found only under the following very restrictive conditions: 1-dimensional, semi-infinite geometry, uniform initial temperature, constant imposed temperature (at the boundary), and thermophysical properties constant in each phase.

Within these confines we present a succession of models of increasingly complicated phase-change processes.

We begin with the simplest possible models, the classical 1-phase Stefan problem (§**2.1**), and 2-phase Stefan Problem (§**2.2**), modeling the most basic aspects of a phase-change process (as discussed in §**1.2**). We present the Neumann similarity solution and familiarize the reader with some of the information it conveys.

Next (§**2.3**) we relax the assumption of constant density by allowing the densities of solid and liquid to be different (but each still a constant), thus bringing density change effects into the picture. We study the effect of volume expansion (no voids), and of shrinkage (causing formation of a void near the wall). In each case we formulate explicitly solvable thermal models (neglecting all mechanical effects) and examine the effect of density change on the Neumann solution. More precise models, which include mechanical effects but don't admit explicit solutions, are derived from first principles in the last subsection.

In §**2.4** we introduce supercooling, thus relaxing the assumption that the phase-change occurs at the melt temperature T_m. We discuss the thermodynamics of phase-coexistence and derive the Laplace-Young, Clausius-Clapeyron and Gibbs-Thomson relations from first principles. The classical Mullins-Sekerka morphological stability analysis is also presented.

In §**2.5** we discuss binary alloy solidification, coupling heat conduction and solute diffusion. We present the classical model of Rubinstein and its explicit solution, as well as various other models of freezing over an extended temperature range.

The introduction of each new physical phenomenon in the simplest possible setting (dictated by the desire to have explicit solutions available) helps us understand the phenomenon more easily and see its effects on the solution.

Similarity solutions in cylindrical and spherical geometries for special problems are the subject of **§2.6**. Finally, in **§2.7** we present a contrived (artificial) multi- dimensional phase-change problem whose explicit solution may serve as benchmark for 2 or 3 dimensional numerical codes. Such a debugging tool becomes necessary because **no** explicitly solvable phase-change problem exists in 2 or 3 dimensions.

Each phase-change process involving *melting* has a counterpart involving *freezing*. For consistency throughout our discussions we will be treating the case of *melting*, unless we are specifically interested in a solidification process (as in **§2.5**). The parallel developments for freezing will be mostly left as exercises for the reader in the PROBLEMS, but the changes needed to turn the solution of the one to the other will be indicated in the text.

2.1. THE ONE-PHASE STEFAN PROBLEM

2.1.A Introduction

The simplest explicitly solvable phase-change problem is the 1-phase Stefan Problem (**§1.2.F**) with **constant** imposed temperature and constant thermophysical properties. Its solution is the classical Neumann similarity solution [CARSLAW-JAEGER], [RUBINSTEIN] involving the error function. As prototype example we treat the **melting** problem leaving the case of freezing for the reader to examine via the Problems.

The term "one-phase" refers to only one of the phases (liquid) being "active", the other phase staying at the melt temperature T_m (**§1.2.F**). Thus the physical situation is the following:

PHYSICAL PROBLEM: Melting of a (semi-infinite) slab, $0 \le x < \infty$, initially **solid** at the melt temperature, T_m, by imposing a **constant** temperature $T_L > T_m$ on the face $x = 0$. Thermophysical parameters: ρ, c_L, k_L, L, $\alpha_L = k_L / \rho c_L$, all constant.

The physical realization of this problem is an insulated pipe, filled with a PCM, and exposed at one face to a heat source, while its length is so great that the second face is not reached by the melting front during the life of the experiment (**Figure 2.1.1**). The experiment begins with the material initially solid and at its melt temperature. The nearby face temperature is raised as quickly as possible to the value T_L and maintained at that value for all time. This may be done by pumping a heat exchange fluid at temperature T_L at very high mass flow rate across the face. The mathematical model of this process leads, as in **§1.2**, to the following:

MATHEMATICAL PROBLEM (1-phase Stefan Problem for a slab melting from the left):

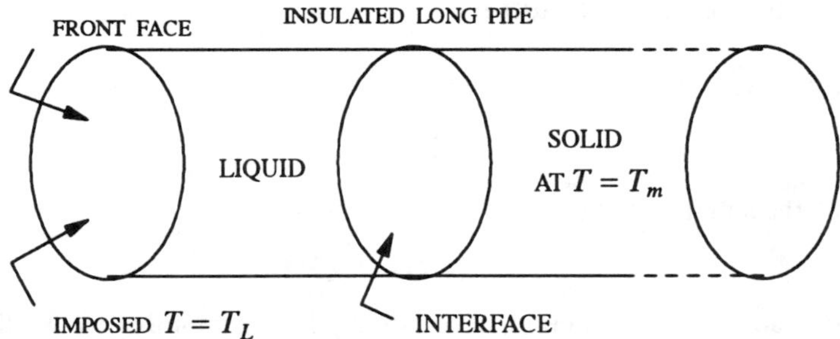

Figure 2.1.1. Physical realization of the One-Phase Stefan Problem.

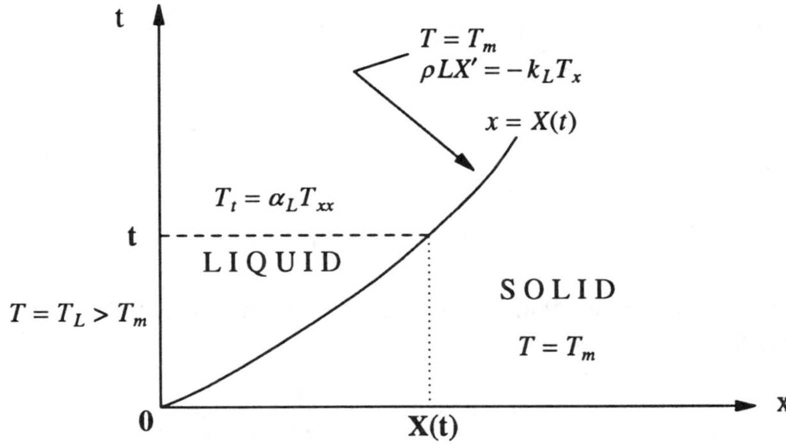

Figure 2.1.2. Space-time diagram for the One-Phase Stefan Problem.

Find $T(x,t)$ and $X(t)$ such that (**Figure 2.1.2**)

$$T_t = \alpha_L T_{xx}, \quad 0 < x < X(t), \quad t > 0 \text{ (liquid)} \tag{1}$$

$$T(X(t),t) = T_m, \quad t \geq 0 \tag{2a}$$

$$\rho L X'(t) = -k_L T_x(X(t),t) \quad , t > 0 \tag{2b}$$

$$X(0) = 0, \quad \text{(material initially completely solid)} \tag{3}$$

$$T(0,t) = T_L > T_m \quad , t > 0 \tag{4}$$

The corresponding problem for a slab **freezing** from the left due to a temperature $T_S < T_m$ being imposed at $x = 0$ is formally obtained by replacing every subscript L by S and the latent heat L by $-L$ in (2b).

2.1.B The Neumann Solution

We introduce the similarity variable

$$\xi = \frac{x}{\sqrt{t}}, \tag{5}$$

and seek the solution in the form

$$T(x,t) = F(\xi), \tag{6}$$

with $F(\xi)$ an unknown function. Accordingly it is natural that we would seek the interface location $X(t)$ to be proportional to $\sqrt{t}$, searching therefore for a constant A for which

$$X(t) = A\sqrt{t}. \tag{7}$$

Substituting into (1) and integrating we obtain

$$F(\xi) = B\int_0^{\xi} e^{-\frac{s^2}{4\alpha_L}}ds + C = B\sqrt{\pi\alpha_L}\,\mathrm{erf}(\frac{\xi}{2\sqrt{\alpha_L}}) + C \tag{8}$$

for B, C constants, where

$$\mathrm{erf}(z) = \frac{2}{\sqrt{\pi}}\int_0^z e^{-s^2}ds \tag{9}$$

denotes the **error function** [ABRAMOWITZ-STEGUN] (see §**1.2**, also (28-35) below). Conditions (4) and (2a) yield

$$C = T_L \text{ and } B = \frac{T_m - T_L}{\sqrt{\pi\alpha_L}\,\mathrm{erf}(A/2\sqrt{\alpha_L})} \tag{10}$$

Set

$$\lambda = \frac{A}{2\sqrt{\alpha_L}}, \qquad \Delta T_L = T_L - T_m, \tag{11}$$

and

$$St_L = \frac{c_L \Delta T_L}{L} = \textbf{Stefan Number}. \tag{12}$$

Then the Stefan condition (2b) leads to an equation for λ:

$$\lambda e^{\lambda^2}\mathrm{erf}(\lambda) = \frac{k_L}{\rho L}\frac{\Delta T_L}{\sqrt{\pi\alpha_L}} = \frac{c_L\Delta T_L}{\sqrt{\pi}L} = \frac{St_L}{\sqrt{\pi}}. \tag{13}$$

Hence it is more convenient to express the solution in terms of λ. From (5-7, 11),

$$X(t) = 2\lambda\sqrt{\alpha_L t}, \tag{14}$$

and from (5-8,10)

$$T(x,t) = T_L - \Delta T_L \frac{\mathrm{erf}(\frac{x}{2\sqrt{\alpha_L t}})}{\mathrm{erf}(\lambda)}, \tag{15}$$

with λ a root of the transcendental equation

$$\lambda e^{\lambda^2}\operatorname{erf}(\lambda) = St_L / \sqrt{\pi}\,. \tag{16}$$

It is easily shown (PROBLEM 2) that the quantity $f(\lambda) = \lambda e^{\lambda^2}\operatorname{erf}\lambda$ is a strictly increasing function of $\lambda \geq 0$, $f(0) = 0$, $\lim_{\lambda\to\infty} f(\lambda) = +\infty$, and therefore the graph of $y = f(\lambda)$ intersects any horizontal line $y = St_L/\sqrt{\pi}$ exactly once. In other words, for each value of $St_L > 0$, there exists a **unique** root, λ, of equation (16), **Figure 2.1.3**. Once λ is found by solving the transcendental equation (16), the solution of the Stefan Problem is given by (14-15). This is the classical **Neumann solution** to the Stefan Problem (after F. Neumann).

Note that the uniqueness of the root λ implies the uniqueness of the *similarity* solution, i.e. that (14-15) is the only solution of the form (6-7). Is this the *only possible* solution of (1-4)? The answer is Yes. The Stefan problem is a well-posed mathematical problem (**§1.2.C**,**§4.5**), so it admits only one solution. Uniqueness of the solution follows from the much more general uniqueness of a weak solution presented in **§4.4** .

2.1.C Dimensionless form

We observe in (16) that the value of the root λ and hence also the solution, depends on a *single dimensionless parameter*, the **Stefan Number**, defined in (12). This is better brought out by undimensionalizing the problem itself. We introduce the dimensionless length and time variables,

$$\zeta = \frac{x}{\hat{x}}, \quad Fo = \frac{\alpha_L}{\hat{x}^2}\, t = \textbf{Fourier Number}, \tag{17}$$

where $\hat{x}$ is any convenient length scale (note that there is no "natural" length in this problem), and the dimensionless interface and temperature

$$\Sigma(Fo) = \frac{X(t)}{\hat{x}}, \quad u(\zeta, Fo) = \frac{T(x,t) - T_m}{\Delta T_L}, \tag{18}$$

where $\Delta T_L = T_L - T_m$ as in (11). Then the Stefan Problem (1-4) takes the form (PROBLEM 6)

$$u_{Fo} = u_{\zeta\zeta}, \quad 0 < \zeta < \Sigma(Fo), \quad Fo > 0 \tag{19}$$

$$u(\Sigma(Fo), Fo) = 0, \quad Fo > 0 \tag{20}$$

$$\Sigma'(Fo) = -\, St_L \cdot u_\zeta(\Sigma(Fo), Fo), \quad Fo > 0 \tag{21}$$

$$\Sigma(0) = 0 \tag{22}$$

$$u(0, Fo) = 1, \quad Fo > 0, \tag{23}$$

containing a *single parameter*, the Stefan number (12). For alternative dimensionless forms see **§3.1**.

The Neumann similarity solution of the dimensionless problem (19-23) is given by (PROBLEM 7)

$$\Sigma(Fo) = 2\lambda\sqrt{Fo}\,, \qquad Fo \geq 0\,, \tag{24}$$

$$u(\zeta, Fo) = 1 - \frac{\operatorname{erf}(\frac{\zeta}{2\sqrt{Fo}})}{\operatorname{erf}\lambda}\,, \qquad 0 \leq \zeta \leq Fo\,, \qquad Fo \geq 0, \tag{25}$$

with λ the root of the same transcendental equation

$$\lambda e^{\lambda^2} \operatorname{erf}\lambda = \frac{St_L}{\sqrt{\pi}}\,. \tag{26}$$

2.1.D The root λ versus the Stefan Number

As the only parameter present in the problem (19-23), the Stefan number St_L completely characterizes the melting process. We may think of it as representing the ratio of the "sensible heat", $c_L \Delta T_L$ to the latent heat L. That this is indeed a correct interpretation will be shown in §**2.2.G**. Note that for a freezing process we define the Stefan number by

$$St_S = \frac{c_S(T_m - T_S)}{L}\,.$$

To gain perspective, let us compute St for some materials in typical phase-change processes.

EXAMPLE 1: *Ice and Water.* Ice and water are the solid and liquid phases of the same material (H_2O). Under ordinary conditions the temperature of ice is less than the value $T_m = 273.15\,K$ ($0°C$); upon warming, ice melts at this temperature with a latent heat $L = 333.4\,kJ/kg$, and water is found at temperatures above T_m. Its specific heat is $c_L = 4.1868\,kJ/kg\,K$. Due to the low value of the ratio c_l/L, the Stefan Number for melting of ice is typically no more than 1; e.g. with $T_L = 37°C$ (body temperature), we have $St_L = 0.46$.

In freezing of water, the specific heat of ice varies strongly with temperature (see (1) §**1.2**), typically in the range of $1. \leq c_S \leq 2.09$, which results in even smaller Stefan numbers. For example, in a food freezing process, we may have $T_S = -20°C$; taking $c_S = 2$ as representative value, we find $St_S = c_S(T_m - T_S)/L \approx 0.12$.

EXAMPLE 2: *Copper.* For copper $T_m = 1356.2\,K$ and $L = 204.9$ kJ/kg. Suppose that copper, initially at the temperature 1470 K, cools down to T_m and solidifies; for liquid copper the average specific heat is $c_L = .51$ kJ/kg K, hence we have $St = .28$. Suppose that the process includes cooling to room temperature; for this temperature range a representative value of the specific heat is $c_S = .45$ kJ/kg K; thus the temperature drop ΔT in the Stefan number is approximately 1200 K and $St = 2.64$.

EXAMPLE 3: *Melting of a Paraffin Wax.* Paraffin waxes have high latent heat. A typical paraffin wax is N- Octadecane for which $\bar{c} = 2.16$ kJ/kg K, $L = 243$ kJ/kg and the melting temperature $T_m = 301.2$ K. Over a range of temperature $\Delta T = 100K$, $\boldsymbol{St} = .89$. Like water, paraffin waxes generally have low Stefan numbers associated with their melting and solidification.

EXAMPLE 4: *Melting of Silicon-Dioxide from Room Temperature.* Silica (silicon dioxide) is a material with a very large ratio of sensible to latent heat. In fact, over a range of temperatures from room temperature to its melting point at $T_m = 1883.15\,K$ the average specific heat is $\bar{c} = 39.23$ kJ/kg K, while its latent heat is $L = 142.35$ kJ/kg. Hence for a melting process from room temperature (300 K), $St = 436.3$.

These examples point to the following rule of thumb. For certain families of non-metallic solids such as waxes, $\boldsymbol{St}$ is small; hence, the bulk of heat stored or released from them is latent heat. For metals $\boldsymbol{St}$ is of the order 1-10 and so the effect of sensible heat is at least as large as that of latent heat. For other materials such as silicates $\boldsymbol{St}$ may be very large; the sensible heat will then dominate the heat transfer process. Of course $\boldsymbol{St}$ depends on the temperature drop ΔT experienced by the material during the heat transfer process (PROBLEM 14).

In general, the size of $\boldsymbol{St}$ will determine the suitability of a particular method for analyzing a given heat transfer process. *For large* $\boldsymbol{St}$ *the process will essentially be one of pure conduction* to which a variety of existing techniques are applicable. For small $\boldsymbol{St}$ the conduction heat transfer process will be dominated by the phase change.

Given a melting or freezing process, hence a Stefan number $\boldsymbol{St}$, the transcendental equation

$$\lambda e^{\lambda^2} \operatorname{erf} \lambda = St/\sqrt{\pi} \tag{26}$$

is easily solvable by the Newton-Raphson iterative method [PRESS et al] using as initial "guess" the value $\sqrt{St/2}$. The latter is the approximate solution to (26) when $\boldsymbol{St} \approx 0$, as we shall show in **§2.1.F**. **Figure 2.1.3** displays the values of the root λ for each $\boldsymbol{St}$ in the range $0 \le \boldsymbol{St} \le 5$, found by the Newton-Raphson method.

For the convenience of the reader and easy reference we list here the basic properties of the **error function** [ABRAMOWITZ-STEGUN]

$$\operatorname{erf}(z) = \frac{2}{\sqrt{\pi}} \int_0^z e^{-s^2} ds \tag{27}$$

$$\operatorname{erf}(0) = 0, \qquad \operatorname{erf}(\infty) = 1, \tag{28}$$

$$\operatorname{erf}(-z) = \operatorname{erf}(z) \tag{29}$$

$$\frac{d}{dz}\operatorname{erf}(z) = \frac{2}{\sqrt{\pi}} e^{-z^2} > 0, \tag{30}$$

$$\operatorname{erf}(z) = \frac{2}{\sqrt{\pi}} \left(z - \frac{z^3}{3 \cdot 1!} + \frac{z^5}{5 \cdot 2!} - \frac{z^7}{7 \cdot 3!} + \cdots\right), \tag{31}$$

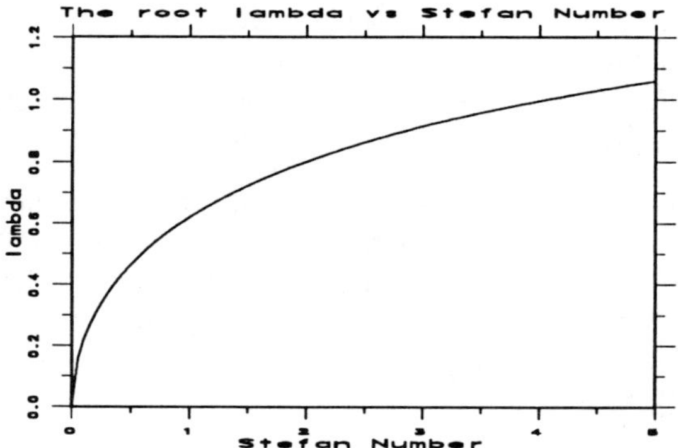

Figure 2.1.3. The root λ of (26) vs the Stefan Number.

$$\operatorname{erf}(z) \approx 1 - \frac{e^{-z^2}}{z\sqrt{\pi}}\left(1 - \frac{1}{2z^2} + \cdots\right) \quad \text{as } z \to \infty, \tag{32}$$

$$\text{complementary error function:} \qquad \operatorname{erfc}(z) = 1 - \operatorname{erf}(z)\,. \tag{33}$$

Extensive tables of values appear in [ABRAMOWITZ-STEGUN], but many Fortran libraries contain erf(z). A useful analytical approximation to the error function is given by the following relation:

$$\operatorname{erf} z = \begin{cases} 1.128z\,, & 0 \le z \le 0.15 \\ -0.0198 + z(1.2911 - 0.4262z)\,, & 0.15 \le z \le 1.5 \\ 0.8814 + 0.0584z\,, & 1.5 \le z \le 2 \\ 1\,, & 2 \le z \end{cases} \tag{34}$$

In the same spirit, an effective approximation to the root λ of (26) is given by the expression

$$\lambda \approx 0.706\sqrt{St}\,\{1 - 0.21\,(0.5642\cdot \mathbf{St})^{0.93-0.15St}\}\,. \tag{35}$$

This relation has less than a 1% relative error for $0 < \mathbf{St} < 0.83$, a relative error below 5% for $0.83 < \mathbf{St} < 4.28$ and below 10% for $\mathbf{St} < 4.86$.

2.1.E Example: Melting a slab of ice

A slab of ice is 10 cm thick. It is initially solid and at its melt temperature of 0°C. One face of the slab is insulated while from the initial moment $t = 0$ of our experiment, the other will be set at the warm temperature of 25 ° C and maintained at this value for all time. We have placed three thermocouples in the slab, at depths of 1 cm, 3 cm and 5 cm. We wish to know the time of melt of the portions up to each thermocouple location, as well as the melting time for the entire slab. We wish also to learn about the appearance of the time-temperature measurements provided by the thermocouples, as well as the appearance of the temperature distribution as a function of position. The information that we seek is provided by the relations (14) (for the melt front), and (15) (for the temperature distribution). Before we hasten to compute, however, let us consider what we need to know and what we are ignoring in our idealized melting model.

Firstly, this problem is not just a "textbook" question, but one that appears in various forms in a variety of applications. Three analogous cases that come to mind are the thawing of food, the freezing or melting of the ground under a highway, and the freezing of ground around an earth-based heat exchanger for a heat pump. Even if the basic geometry of the process is not slab-like, the slab geometry may be a good approximation to it. Thus for a large portion of the melting process of a rectangular region, the corners do not effect the process very much and it may be considered as if uncoupled melting or freezing processes are taking place at each face. Similarly unless the pipe radius is very small the freezing or melting around a pipe is roughly speaking, slab-like.

The words "roughly speaking" as used above are meaningful. While one may strive for unlimited accuracy (and indeed, in the absence of that goal, a "rule of thumb approach" will be questionable), nevertheless heat transfer process simulation carries with it the burden of many sources of inaccuracy. These include the lack of accurate thermophysical parameter values [TOULOUKIAN], the simplifications needed to apply the tools of mathematics to the goal of simulation, and the simplifications needed to carry out experiments. The latter may arise, for example, from the changes of density of a material under a change of phase, etc. (see PROBLEMS 11-13). Besides, rough, first-cut approximations are informative, and, possibly, sufficient in some circumstances.

In our example we are ignoring the thermal effects of the change of density, which is reasonable for small temperature gradients and small volumes. For the only effect of a density change is to replace less dense ice by denser water, thus in effect "pulling" the material towards the heating face. This action would induce convection in the liquid region (which is negligible due to the smallness of the region involved), while the solid remains at the melt temperature for all time. The mechanical effect of this action would, of course, be to buckle the container (if it is tightly sealed) at the far end, (.1 m), an effect with which we are not concerned ! In any case, density change effects will be discussed in §**2.3**.

The assumption of the initial temperature being at the melting point is difficult to attain in practice but may be "nearly" reached. As we will see in §**2.2**, the estimate of the melt-depth that we will obtain will be greater than that obtained when the initial subcooling is indeed taken into account (see §**2.2**).

Since our process is only "one-phase", we only need the relevant properties of water which are: melt temperature = $T_m = 0\ °C$, density = $\rho = 1\ g/cm^3$, specific heat = $c_L = 4.1868\ J/g°C$, conductivity = $k_L = 0.564 \times 10^{-2}\ J/cm\,s\,°C$, thermal diffusivity $= \alpha_L = k_L / \rho c_L = 1.347 \times 10^{-4}\ cm^2/s$, latent heat $= L = 333.4\ J/g$. We then compute the Stefan number $\boldsymbol{St}$ and obtain the melt front history $X(t)$ and the temperature distribution $T(x,t)$ from (14-16).

The temperature drop for our process is $\Delta T_L = T_L - T_m = 25°C$, and thus the Stefan number is $St = c_L \, \Delta T_L/L = 0.314$. Using the Newton-Raphson method of PROBLEM 9 we find the root λ of (16) to be $\lambda = 0.3777$. A much simpler method for solving the transcendental equation is to use relation (35), giving us the value $\lambda = 0.3776$. If you are really "in a hurry" to obtain a "back of the envelope" estimate, the value of $\sqrt{St/2}$ is 0.3962 (see **§2.1.F**) with a relative error of 4.9%, which is well within the needs of a reasonable "sizing" estimate.

From (14), the time needed for the melting front to reach a given depth X is

$$t_{melt} = X^2 \; / \; (4\,\lambda^2\,\alpha_L) \;. \tag{36}$$

Let t^1_{melt} , t^2_{melt} , t^3_{melt} and t^4_{melt} be the times needed for the melt front $X(t)$ to reach the thermocouples at depths $X = 1$, 3 , 5 centimeters, and the right hand face at 10 centimeters, respectively. Substitution into (36) yields

$$\begin{aligned} t^1_{melt} &= 1301.44 \text{ sec} = 0.36 \text{ hr} \\ t^2_{melt} &= 11713 \text{ sec} \;\;= 3.25 \text{ hr} \\ t^3_{melt} &= 32536 \text{ sec} \;\;= 9.04 \text{ hr} \\ t^4_{melt} &= 130144 \text{ sec} \;= 36.15 \text{ hr} \end{aligned}$$

In **Figure 2.1.4** we see the simulated thermocouple readings at the three depths where they were assumed placed. Note that the curves are all convex downward, and heading asymptotically to the wall value of 25 °C. **Figure 2.1.5** is interesting. It shows temperature distributions in the liquid at the times t^1_{melt}, t^2_{melt} and t^3_{melt}, that to all intents and purposes are linear in the spatial variable. This is a particular case of the general "rule of thumb" that for small values of the Stefan number St, the temperature in the phase change process is at any time essentially at its steady state, (quasistationary, see (38) and **§3.1**) because of the quickness of the response of temperature relative to the movement of the phase change front. In **Figure 2.1.6** we see the moving front as a function of time.

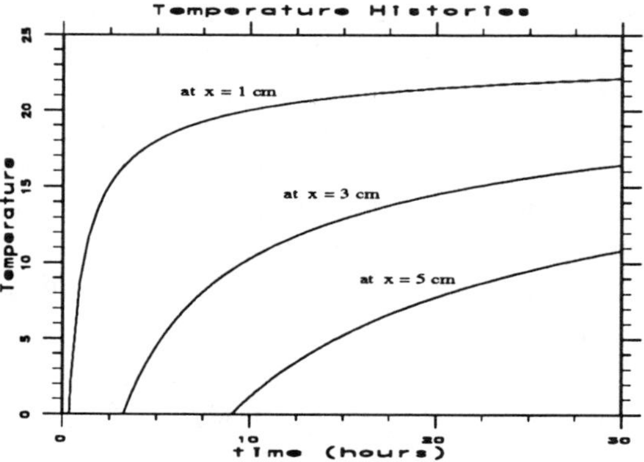

Figure 2.1.4. Melting of ice: temperature histories at three depths.

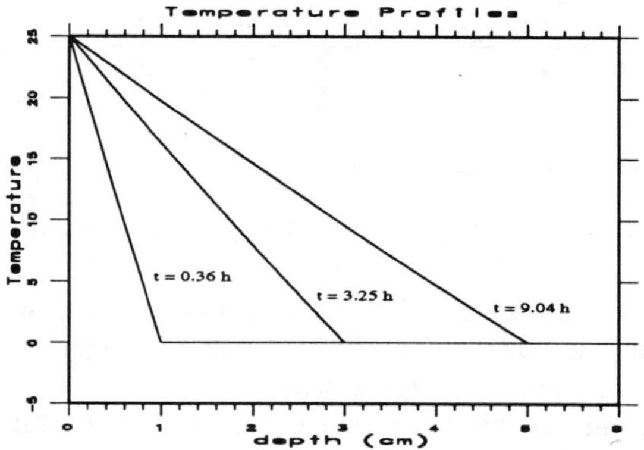

Figure 2.1.5. Melting of ice: temperature profiles at three times.

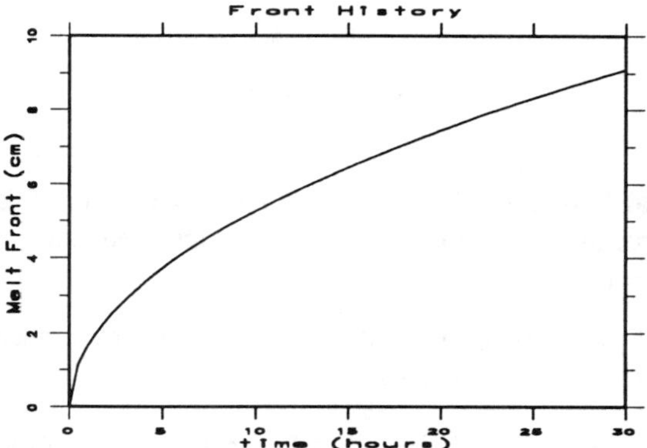

Figure 2.1.6. Melting of ice: interface location.

2.1.F The case of small Stefan Number

For many materials of interest the specific heat is considerably smaller than the latent heat so that the Stefan number for processes with moderate ΔT is of the order 10^{-1} or less. Consider equation (26). For $\boldsymbol{St} \approx 0$, λ must also be small, and by (31) $\lambda\, e^{\lambda^2} \text{erf}\lambda \approx \lambda \cdot 1 \cdot \frac{2}{\sqrt{\pi}} \lambda$; hence (26) is approximately $\frac{2}{\sqrt{\pi}} \lambda^2 = \frac{\boldsymbol{St}}{\sqrt{\pi}}$, or

$$\lambda \approx \sqrt{\frac{\boldsymbol{St}}{2}} \quad \text{for} \quad \boldsymbol{St} \approx 0. \tag{37}$$

It also follows that for $0 \le x \le X(t) = 2\lambda\sqrt{\alpha_L t}$ the quantity $\frac{x}{2\sqrt{\alpha t}} \le \lambda \ll 1$, whence, by (31),

$$\frac{\operatorname{erf}(\frac{x}{2\sqrt{\alpha_L t}})}{\operatorname{erf}\lambda} \approx \frac{\frac{2}{\sqrt{\pi}}\frac{x}{2\sqrt{\alpha_L t}}}{\frac{2}{\sqrt{\pi}}\lambda} = \frac{x}{X(t)},$$

and the Neumann temperature (15) becomes

$$T(x,t) \approx T_L - \Delta T_L \frac{x}{X(t)}, \quad 0 \le x \le X(t), \quad t \ge 0. \tag{38}$$

For each $t > 0$ this is linear in x, i.e. the temperature profile at each time is a straight line joining the point $(x = 0, T = T_L)$ with $(x = X(t), T = T_m)$. This is the reason for the linear profile in **Figure 2.1.5**.

Note that (38) satisfies the steady-state equation $T_{xx} = 0$, while the Neumann temperature satisfies $T_{xx} = \frac{1}{\alpha_L} T_t$. It is an approximate solution to (1-4), valid when $St_L \approx 0$ and it is called the *quasistationary approximation*, the subject of **§3.1, 3.2**.

PROBLEMS

PROBLEM 1. (a) Formulate the 1-phase Stefan problem for a slab initially liquid at T_m, freezing from the left due to an imposed constant temperature $T_S < T_m$ at $x = 0$.

(b) Verify that the freezing problem results formally by replacing every subscript L by S and L by $-L$ in (1-4).

PROBLEM 2. Show that $f(\lambda) = \lambda e^{\lambda^2} \operatorname{erf}\lambda$, $\lambda > 0$, is strictly increasing $[f'(\lambda) > 0$ for any $\lambda > 0]$.

PROBLEM 3. Verify the Neumann solution, i.e. that (14-16) satisfy (1-4).

PROBLEM 4. For the case of freezing, in PROBLEM 1,

(a) Seek the similarity solution in the form: $X(t) = 2\lambda\sqrt{\alpha_S t}$, $T(x,t) = F(\xi)$, $\xi = x / \sqrt{t}$, and show that, with λ satisfying $\lambda e^{\lambda^2} \operatorname{erf}\lambda = St_S / \sqrt{\pi}$,

$$T(x,t) = T_S + [T_m - T_S] \operatorname{erf}(\frac{x}{2\sqrt{\alpha_S t}}) / \operatorname{erf}\lambda, \quad 0 \le x \le X(t), \quad t \ge 0,$$

(b) Verify that the solution for freezing results formally from the solution for melting by the formal substitutions mentioned in PROBLEM 1 (b). Note in particular, that the equation for λ is the same.

PROBLEM 5. Show that (5) is the only possible similarity variable for the heat equation (1) of the form $\xi = x^\gamma t^\delta$.

PROBLEM 6. Derive the dimensionless form (19-23) of the 1-phase Stefan Problem (1-4) for the variables (17-18).

PROBLEM 7. (a) Seek the similarity solution of (19-23) in the form $\Sigma(Fo) = 2\lambda\sqrt{Fo}$, $u(\zeta, Fo) = F(\xi)$, $\xi = \frac{\zeta}{\sqrt{Fo}}$, and show that this leads to (24-26).

(b) By direct change of variables, obtain the solution in physical variables (14-15) from the dimensionless solution (24-25).

PROBLEM 8. Repeat PROBLEMS 6,7 for the freezing case (see PROBLEMS 1,4).

PROBLEM 9. Write and implement a numerical scheme for solving (26) based on the Newton-Raphson method

$$\lambda_{n+1} = \lambda_n - \frac{f(\lambda_n)}{f'(\lambda_n)}, \quad \lambda_0 = \sqrt{St/2},$$

where $f(\lambda) = \lambda e^{\lambda^2} \operatorname{erf} \lambda - \frac{St}{\sqrt{\pi}}$, and produce a table of values of the root λ for $0 < St < 5$, $St = .01, .02, \cdots$.

PROBLEM 10. Freezing of water: Repeat the work of **§2.1.E** for the freezing of water initially at its melting point and subject to a face temperature of -25 ° C. In this range, the properties of ice may be taken as: $\rho_S = 916.8\ kg/m^3$, $c_S \approx 2\ kJ/kg\ K$, $k_S \approx .58\ kJ/m\ s\ K$, hence $\alpha_S = 3.16 \times 10^{-4}\ m^2/s$. You may use the approximation (35) for the root of equation (26), and the approximation of (34) in the evaluation of the temperature function.

PROBLEM 11. A box has been constructed to house an experiment in melting and freezing of materials. The experiments are to be "one-dimensional", in the sense that all but one of the box's faces are insulated; the non-insulated face is Aluminum through which Copper tubes carrying a cooling/heating fluid pass. What problems can arise from the change of density of the material that takes place when it melts or freezes? How would you deal with these problems? What difficulties would arise in your modeling efforts as a result of your handling of these problems.

PROBLEM 12. In the course of doing the experiments of the last PROBLEM you encounter a material which in its liquid phase dissolves large amounts of air. What may happen to your experiment?

PROBLEM 13. In the course of performing a melting experiment and recording the temperature values read by a thermocouple held along a thin wire across the box of PROBLEM 11, you find that the temperature value "jumped" through a short temperature range discontinuously. The range begins at the melt temperature, the density of the solid is greater than that of the liquid, and the thermocouple and recording equipment are in order. What could have caused the jump?.

PROBLEM 14. Let us stretch our imagination to the following (physically) imaginary case: A slab of ice at the melt temperature is to be melted via an imposed face temperature of 36000°C. How much will melt in 10 minutes? Explain your result.

2.2. THE TWO-PHASE PROBLEM ON A SEMI-INFINITE SLAB

In §**2.1** we examined the explicit solution to the one-phase problem on a semi-infinite slab corresponding to a uniform initial temperature T_m and an initially solid phase (for a melting problem), or liquid phase (for a freezing problem). A more realistic scenario is one in which the initial state of the PCM, say for a melting process, is solid, but its initial temperature is some value T_S below T_m. This is the case discussed in §**1.2** and the subject of this section. For the problem to be *explicitly* solvable, it is necessary to assume that the slab is semi-infinite. While this would seem to rule out its utility for problems on a finite slab, the slow heat conduction and phase change process found in most actual melting, freezing and casting processes make the semi-infinite case a reasonable approximation to that of the finite interval case (see §**2.2.C**). The assumptions listed in §**1.2.B** are assumed to hold.

2.2.A Problem statement and solution

As prototype 2-phase process we consider the following

PHYSICAL PROBLEM: Melting of a semi-infinite slab, $0 \le x < \infty$, initially **solid** at a **uniform** temperature $T_S \le T_m$, by imposing a **constant** temperature $T_L > T_m$ on the face $x = 0$. Thermophysical parameters: ρ, c_L, c_S, k_L, k_S, L, $\alpha_L = k_L/\rho c_L$, $\alpha_S = k_S/\rho c_S$, all constant (see §**1.2**).

The mathematical model of this process, derived in §**1.2** is the following:

Two-phase Stefan Problem (for a semi-infinite slab melting from the left):

Find a temperature distribution $T(x, t)$ and an interface function $X(t)$ satisfying the following conditions (**Figure 2.2.1**):

Heat equation in melt region

$$T_t = \alpha_L T_{xx}, \quad 0 < x < X(t), \quad t > 0, \tag{1a}$$

Heat equation in solid region

$$T_t = \alpha_S T_{xx}, \quad X(t) < x, \quad t > 0, \tag{1b}$$

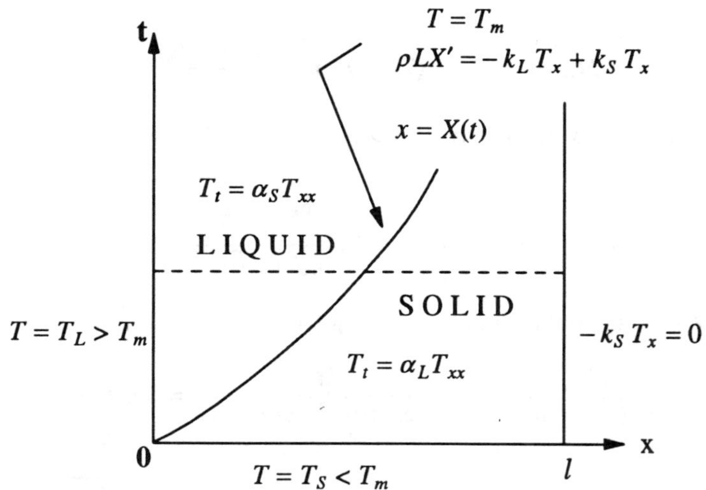

Figure 2.2.1. Space-time diagram for the Two-Phase Stefan Problem.

Interface temperature

$$T(X(t), t) = T_m, \qquad t > 0, \tag{1c}$$

Stefan condition

$$\rho L X'(t) = -k_L T_x(X(t)^-, t) + k_S T_x(X(t)^+, t), \quad t > 0, \tag{1d}$$

Initial conditions

$$T(x, 0) = T_S < T_m, \quad x > 0, \qquad X(0) = 0, \tag{1e}$$

Boundary conditions

$$T(0, t) = T_L > T_m, \qquad \lim_{x \to \infty} T(x, t) = T_S, \quad t > 0. \tag{1f}$$

Recall (§1.2) that the right-hand side of the Stefan condition (1d) is the flux jump $q_L - q_S$ at the interface $x = X(t)$. The notation $T_x(X(t)^{\mp}, t)$ serves to remind us that these are *limiting* values of $T_x(x, t)$ as $x \to X(t)^{\mp}$ (from the left (liquid) and from the right (solid)).

Because of the structure of the problem we can again find a solution in terms of the similarity variable $\xi = x/\sqrt{t}$. Guided by the 1-phase case, we seek the solution in the form $X(t) = 2\lambda\sqrt{\alpha_L t}$, $T(x,t) = F_L(\xi)$ in the liquid and $T(x, t) = F_S(\xi)$ in the solid, with λ an unknown constant and F_L, F_S unknown functions of the similarity variable ξ. Using the procedure of §2.1.B, we obtain (PROBLEM 2) the

Neumann solution of the 2-phase Stefan Problem (1):

Interface location

$$X(t) = 2\lambda\sqrt{\alpha_L t}, \qquad t > 0 \tag{2a}$$

Temperature in the liquid region $\quad 0 < x < X(t), \qquad t > 0:$

$$T(x,t) \;=\; T_L - (T_L - T_m)\,\frac{\operatorname{erf}\!\left(\dfrac{x}{2\sqrt{\alpha_L t}}\right)}{\operatorname{erf}\lambda}\,, \tag{2b}$$

Temperature in the solid region $\quad x > X(t), \qquad t > 0:$

$$T(x,t) \;=\; T_S + (T_m - T_S)\;\frac{\operatorname{erfc}\!\left(\dfrac{x}{2\sqrt{\alpha_S t}}\right)}{\operatorname{erfc}(\lambda\sqrt{\alpha_L/\alpha_S})}\,. \tag{2c}$$

Here λ is the solution to the transcendental equation

$$\frac{\mathbf{St}_L}{\exp(\lambda^2)\ \operatorname{erf}(\lambda)} \;-\; \frac{\mathbf{St}_S}{\nu\ \exp(\nu^2\lambda^2)\ \operatorname{erfc}(\nu\,\lambda)} \;=\; \lambda\sqrt{\pi}, \tag{2d}$$

with

$$\mathbf{St}_L = \frac{c_L(T_L - T_m)}{L}\,, \qquad \mathbf{St}_S = \frac{c_S(T_m - T_S)}{L}\,, \qquad \nu = \sqrt{\frac{\alpha_L}{\alpha_S}}\;. \tag{3}$$

By PROBLEM 5, the transcendental equation (2d) has exactly one root $\lambda > 0$, and therefore the similarity solution (2) is unique for each $\mathbf{St}_L > 0$, $\mathbf{St}_S \geq 0$, $\nu > 0$. The fact that this is the *only* solution follows from the general uniqueness theory (see **§4.4**) or it may be proved directly [RUBINSTEIN].

Note that when $T_S = T_m$, we have $\mathbf{St}_S = 0$ and (2) reduces to the similarity solution of the 1-phase problem (**§2.1.B**), as expected. The presence of the term containing $\mathbf{St}_S$ in (2d) simply reduces the magnitude of the root and therefore, for any $\mathbf{St}_S > 0$, we have

$$\lambda_{2-phase} < \lambda_{1-phase}\ . \tag{4}$$

This of course is expected for it says merely that the presence of initial subcooling in the solid will slow down the melting process (given by $X(t) = 2\,\lambda\sqrt{\alpha_L t}$), since some heat must go to raising the temperature of the solid to T_m before it can melt.

The Neumann solution for the case of freezing (PROBLEM 1) may be formally obtained from (2) by simply interchanging the subscripts L and S and replacing the latent heat L by $-L$ (PROBLEM 7).

2.2.B Dimensionless form

The 2-phase problem in physical variables, (1), contains nine parameters, namely, ρ, c_L, c_S, k_L, k_S, L, T_m, T_L, T_S. Undimensionalization reduces the number to four, the minimum necessary to specify the problem. Indeed, set (see (3))

$$\mathbf{St}_L = \frac{c_L(T_L - T_m)}{L}\,, \quad \mathbf{St}_S = \frac{c_S(T_m - T_S)}{L}\,, \quad \nu = \sqrt{\frac{\alpha_L}{\alpha_S}}\,, \tag{5}$$

and with $\hat{x}$ being any convenient length (no *natural* length scale is present here), introduce the dimensionless variables: $\zeta = x/\hat{x}$, $Fo = \alpha_L t/\hat{x}^2$, and

$$\Sigma(Fo) = \frac{X(t)}{\hat{x}}, \quad u^L(\zeta, Fo) = \frac{T(x,t) - T_m}{T_L - T_m}, \quad u^S(\zeta, Fo) = \frac{T(x,t) - T_m}{T_m - T_S}. \tag{6}$$

Then, problem (1) takes the form (PROBLEM 8)

$$u^L_{Fo} = u^L_{\zeta\zeta}, \qquad 0 < \zeta < \Sigma(Fo), \quad Fo > 0 \text{ (liquid)}, \tag{7a}$$

$$\nu^2 u^S_{Fo} = u^S_{\zeta\zeta}, \quad \Sigma(Fo) < \zeta < \infty, \quad Fo > 0 \text{ (solid)}, \tag{7b}$$

$$u^L(\Sigma(Fo), Fo) = u^S(\Sigma(Fo), Fo) = 0, \qquad Fo > 0, \tag{7c}$$

$$\Sigma'(Fo) = -\mathbf{St}_L \cdot u^L_\zeta + (\mathbf{St}_S/\nu^2)\cdot u^S_\zeta, \quad \zeta = \Sigma(Fo), \quad Fo > 0, \tag{7d}$$

$$u^S(\zeta, 0) = -1, \quad 0 < \zeta < \infty, \qquad \Sigma(0) = 0 \tag{7e}$$

$$u^L(0, Fo) = +1, \qquad \lim_{\zeta\to\infty} u^S(\zeta, Fo) = -1, \qquad Fo > 0, \tag{7f}$$

which contains only the three parameters defined in (5).

Its similarity solution is easily found to be (PROBLEM 10)

$$\Sigma(Fo) = 2\lambda\sqrt{Fo} \tag{8a}$$

$$u^L(\zeta, Fo) = +1 - \frac{\operatorname{erf}\left(\frac{\zeta}{2\sqrt{Fo}}\right)}{\operatorname{erf}\lambda}, \quad 0 \le \zeta \le \Sigma(Fo), \quad Fo > 0, \text{ (liquid)} \tag{8b}$$

$$u^S(\zeta, Fo) = -1 + \frac{\operatorname{erfc}\left(\nu\frac{\zeta}{2\sqrt{Fo}}\right)}{\operatorname{erfc}(\nu\lambda)}, \quad \Sigma(Fo) \le \zeta, \quad Fo > 0, \text{ (solid)} \tag{8c}$$

with λ the unique root of the (already dimensionless) equation (2d).

2.2.C Approximations to the root λ

We know from **§2.1.F** that in the 1-phase case, $\mathbf{St}_L \approx 0$ implies $\lambda_{1\text{-}phase} \approx \sqrt{\mathbf{St}_L/2}$. Since, by (4), $\lambda_{2\text{-}phase} < \lambda_{1\text{-}phase}$ always holds, $\mathbf{St}_L \approx 0$ implies $\lambda_{2\text{-}phase} \approx 0$. To the lowest order, equation (2d) is approximately $\mathbf{St}_L/(2\lambda^2) - \mathbf{St}_S/(\nu\lambda\sqrt{\pi}) = 1$, whence

$$\lambda_{2\text{-}phase} \approx \frac{1}{2}\left[-\frac{\mathbf{St}_S}{\nu\sqrt{\pi}} + \sqrt{2\mathbf{St}_L + \left(\frac{\mathbf{St}_S}{\nu\sqrt{\pi}}\right)^2}\,\right] \quad \text{for} \quad \mathbf{St}_L \approx 0. \tag{9a}$$

If, in addition, $\mathbf{St}_S >> \mathbf{St}_L \approx 0$, this simplifies to

$$\lambda_{2\text{-}phase} \approx \frac{\nu\sqrt{\pi}}{2}\,\frac{\mathbf{St}_L}{\mathbf{St}_S}. \tag{9b}$$

An approximation, analogous to (35) §**2.1**, applicable to a narrow but useful range of situations is the following:

$$\lambda_{2-phase} \approx 0.706\sqrt{St}\,\{1 - [0.21 + U(0.51 - 0.169\,St)]\cdot(0.5642\,St)^B\} \quad (9c)$$

which is valid when $\alpha_L = \alpha_S$ for $St + 0.8\,U \le 2$, where $St := \mathbf{St}_L$, $U = (T_m - T_S)/(T_L - T_m)$, and $B = \dfrac{0.93}{1 + 0.69\,U^{0.7}} - 0.15\,St$. Its error is less than 10% and usually less than 3-5%.

Highly accurate values may be found numerically, using, for example, Brent's method [PRESS et al] or even plain bisection. Note that most FORTRAN libraries already contain the error function.

Some materials have extremely small latent heat of melting. Considering $L = 0$ as an approximation to such a case, (2d) reduces to

$$\frac{c_L(T_L - T_m)}{e^{\lambda^2}\,\mathrm{erf}\,\lambda} = \frac{c_S(T_m - T_S)}{\nu\, e^{\nu^2\lambda^2}\,\mathrm{erfc}(\nu\lambda)} . \quad (9d)$$

Note that then by (1d), there will be no flux jump at the interface, only the specific heat and conductivity may have jumps. The interface will simply be the isotherm $T = T_m$.

2.2.D Approximating the finite slab case

It is of interest and useful to know when we may consider a finite slab, $0 \le x \le l$, as being semi-infinite. Clearly, if the back face, $x = l$, is anything but insulated then there is an active boundary condition there which influences the temperature throughout the slab immediately (§**1.2**) and no semi-infinite approximation is possible. With $q(l, t) = 0$, the question is up to what time will the Neumann solution approximately satisfy this boundary condition? In other words, given $\varepsilon > 0$, we want the time up to which

$$q(l, t) = -k_S T_x(l, t) = \frac{k_S\,\Delta T_S}{\sqrt{\pi\alpha_S t}\; e^{l^2/4\alpha_S t}\,\mathrm{erfc}(\nu\lambda)} < \varepsilon , \quad (10)$$

$\Delta T_S = T_m - T_S$. Using $e^{l^2/4\alpha_S t} > l^2/4\alpha_S t$, (10) will certainly hold up to time t^* given by

$$t < t^* := \varepsilon^2\left(\frac{\pi}{\alpha_S}\right)\left(\frac{l^2\,\mathrm{erfc}(\nu\lambda)}{4k_S\,\Delta T_S}\right)^2 . \quad (11)$$

EXAMPLE 1 : In melting a slab of ice of thickness $l = 1\ m$, initially at $T_S = -10°C$, via $T_L = 25°C$ at $x = 0$, an estimate of the time up to which the Neumann solution flux at $x = l$ remains less than $\varepsilon = 10^{-3} kJ/m^2\,s$ is $t^* \approx 20 \times 10^{-6}$ seconds ! (PROBLEM 13).

An alternative to keeping the flux at $x = l$ small is to ask for how long does the (solid) flux at $x = l$, $q_S(l,t)$, stay a small percentage of the incoming flux, $q_L(0,t)$, namely

$$\left| \frac{q_S(l,t)}{q_L(0,t)} \right| \leq \varepsilon \,. \tag{12}$$

We find that this will happen up to time (PROBLEM 12)

$$t \leq t^{**} := \frac{l^2}{4\alpha_S \ln \frac{A}{\varepsilon}} \quad \text{with } A = \nu \frac{k_S \Delta T_S}{k_L \Delta T_L} \frac{\operatorname{erf} \lambda}{\operatorname{erfc}(\nu\lambda)} \,, \tag{13a}$$

which may be roughly estimated, using $\operatorname{erf} \lambda > \frac{2}{\sqrt{\pi}}(\lambda - \frac{\lambda^3}{3})$, $\operatorname{erfc}(\nu\lambda) < 1$, by

$$t \leq \frac{l^2}{4\alpha_S \ln \frac{B}{\varepsilon}} \,, \qquad B = \nu \frac{k_S \Delta T_S}{k_L \Delta T_L} \frac{2}{\sqrt{\pi}} (\lambda - \frac{\lambda^3}{3}) \,. \tag{13b}$$

EXAMPLE 2: For the situation of EXAMPLE 1, we have $B \approx 0.94$, so (12) will hold with $\varepsilon = 1\%$ at least up to time $t^{**} \approx 186\ s$, and with $\varepsilon = 10^{-6}$ up to 61 seconds (PROBLEM 13). Yet another possibility is described in PROBLEM 15.

In order to use the Neumann solution as a debugging tool for a numerical simulation of phase-change in a finite slab, we do not have to rely to such approximations. Instead, one may impose at the back face $x = l$ the Neumann temperature itself. This neutralizes the effect of the back face and direct comparison of the computed temperatures and front location with the Neumann solution is meaningful. The same could be done experimentally if one had the means to exactly control the time varying back-face temperature, which is difficult to achieve.

2.2.E Energy content and Stefan numbers

In the 2-phase Stefan Problem of §2.2.A, at any $t > 0$, the interval $[0, X(t))$ is occupied by liquid and $[X(t), \infty)$ by solid. The total energy (heat) in the system consists of the sensible heat of the solid and of the liquid and the latent heat of liquid. Taking T_S as the (reference) temperature of zero energy (in order to have zero energy at $x = \infty$), we have (per unit crossectional area)

$$\text{sensible heat of liquid:} \quad E_L^{sens}(t) = \int_0^{X(t)} \rho c_S [T_m - T_S] dx + \int_0^{X(t)} \rho c_L [T(x,t) - T_m] dx \,, \tag{14a}$$

$$\text{sensible heat of solid:} \quad E_S^{sens}(t) = \int_{X(t)}^{\infty} \rho c_S [T(x,t) - T_S] dx \,, \tag{14b}$$

Latent heat of liquid: $\quad E^{lat}(t) = \rho L X(t)\,.$ (14c)

For the Neumann solution, (2), these turn out to be (PROBLEM 18)

$$E_L^{sens}(t) = \rho\, L\, \mathbf{St}_S\, X(t) + \rho\, L\, \mathbf{St}_L\, X(t)\,\frac{1 - e^{-\lambda^2}}{\sqrt{\pi}\,\lambda\,\mathrm{erf}\,\lambda}\,, \tag{15a}$$

$$E_S^{sens}(t) = \rho\, L\, \mathbf{St}_S\, X(t)\left[\frac{1}{\sqrt{\pi}\,\nu\,\lambda\, e^{(\nu\lambda)^2}\mathrm{erfc}(\nu\lambda)} - 1\right]. \tag{15b}$$

At time $t = 0$, the system was solid at T_S, so had energy zero (by our choice of T_S as the reference temperature); the only heat that came in up to time $t > 0$ is (PROBLEM 19)

$$Q(t) := \int_0^t q(0, s)ds = \int_0^t -k_L T_x(0, s)ds = \frac{\rho\, L\, \mathbf{St}_L X(t)}{\sqrt{\pi}\,\lambda\,\mathrm{erf}\,\lambda}\,. \tag{16}$$

One may easily verify the heat balance (PROBLEM 20)

$$E_L^{sens} + E_S^{sens} + E^{lat} = Q\,. \tag{17}$$

It is interesting to note that the ratios of sensible to latent heats are constants (independent of time). Indeed, from (15) and (2d), we find, for example,

$$\frac{E_L^{sens}(t)}{E^{lat}(t)} = \mathbf{St}_S + \mathbf{St}_L\,\frac{1 - e^{-\lambda^2}}{\sqrt{\pi}\,\lambda\,\mathrm{erf}\,\lambda}\,, \tag{18a}$$

$$\frac{E_L^{sens} + E_S^{sen}}{E^{lat}} = \frac{\mathbf{St}_L}{\sqrt{\pi}\,\lambda\,\mathrm{erf}\,\lambda} - 1. \tag{18b}$$

When $\mathbf{St}_L \approx 0$, the right-hand side of (18a) is $\approx \mathbf{St}_S + \frac{\mathbf{St}_L}{2}$, (PROBLEM 22), and therefore *in the 1-phase case (i.e. $\mathbf{St}_S = 0$), the Stefan number represents twice the ratio of sensible to latent heat* (PROBLEM 23).

The above relationships may be used as simple checks on the validity of computer codes as well as in finding the heat stored in the system, see **§2.2.G**.

2.2.F Shape of melting and cooling curves

Placing a thermocouple at a location x^*, an experimenter records the temperature of that location over time. Plotting these temperatures against time produces a *melting* or *cooling curve*, $T = T^*(t)$. This is the most precisely and easily measurable quantity in a phase-change experiment, so it is important to know what to expect. The theoretical melting curve, corresponding to the experimental one, is the curve $T = T(x^*, t)$, with x^* the fixed thermocouple location. We are interested in its shape and qualitative features.

As it is easier to work in dimensionless variables, we set

$$\zeta^* = \frac{x^*}{\hat{x}} \quad \text{and} \quad \tau^* = \left(\frac{\zeta^*}{2\lambda}\right)^2 = \text{melt time of } \zeta^* , \tag{19}$$

and examine the Neumann solution (8), of our melting problem (1). For notational convenience we denote dimensionless time (Fourier Number) by τ.

During $0 \le \tau \le \tau^*$, the point $\zeta = \zeta^*$ is solid with temperature, (8c),

$$u(\zeta^*, \tau) = -1 + \frac{\operatorname{erfc}(\nu \frac{\zeta^*}{2\sqrt{\tau}})}{\operatorname{erfc}(\nu\lambda)} , \quad 0 \le \tau \le \tau^* , \tag{20}$$

where λ is the root of (2d). Computing $u_\tau(\zeta^*, \tau)$ and $u_{\tau\tau}(\zeta^*, \tau)$, we see that (PROBLEM 24) $u_\tau(\zeta^*, \tau) > 0$, hence $u(\zeta^*, \tau)$ is increasing for $0 \le \tau \le \tau^*$, and

$$u_{\tau\tau}(\zeta^*, \tau) \ge 0 \quad \text{for} \quad \tau \le \hat{\tau} = \frac{\nu^2 \zeta^{*2}}{6} , \tag{21}$$

hence $u(\zeta^*, \tau)$ is a convex curve up to $\tau \le \hat{\tau}$. Thus, we have two possibilities. If $\hat{\tau} > \tau^*$, i.e. if $\nu\lambda > \sqrt{3/2}$, then the melting curve stays convex during $0 \le \tau \le \tau^*$, **Figure 2.2.2(a)**. Otherwise, it changes its concavity at $\tau = \hat{\tau} < \tau^*$, **Figure 2.2.2(b)**.

After ζ^* melts, its temperature follows (8b). Again, we see that (PROBLEM 25) $u_\tau(\zeta^*, \tau) > 0$ and

$$u_{\tau\tau}(\zeta^*, \tau) \le 0 \quad \text{for} \quad \tau \ge \hat{\hat{\tau}} := \frac{\zeta^{*2}}{6} . \tag{22}$$

Hence, if $\hat{\hat{\tau}} < \tau^*$, i.e. if $\lambda < \sqrt{3/2}$, then by (22) the melting curve is concave forever, **Figure 2.2.3(a)**. Otherwise, $u(\zeta^*, \tau)$ is convex during $\tau^* \le \tau \le \hat{\hat{\tau}}$ and concave everafter, **Figure 2.2.3(b)**.

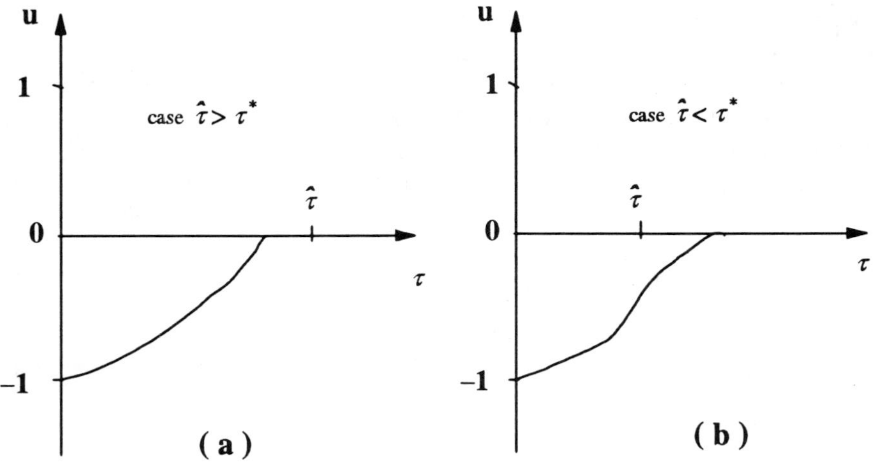

Figure 2.2.2. Shape of melting curve before ζ^* melts.

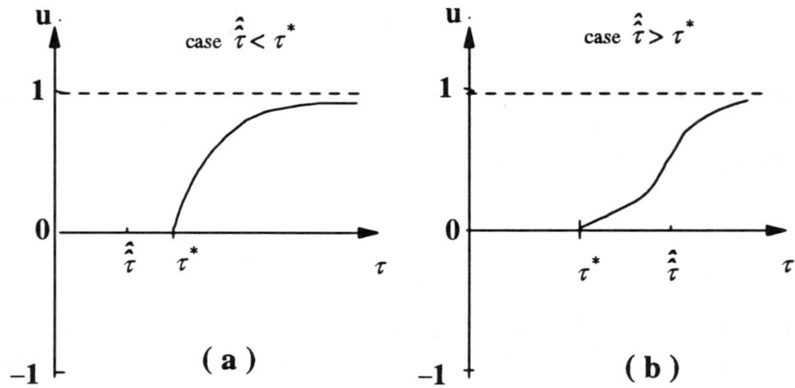

Figure 2.2.3. Shape of melting curve after ζ^* melts.

It follows that the overall melting curve will consist of an appropriate combination of convex and concave time-histories. Clearly, cooling curves are similar in shape but, of course, decreasing.

EXAMPLE 3 : Melting of ice, initially at $T_S = -10°C$, via $T_L = 25°C$ at $x = 0$. With parameter values given in **§2.1.E** and PROBLEM 10 **§2.1** the dimensionless constants are: $\boldsymbol{St}_L = 0.314$, $\boldsymbol{St}_S = 0.06$, $\nu = 0.02137$, and the corresponding root is $\lambda = 0.09178$. Let us imagine a thermocouple at location $\zeta^* = 0.2$. Then, from (19, 21-22) we find $\tau^* = 1.187$, $\hat{\tau} = 3 \times 10^{-6}$, $\hat{\hat{\tau}} = 0.0067$. Since $\hat{\tau} \ll \tau^*$, the melting curve at $\zeta^* = 0.2$ would not exhibit any change in concavity.

When an actual experiment or a computer simulation exhibit qualitatively different melting curves than expected from the above, then one must question the simplifying features of the model used (constant thermophysical properties, neglecting various effects, like supercooling or convection, etc).

On the other hand, if the expected behavior occurs then such curves may be used to check the accuracy of parameter values. For example, the jump in the slope of a melting curve at the melt time is a measurable quantity (with a differential thermal analyzer). For the Neumann solution this jump is, (from PROBLEMS 24-25)

$$[\![u_\tau(\zeta^*, \tau^*)]\!]_{solid}^{liquid} = \frac{\lambda}{\sqrt{\pi}\,\tau^*}\left[\frac{e^{-\lambda^2}}{\text{erf}\,\lambda} - \frac{\nu e^{-\nu^2\lambda^2}}{\text{erfc}(\nu\lambda)}\right]. \tag{23}$$

For $\boldsymbol{St}_L \approx 0$ and $\boldsymbol{St}_S > \boldsymbol{St}_L$ we know that $\lambda \approx \dfrac{\nu\sqrt{\pi}}{2}\dfrac{\boldsymbol{St}_L}{\boldsymbol{St}_S}$ (see (9b)) and, expanding erf to first order in λ, (23) yields

$$[\![u_\tau(\zeta^*, \tau^*)]\!]_{solid}^{liquid} \approx \frac{1}{2\tau^*}\left[1 - \frac{1}{\dfrac{k_S}{k_L}\dfrac{\Delta T_S}{\Delta T_L} - 1}\right], \tag{24}$$

with $\Delta T_S = T_m - T_S$, $\Delta T_L = T_L - T_m$. By measuring the left-hand side, (24) could be used to check the correctness of data values for the ratio of conductivities k_S / k_L.

2.2.G An example

For many years effective means have been sought for storing heat as the latent heat of melting of a material. The prime source of such energy is solar, which is intermittent, and whose energy, derived during sunlit periods, is needed at other times. A material under intense study as a candidate for such a role is Glauber's salt (sodium sulfate decahydrate). Its thermal properties are as follows:
$\rho = 1460\ kg/m^3$, $T_m = 32\,°C$, $L = 251.21\ kJ/kg$, $c_L = 3.31$, $c_S = 1.76$ $(kJ/kg°C)$, $k_L = 0.59 \times 10^{-3}$, $k_S = 2.16 \times 10^{-3}$ $(kJ/m\,s\,°C)$ whence $\alpha_L = 1.22 \times 10^{-7}$, $\alpha_S = 8.4 \times 10^{-7} (m^2/s)$.

Let the face of a long can of Glauber's salt be exposed to a warm temperature $T_L = 90°C$. Initially it is solid at the temperature $T_S = 25°C$. We wish to describe the resulting melting process. Here we have $\mathbf{St}_L = 0.76422$, $\mathbf{St}_S = 0.049$, $\nu = 0.3811$, and solving (2d) numerically, $\lambda = 0.520815$. Hence, the location of the phase change front at any time is $X(t) = 3.64 \times 10^{-4}\sqrt{t}$ meters.

In **Figure 2.2.4** we see the temperature profiles at three times. We note that the temperature is very nearly linear in space with a jump in slope at the interface $x = X(t)$. The variation of the temperature in time at three depths in the material is shown in **Figure 2.2.5**. Again we see a jump in the slope at the time when the phase change takes place. We note the extreme flatness of the curves as we move to greater depth, something that is typical of actual processes. A simple calculation shows that for any x we have $T_t(x,0) = 0$ so, all melting curves start with zero slope; since $(\nu\lambda)^2 < 3/2$ here (**§2.2.F**), there is a concavity change before the melt time, as in **Figure 2.2.2(b)**, but due to scaling this effect cannot be seen in **Figure 2.2.5**.

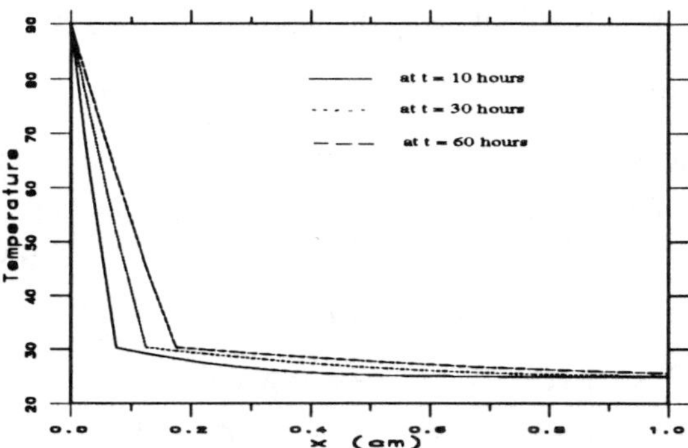

Figure 2.2.4. Temperature profiles at three times.

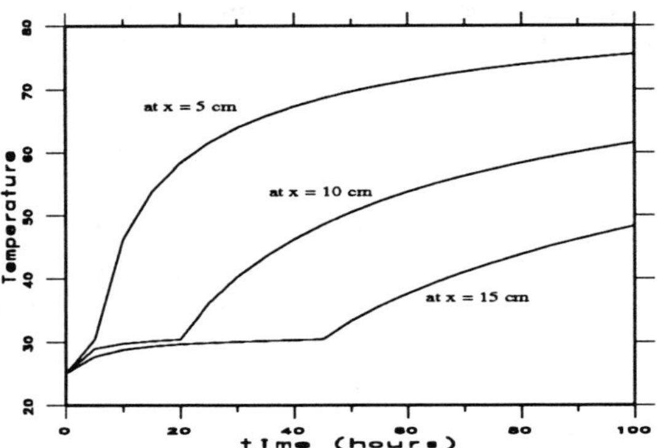

Figure 2.2.5. Temperature histories at three points.

At this point we have "modeled" the process, meaning that the temperature (given by (2b,c) is related in a known way to the thermal properties and the initial and boundary temperature. Of primary interest in heat storage is the "heat inventory" at any time. For example, after 1 hour the melt-depth is $X(3600s) = 0.0218\ m$, and the amount of heat that has entered the system is (from (16)) $Q(3600s) = 12306\ kJ/m^2$. Of this amount, $E^{lat} = \rho LX = 7995.5\ kJ/m^2$ is stored as latent heat and the rest, $4310.5\ kJ/m^2$, as sensible heat (of which, $E_L^{sens} = 3308$ in the liquid and only $E_S^{sens} = 1002.5$ in the semi-infinite solid, by (15)).

PROBLEMS

PROBLEM 1. State precisely the 2-phase Stefan Problem for **freezing** from the left with initial temperature $T_L > T_m$ and imposed temperature $T_S < T_m$ at $x = 0$. Compare with (1).

PROBLEM 2. Derive (2) by following the procedure of **§2.1.B**.

PROBLEM 3. Verify that (2) solves (1).

PROBLEM 4. The function $g(x) := e^{x^2}\operatorname{erfc}(x)$ is *decreasing* while the function $h(x) := x\,e^{x^2}\operatorname{erfc}(x)$ is *increasing*, for $x > 0$. These functions arise in the similarity solutions in **§2.2, 2.3** and **2.4**. Their monotonicity properties may be proved as follows: (a) Set $\phi(x) := e^{-x^2} g'(x) = 2\,x \operatorname{erfc} x - \frac{2}{\sqrt{\pi}} e^{-x^2}$, $x > 0$. Show that $\phi(0) < 0$, $\phi(x) \to 0$ as $x \to \infty$ and $\phi'(x) > 0$ for $x > 0$. Conclude that $\phi(x) < 0$, whence $g'(x) = e^{x^2}\phi(x) < 0$, for $x > 0$.
(b) Set $\psi(x) := e^{-x^2} h'(x)$, $x > 0$. Show that $\psi(0) = 1$, $\psi(x) \to 0$ as $x \to \infty$

and $\psi'(x) = 2e^{-x^2}g'(x) < 0$ (from part (a)). Conclude that $\psi(x) > 0$, so also $h'(x) = \psi(x)e^{x^2}(x) > 0$ for $x > 0$.

PROBLEM 5. To prove that, for any $St_L > 0$, $St_S \geq 0$, $\nu > 0$ equation (2d) has a unique root $\lambda > 0$, set $f(\lambda)$ equal to the left-hand side of (2d), $\lambda > 0$. Use (28-33) of **§2.1** and the previous PROBLEM to show that

(a) $f(0) = +\infty$, $f(\infty) = -\infty$; conclude that the equation $f(\lambda) = \lambda\sqrt{\pi}$ has at least one (positive) solution.

(b) $f(\lambda)$ is strictly decreasing for $\lambda > 0$; conclude that $f(\lambda) = \lambda\sqrt{\pi}$ has exactly one (positive) solution.

PROBLEM 6. Seek an alternative form of the similarity solution by setting $X(t) = 2\lambda\sqrt{\alpha_S t}$ with $T(x,t)$ as before. Compare with (2).

PROBLEM 7. Derive the Neumann solution for the freezing case described in PROBLEM 1.

PROBLEM 8. Derive the dimensionless form (7).

PROBLEM 9. Corresponding to the alternative choice suggested in PROBLEM 6, one may choose as Fourier number (dimensionless time) $Fo = \alpha_S t / \hat{x}^2$. Derive the dimensionless form of (1) with this choice of Fo. Compare with PROBLEM 8.

PROBLEM 10. Seek the similarity solution of (7) in the form $\Sigma = 2\lambda\sqrt{Fo}$, $u(\zeta, Fo) = F(\xi)$, $\xi = \zeta / \sqrt{Fo}$, to obtain (8).

PROBLEM 11. Prove that equation (9d) has unique solution. Using other sources find a material for which a phase change with an ignorable latent heat is of interest and for which the relation is relevant.

PROBLEM 12. Derive (13).

PROBLEM 13. In the situation of Example 1, **§2.2.D**, estimate the time up to which the Neumann solution will be a reasonable approximation to this finite-slab problem, according to the criterion (10) with $\varepsilon = 10^{-3}$, and according to criterion (12) with $\varepsilon = 1\%$.

PROBLEM 14. Repeat PROBLEM 13 with the same ε, but using the approximation (9c) for λ and the first term of (32) **§2.1** for $\mathrm{erfc}(\nu\lambda)$.

PROBLEM 15. As another alternative in **§2.2.D**, consider the problem of **§2.2.C** for the condition that the temperature at the right hand side of a finite slab be closer to the initial temperature T_S than some prescribed tolerance ε. Derive an estimate similar to (10) for the condition that $|T(l,t) - T_S| < \varepsilon$.

PROBLEM 16. Apply the results of (10), of (12), and of PROBLEM 15 to the case of Glauber's salt of **§2.2.G** for $l = 0.5m$. Feel free to use the approximation (34) **§2.1** for the error function.

PROBLEM 17. Integral of the error function. By interchanging the order of integration show that

$$\int_0^X \operatorname{erf}\left(\frac{x}{2\sqrt{\alpha_L t}}\right)dx = \frac{2}{\sqrt{\pi}}\int_0^X \int_0^{\frac{x}{2\sqrt{\alpha_L t}}} e^{-s^2}\,ds\,dx = X \operatorname{erf}\lambda - \frac{2}{\sqrt{\pi}}(1 - e^{-\lambda^2})\sqrt{\alpha_L t}\,,$$

$$\int_X^\infty \operatorname{erfc}\left(\frac{x}{2\sqrt{\alpha_S t}}\right)dx = \frac{2}{\sqrt{\pi}}\int_X^\infty \int_{\frac{x}{2\sqrt{\alpha_S t}}}^\infty e^{-s^2}\,ds\,dx = \frac{2\sqrt{\alpha_S t}}{\sqrt{\pi}}\,e^{-(\nu\lambda)^2} - X \operatorname{erfc}(\nu\lambda)\,.$$

PROBLEM 18. Using the previous PROBLEM, derive relations (15).

PROBLEM 19. Derive relation (16).

PROBLEM 20. Verify the heat balance (17).

PROBLEM 21. Derive relations (18).

PROBLEM 22. Show that when $\lambda \approx 0$, the last term in (18a) to lowest order is $St_L/2$.

PROBLEM 23. In the 1-phase case, approximate the temperature by the average of T_m and T_L to show that the ratio of sensible to latent heat is half the Stefan number, in agreement with the result of PROBLEM 18.

PROBLEM 24. Compute the time derivatives $u_\tau(\zeta^*, \tau)$, $u_{\tau\tau}(\zeta^*, \tau)$ of (20) and verify (21).

PROBLEM 25. Compute the time derivatives $u_\tau(\zeta^*, \tau)$, $u_{\tau\tau}(\zeta^*, \tau)$ of (8b) and verify (22).

PROBLEM 26. Verify the work in the Example of **§2.2.F**.

PROBLEM 27. Derive relation (23) and its approximation (24).

PROBLEM 28. Check the work in **§2.2.G**.

2.3. THE EFFECT OF DENSITY CHANGE

Melting and solidification are generally accompanied by changes in density (**§1.1**). These inescapably induce movement in the material and complicate the phase-change process considerably.

It should be recalled (**§1.2.B**) that one of the basic assumptions in the Stefan Problem is that of constant density, which we now wish to relax. The physical consequences of density variation were briefly mentioned in **§1.1**, and illustrative experiments are described in **§2.3.A**. In the rest of the section we concentrate on the expansion/shrinkage effect due to $\rho_L \neq \rho_S$. All the other assumptions of the Stefan Problem (**§1.2**) are retained and the effect this density change has on the Neumann solution is examined.

Most materials have $\rho_L < \rho_S$, so they *expand* upon melting creating containment problems (the container will burst unless the excess volume is accommodated somehow, see **§2.3.A**). In order to retain the explicit solvability of the problem, we shall assume that the expansion is accommodated by *bulk movement* of the existing (semi-infinite) phase and ignore all mechanical effects. Then, the resulting purely thermal model still admits similarity solution, to be presented in **§2.3.B**. The analogous case of shrinkage with void formation is studied in **§2.3.D**. However, having neglected all mechanical effects, these models must be considered as approximate ones. More precise models, which include mechanical effects but don't admit explicit solutions, are derived in **§2.3.E** from first principles.

2.3.A Physical effects

Two types of density change are relevant in phase-change processes. One is due to dependence of density on temperature, arising in any heat transfer process, the other is due to the difference between the solid and liquid densities at the melt temperature. Let us describe, qualitatively, their physical consequences.

The density of a liquid may vary considerably with temperature. Typically, the density decreases with rising temperature, so the warmer the liquid the more volume it occupies (per unit mass). In the presence of gravity, a temperature gradient induces flow due to buoyancy, known as natural (or free) convection [BIRD et al]. Consider a box with a vertical heated face, the other faces being insulated. The box is filled with a solid PCM, but not all the way to the top, to allow for expansion. As liquid forms at the heated face, natural convection sets in. The convection current carries more heat to the upper part of the melt front (**Figure 2.3.1**), resulting in a "tilted" interface. The effect here is (at least) two-dimensional, no simple modeling is possible, [VISKANTA], [TIEN-CAREY-FERRELL], and we shall not consider it any further.

The density change we are concerned with here is the sudden change occurring at the melt temperature itself. From now on we assume that ρ_L and ρ_S are constants but $\rho_L \neq \rho_S$. For most materials they differ by up to 10%, in extreme

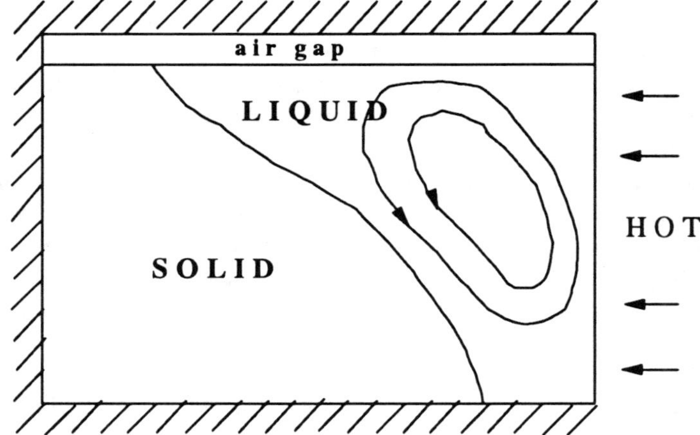

Figure 2.3.1. Phase-change with convective effects (schematic).

cases by up to 30%. Usually $\rho_L < \rho_S$, so the volume expands upon melting.

What would happen if the box, in the example above, were completely filled with solid initially? With $\rho_L < \rho_S$, the volume expansion due to melting will burst the container at its weakest point. To avoid this, one must place an "expansion tube" near the top of the heated face to allow (the hottest) liquid to escape. The expansion tube itself must be heated or else the liquid in it may freeze and clog it.

Now consider the case of shrinkage in a closed container. Assume the box is completely filled with liquid, $\rho_L < \rho_S$, and we cool the non-insulated face. Now, the forming solid occupies *less* volume, the interior becomes increasingly underpressurized and something must "give" sooner or later. Either the container will buckle or the PCM will "tear". If the container is sufficiently strong, then "voids" will form in the PCM (bubbles or gaps filled with vapor of the material). Where these form depends on the relative strength of the adhesion forces between wall and liquid, wall and solid, liquid and solid, or within the liquid or the solid. Usually the weakest forces are between wall and PCM, so a void is most likely to form there. If, on the other hand, microbubbles of trapped air (or other foreign gas) pre-exist in the liquid then that is from where the voids will start growing. The growing voids, being buoyant, will float toward the top. Even in zero-gravity there will be convection, the so called Marangoni flow, due to surface tension (capillary) effects at the void-liquid interfaces (driving the void towards the hot spot) [MYSHKIS et al]. Modeling of such complicated processes, in the context of energy storage in space systems (§5.3), is an active area of research [WICHNER et al], [DRAKE, 1991].

On the other hand, there are situations where the volume change may be accommodated simply by *bulk displacement* of the existing phase. For example, consider a long insulated tube in vertical position, closed at the bottom and open at the top, filled with a liquid having $\rho_L < \rho_S$. If we freeze from below then, as solid forms, the liquid will simply be "pulled down" to take up the reduced volume. The solid remains stationary, growing into the liquid which may be considered as "moving down" with uniform velocity (but, of course, time-dependent). This is a case where the density change has "minimal" thermal effect. A purely thermal (approximate) model admitting explicit solution will be presented in **§2.3.B** and applied to a cryosurgery scenario in **§2.3.C** (where we will see that the effect may still be significant).

A simple case of void formation may arise as follows: A long vertical tube filled with ice (now $\rho_S < \rho_L$) is heated from above. The water forming at the top occupies less volume than the melted ice so a vapor gap (void) appears between it and the heated upper end of the tube. As the water layer grows into the ice (which does not move), the void increases, "pushing" the water down with uniform (but time-dependent) speed. Contrary to the previous case, the density-change here is expected to affect the phase-change process considerably (the vapor, being a poor conductor, inhibits heat transfer). Such a situation will be dealt with in **§2.3.D**, again by an approximate model with explicit solution.

2.3.B Expansion with bulk movement due to $\rho_L \leq \rho_S$

As a model situation we consider the following

PHYSICAL PROBLEM: A semi-infinite slab, $x \geq 0$, initially solid at $T_S < T_m$, is melted by imposing a constant temperature, $T_L > T_m$, at $x = 0$. We assume $\rho_L \leq \rho_S$, c_L, c_S, k_L, k_S, L, T_m are (positive) constants. On melting, the liquid expands and pushes the solid to the right (without friction!) with uniform speed $\mathrm{v} = \mathrm{v}(t)$, while the liquid itself does not move. We expect a melt front, $x = X(t)$, starting at $X(0) = 0$ and advancing to the right.

We develop the model more or less "intuitively" here, while a more complete discussion of "first principles" is presented in **§2.3.E**. Clearly, for the liquid the situation is as in the Stefan Problem of **§2.2.A**. On the other hand, since the solid is moving with speed v, the heat flux there consist of the *sum* of the *conductive* and *convective* fluxes:

$$q_S(x,t) := -k_S T_x(x,t) + \rho_S c_S [T(x,t) - T_m] \mathrm{v}(t) \ . \tag{1}$$

The latter represents the (sensible) heat crossing a unit area per unit time due to movement of the solid with speed $\mathrm{v}(t)$. Thus, the heat conduction equation in the solid takes the form

$$\rho_S c_S [T_t + \mathrm{v}(t) T_x] = k_S T_{xx} \ , \tag{2a}$$

or, dividing by $\rho_S c_S$,

$$T_t + \mathrm{v}(t) T_x = \alpha_S T_{xx} \quad \text{in the solid} \ . \tag{2b}$$

We see that the motion of the solid introduces the **convective** term, $\mathrm{v} T_x$, in the heat equation, and the additional unknown $\mathrm{v}(t)$.

The additional equation needed to determine $v(t)$ comes from the requirement that *mass be conserved.* Let $x = \xi(t)$ be the location, at time t, of a *fixed material point* far into the solid. It moves to the right with speed $\xi'(t) = v(t)$, (until it melts much later). The total mass (per unit cross-sectional area) inside the interval $0 \le x \le \xi(t)$ is $M(t) = \rho_L[X(t) - 0] + \rho_S[\xi(t) - X(t)]$, $t \ge 0$, and mass conservation requires $\frac{dM}{dt} = 0$, whence

$$v(t) = (1 - \frac{\rho_L}{\rho_S})\, X'(t), \quad t \ge 0\,, \tag{3a}$$

(another derivation is presented in **§2.3.E**). We see that the solid speed, $v(t)$, is proportional to the melt-front speed, $X'(t)$, with proportionality constant $0 \le 1 - (\rho_L / \rho_S) < 1$ (we are assuming $\rho_L \le \rho_S$), so the solid is moving slower than the melt front.

What is the interface condition (assuring energy conservation) here? It is easy to fall in a trap and simply apply the Stefan condition ((41)**§1.2**) $[\![\rho e]\!]_S^L X' = [\![q]\!]_S^L$ on $x = X(t)$, which would say

$$\rho_L L X'(t) = -k_L T_x(X(t)^-, t) + k_S T_x(X(t)^+, t)\,, \tag{3b}$$

(because $[\![\rho e]\!]_S^L = \rho_L c_L [T(X(t),t) - T_m] + \rho_L L - \rho_S c_S$, $[T(X(t),t) - T_m] = \rho_L L$, and $[\![q]\!]_S^L = q_L - q_S = -k_L T_x(X(t)^-,t) + k_S T_x(X(t)^+,t) - \rho_S c_S [T(X(t),t) - T_m] v(t)$ = right-hand side of (3b), see also PROBLEM 1). What is wrong with this? It is that it expresses *conservation of enthalpy only*, which is *no longer* the total energy in our situation here (due to the volume change)! The correct conservation condition of *total* energy will be derived in **§2.3.E** (see (36) below); it contains an additional term which, being proportional to X'^3, destroys the similarity solution unfortunately. Except at very small times however, this term will is expected to be ignorably small compared to $\rho_L L X'$ [ALEXIADES-DRAKE]. By dropping it, we effectively reduce the left-hand side thus speeding up the melting process. Under such an *approximation*, the interface condition reduces to (3b) and we are led to the following

MATHEMATICAL PROBLEM *(2-phase semi-infinite slab with $\rho_L \le \rho_S$, melting from the left):*

Find $T(x, t)$, $X(t)$, $v(t)$ such that

$$T_t = \alpha_L T_{xx}\,, \qquad 0 < x < X(t), \qquad t > 0 \ \ (liquid) \tag{4a}$$

$$T_t + v(t) T_x = \alpha_S T_{xx}\,, \qquad X(t) < x < \infty\,, \qquad t > 0 \ \ (solid) \tag{4b}$$

$$v(t) = [1 - \frac{\rho_L}{\rho_S}]\, X'(t), \qquad t > 0 \tag{4c}$$

$$T(X(t), t) = T_m, \qquad t > 0 \tag{4d}$$

$$\rho_L L X'(t) = -k_L T_x(X(t)^-, t) + k_S T_x(X(t)^+, t), \quad t > 0 \ \ (approximate) \tag{4e}$$

$$X(0) = 0, \qquad T(x, 0) = T_S < T_m, \qquad 0 < x < \infty \tag{4f}$$

$$T(0,t) = T_L > T_m, \qquad \lim_{x\to\infty} T(x,t) = T_S, \qquad t>0\ . \tag{4g}$$

Thanks to (4c), a similarity solution is still possible [CARSLAW-JAEGER]. We introduce the similarity variable $\xi = \frac{x}{\sqrt{t}}$, the notation $\mu = \frac{\rho_L}{\rho_S}$, and seek the three unknowns in the forms

$$X(t) = 2\lambda\sqrt{\alpha_L t}, \quad \mathrm{v}(t) = (1-\mu)\lambda\sqrt{\alpha_L/t}, \quad T(x,t) = F(\xi)\ . \tag{5}$$

We find (PROBLEM 2)

$$X(t) = 2\lambda\sqrt{\alpha_L t}\ , \tag{6a}$$

$$\mathrm{v}(t) = (1-\mu)\lambda\sqrt{\alpha_L/t} = (1-\mu)\nu\lambda\sqrt{\alpha_S/t}\ , \tag{6b}$$

$$T(x,t) = T_L - (T_L - T_m)\frac{\operatorname{erf}\left(\frac{x}{2\sqrt{\alpha_L t}}\right)}{\operatorname{erf}\lambda}, \quad 0 \le x \le X(t),\ t>0 \ (liquid) \tag{6c}$$

$$T(x,t) = T_S + (T_m - T_S)\frac{\operatorname{erfc}\left[\frac{x}{2\sqrt{\alpha_S t}} - (1-\mu)\nu\lambda\right]}{\operatorname{erfc}(\mu\nu\lambda)}, \quad X(t) \le x,\ t>0 \ (solid) \tag{6d}$$

with λ the root of the transcendental equation

$$\frac{\mathbf{St}_L}{\lambda e^{\lambda^2}\operatorname{erf}\lambda} - \frac{\mathbf{St}_S}{(\mu\nu\lambda)e^{(\mu\nu\lambda)^2}\operatorname{erfc}(\mu\nu\lambda)} = \sqrt{\pi}\ , \tag{6e}$$

where

$$\mathbf{St}_L = \frac{c_L(T_L - T_m)}{L}, \quad \mathbf{St}_S = \frac{c_S(T_m - T_S)}{L}, \quad \nu = \sqrt{\frac{\alpha_L}{\alpha_S}}, \quad \mu = \frac{\rho_L}{\rho_S}\ . \tag{6f}$$

The existence and uniqueness of the root λ is easily shown, as in **§2.2**. We observe that the root λ depends only on the parameters $\mathbf{St}_L$, $\mathbf{St}_S$, and $\mu\nu$ (see (6f)). In fact, the only change from the Stefan Problem (§**2.2**) comes from replacing ν by $\mu\nu$, (see (2d) **§2.2**) so the densities enter only via their ratio $\mu = \frac{\rho_L}{\rho_S}$. When $\rho_L = \rho_S$, $\mu = 1$ and we recover the Neumann solution of the classical Stefan Problem ((2) **§2.2**). For most materials the densities differ by up to 10%, so μ is usually between 0.9 and 1.1 (e.g. for water-ice, $\mu \approx 1.09$), but there are extreme cases with density variation up to 30% (e.g. LiF has $\mu \approx 0.8$).

To see the dependence of the root λ on the density ratio, μ, one may fix $\mathbf{St}_L$, $\mathbf{St}_S$ and ν and solve (6e) numerically for a range of μ's. Sample values are shown in **Table 2.3.1**. For combinations of $\mathbf{St}_L$, $\mathbf{St}_S$, ν within the range 0.1 to 2 (which is of practical interest), the dependence of λ on μ is linear with small slope, typically changing by ±3% (relative to the $\mu = 1$ case) when μ changes by ±30%. Since for most materials μ changes only by ±10%, one expects only 1% change in λ in most cases. Therefore, the effect of $\rho_L \neq \rho_S$ on any quantity that is proportional to λ (e.g. the front location) is usually (ignorably) small. The melt time, $t = x^2/4\lambda^2\alpha_L$, is a little more sensitive, a 1% change in λ producting a, roughly, 2% change in the

Table 2.3.1. Sample values of λ, root of (6e) for various values of St_L, St_S, ν, μ of (6f).

St_L	St_S	ν	μ	λ
0.01	0.01	0.1	0.7	0.060598
"	"	"	1	0.062087
"	"	"	1.3	0.063046
"	2.	1.	0.7	0.003684
"	"	"	1	0.004392
"	"	"	1.3	0.004995
1.	0.01	1.	0.7	0.615709
"	"	"	1	0.616112
"	"	"	1.3	0.616360
"	2.	5.	0.7	0.330203
"	"	"	1	0.342473
"	"	"	1.3	0.350172
2.	0.01	1.	0.7	0.796339
"	"	"	1	0.796676
"	"	"	1.3	0.796883
"	2.	5.	0.7	0.477778
"	"	"	1	0.488889
"	"	"	1.3	0.495649

melt time. Finally, notice that the temperature in the solid is substantially altered in form from the $\rho_L = \rho_S$ case, since it includes the effect of displacement (see (11) below).

Let us remember, however, that (4e), hence the model and its solution, is only approximately valid since it only balances "thermal energy" (enthalpy) across the interface (see §2.3.E). As we have remarked already, we expect (6a) to *overestimate* the true interface, especially at early times.

Similarly to the case of "expansion on melting" just described, one may consider expansion on freezing, as well as shrinkage on melting or on freezing (without void), (PROBLEMS 10-12).

2.3.C Application to cryosurgery

Cryosurgery is a means of destroying unwanted cells (tumors, etc.) by freezing and thawing. It is performed by bringing the tissue to be killed into contact with a cold source, ordinarily cooled by liquid Nitrogen. While the objective is to kill certain cells, it is equally important that the treatment not be lethal to neighboring, healthy cells. Thus the extent of freezing and lethality conditions are of primary interest.

We shall apply the above results to an idealized situation brought about by placing a flat cryoprobe against a body of tissue with the resulting freezing front essentially one-dimensional during the period of the operation. We wish to take the density change of the tissue into account in order, in particular, to question the effect of the resulting movement upon the readings of thermocouples placed at various depths in the tissue.

Consider the case of a one-dimensional freezing process for a semi-infinite tissue slab ($x > 0$) in which a constant temperature T_S is imposed by a (flat) cryoprobe at $x = 0$ and the following conditions are assumed to hold:

a) Constant thermophysical parameters $\rho_S, \rho_L, c_S, c_L, k_S, k_L, \alpha_S, \alpha_L$.

b) Uniform initial tissue temperature T_L.

This model corresponds best to freezing via a flat plate. The assumption that the specific heat and conductivity are independent of temperature is a very severe one, imposing a significant limitation on the model. Moreover the model is highly idealized, not taking into account heat generation by metabolism, the composition of tissue, and blood flow. Nevertheless from a cryosurgical point of view the model can be of use in a number of ways, for example, to "bracket" the system behavior using extreme values of the parameters; to test the possible effects of density change; to obtain simple "ballpark" estimates of the system behavior.

Since a significant percentage of most biological tissue is water, the thermal properties of the tissue will be determined to a large extent by those of water and ice. Over the range from body temperature (37°C = 310 K) to that in liquid Nitrogen cryoprobes (–200 °C = 73 K) the conductivity k, specific heat c, density ρ and thermal diffusivity α, vary strongly. Using standard data fitting techniques we have obtained expressions of their temperature dependence over this range (**§1.2.B**). Let us recall them for convenience. If T denotes the temperature in degrees-Kelvin, then for $73\,K \le T \le 310\,K$,

$$c = \begin{cases} 7.16 \times 10^{-3}\, T + 0.138\,, & T \le 273\,K \text{ (ice)} \\ 4.1868\,, & T \ge 273\,K \text{ (water)},\quad (kJ/kg\,K), \end{cases} \tag{7a}$$

$$k = \begin{cases} 2.24 \times 10^{-3} + 5.975 \times 10^{-6}(273 - T)^{1.156}\,, & T \le 273\,K\,, \\ 1.017 \times 10^{-4} + 1.695 \times 10^{-6}\, T\,, & T \ge 273\,K \quad (kJ/m\,s\,K), \end{cases} \tag{7b}$$

$$\rho = \begin{cases} 920\,, & T \le 273\,K\,, \\ 1000\,, & T \ge 273\,K \quad (kg/m^3), \end{cases} \tag{7c}$$

and $L = 333.73$ (kJ/kg). In our example we will use the minimum, average and maximum values of c and k.

The idealized situation may be represented by the freezing counterpart to the melting model of **§2.3.B**. Note that we still have expansion (since $\rho_L > \rho_S$ now, see **Figure 2.3.2**), and the solution is obtained from (6) by simply interchanging the subscripts "L" and "s" (and replacing the latent heat L by $-L$ in (6f)) (PROBLEM 13).

With initial uniform tissue temperature $T_L = 37°C$ and imposed temperature $T_S = -200°C$, the minimum, average, and maximum values of the parameters c, k, α, $\boldsymbol{St}$, $\nu = \sqrt{\alpha_S/\alpha_L}$ and the corresponding root λ are shown in **Table 2.3.2.** Hence, the extreme values for the interface are

$$0.59\sqrt{t} \;\le\; X(t) \;\le\; 2.02\sqrt{t}\,, \tag{8}$$

with t in *minutes*, X in *centimeters*, and average value $X(t) = 1.06\sqrt{t}$. The extreme freeze-fronts of (8) are plotted in **Figure 2.3.3**.

Question: How is a marker particle displaced due to freezing-induced expansion?

Letting $x_0(t)$ denote the location, at time t, of a material point in the liquid (which moves with velocity $\mathrm{v}(t)$ until it freezes), by integration of (3) we find

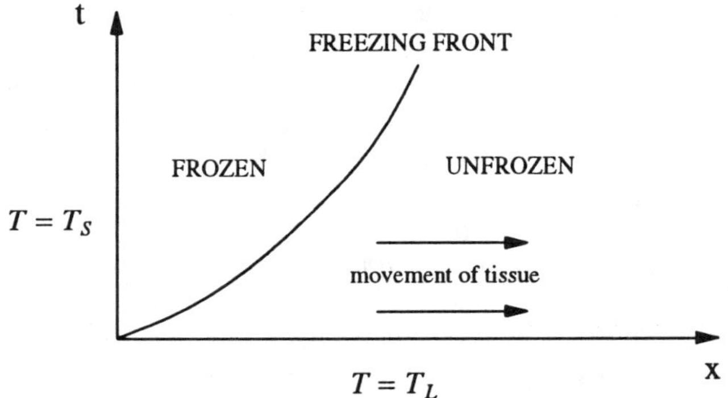

Figure 2.3.2. Space-time diagram for the cryosurgery process.

Table 2.3.2. Extreme and average values of parameters in cryosurgery model

parameter	minimum value	average value	maximum value
k_L	5.64×10^{-4}	6.0×10^{-4}	6.27×10^{-4}
k_S	2.24×10^{-3}	2.66×10^{-3}	3.09×10^{-3}
c_L	4.1868	4.1868	4.1868
c_S	0.661	1.7	2.09
α_S	1.16×10^{-6}	1.7×10^{-6}	5.08×10^{-6}
α_L	1.35×10^{-7}	1.43×10^{-7}	1.5×10^{-7}
$\boldsymbol{St}_S$	0.396	1.02	1.25
$\boldsymbol{St}_L$	0.464	0.464	0.464
ν	2.94	3.45	5.82
λ	0.3551	0.5257	0.5786

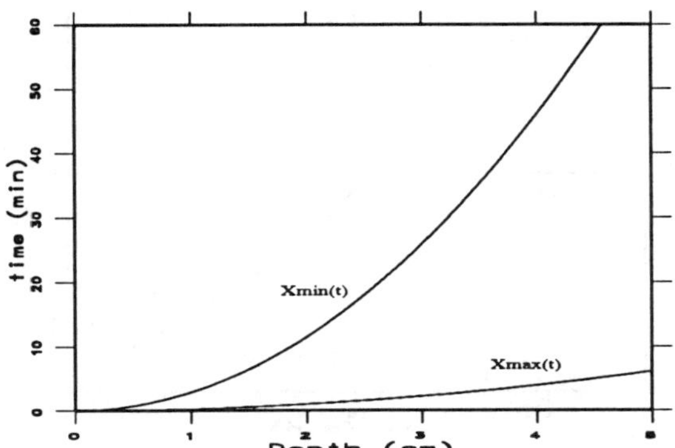

Figure 2.3.3. The extreme freezing fronts of (8).

$$x_0(t) = x_0(0) + (1 - \frac{\rho_S}{\rho_L})X(t), \qquad 0 \leq t \leq t_{freeze} \ . \tag{9}$$

When the freeze-front reaches it, $X(t) = x_0(t)$, whence (PROBLEM 14).

$$x_0(t_{freeze}) = \frac{\rho_L}{\rho_S}\, x_0(0) \quad \text{and} \quad t_{freeze} = \left(\frac{\rho_L}{\rho_S}\frac{x_0(0)}{2\lambda}\right)^2 / \alpha_S \ . \tag{10}$$

For example, a particle initially located at $x_0(0) = 4\,cm$ will freeze at location $4.348\,cm$, at time $t_{freeze} \approx 1006\,s \approx 17\ min$ (assuming the average parameter values), where it will stay ever after.

Question: To what extent are temperature measurements affected by the expansion?

To be specific, consider a thermocouple initially placed $x_0 = 4\,cm$ away from the cryoprobe into the tissue, and assume the average parameter values. If it is rigidly held there, it will freeze at time $t_1 \approx 851.4\,s$ and up to that time its temperature reading will be

$$T(x_0, t) = T_L - (T_L - T_m)\frac{\operatorname{erfc}\left[\dfrac{x_o}{2\sqrt{\alpha_L t}} - \left(1 - \dfrac{\rho_S}{\rho_L}\right)\lambda\sqrt{\dfrac{\alpha_S}{\alpha_L}}\right]}{\operatorname{erfc}\left(\dfrac{\rho_S}{\rho_L}\lambda\sqrt{\dfrac{\alpha_S}{\alpha_L}}\right)}, \quad 0 \leq t \leq t_1 . \tag{11a}$$

If, on the other hand, it moves freely, then it will freeze at time $t_2 \approx 1006\,s$, as we just saw above, and its temperature will be

$$T(x_0, t) = T_L - (T_L - T_m)\frac{\operatorname{erfc}\left[\dfrac{x_o}{2\sqrt{\alpha_L t}}\right]}{\operatorname{erfc}\left(\dfrac{\rho_S}{\rho_L}\lambda\sqrt{\dfrac{\alpha_S}{\alpha_L}}\right)}, \quad 0 \leq t \leq t_2 , \tag{11b}$$

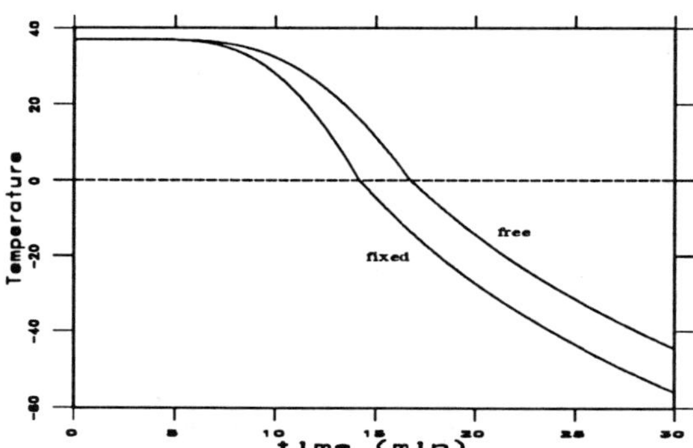

Figure 2.3.4. Cooling curves at fixed and moving point.

(PROBLEM 13). These cooling curves are plotted in **Figure 2.3.4**. Note that at $t = t_1 \approx 851.4\,s$, the moving thermocouple will read about $16.2°C$ as opposed to $0°C$ for the fixed one. Conversely, at $t = t_2 \approx 1006\,s$, the fixed thermocouple will show $-13.6°C$. Clearly, the displacement has a very noticeable effect on the temperature readings and therefore the density change can not be ignored in this situation.

2.3.D Void formation

Now we consider the case in which the shrinkage due to the change of density results in the formation of a void at the heat-transfer face. A situation where this may arise was described at the end of §**2.3.A**.

Traditionally, the formation of voids has been largely ignored by modelers of phase-change processes. This may be because voids often form within the material (at pre-existing, trapped air microbubbles), may be small (due to small density-change and/or dispersion) thus expected to have small thermal effect, and/or because taking them into account introduces major complications. Clearly, however, there are also many situations in which the appearance of a void should not be ignored, for example when the density-change of the material is substantial and/or the container is small. As mentioned in §**2.3.A**, the rigorous treatment of void formation requires the full Navier-Stokes equations and knowledge of the thermodynamics of interfaces, the combination of which is not yet clear! In the spirit of exploration of major effects only, we treat here a case simple enough to be amenable to analysis and realistic enough to be useful.

As a model situation we consider the following

PHYSICAL PROBLEM: A semi-infinite slab, $x \geq 0$, initially liquid at $T_L > T_m$, is solidified by imposing a constant temperature $T_S < T_m$ at $x = 0$. The forming solid occupies less volume and the shrinkage is accommodated by the creation of a void (vapour) between the wall and the solid. As freezing proceeds, the void grows "pushing" the forming solid to the right with uniform speed v(t) (but the liquid does not move, the solid simply grows into it). There are two interfaces, both assumed planar and sharp: $x = X(t)$ between solid and liquid and $x = Y(t)$ between void and solid. We assume the void has negligible density ($\rho_V \approx 0$, since it is about three orders of magnitude smaller than ρ_L), hence negligible mass and energy, but a non-zero conductivity k_V. Also, $\rho_L \leq \rho_S$, c_L, c_S, k_V, k_L, k_S, L, T_m are assumed constants.

As in **§2.3.B**, conservation of mass yields $v(t) = (1 - \rho_L / \rho_S)X'(t)$, (see (3a)), and since $v(t) = Y'(t)$, we have

$$Y(t) = (1 - \frac{\rho_L}{\rho_S})X(t), \qquad t > 0 \ . \tag{12}$$

Clearly, equation (2b) is valid in the solid $(Y(t) < x < X(t),\ t > 0)$, and $T_t = \alpha_L T_{xx}$ is valid in the liquid $(X(t) < x < \infty\ , t > 0)$.

How is heat transferred across the void? We have assumed that the void has negligible mass and energy, so all the heat that enters it must leave it, i.e. the fluxes at $x = 0$ and $x = Y(t)^-$ must be equal. Hence, it suffices to define the flux $q(Y(t)^-, t)$ at $x = Y(t)$. Two assumptions suggest themselves. First, we could assume that the void is a vacuum with radiation the only mode of heat transfer, and the flux would have the form of (34) **§1.2.D**; the resulting problem is interesting, but it does *not* admit an explicit solution, so we pass to an alternative assumption. We assume that the void acts simply as a *thermal resistance layer* with conductivity k_V, so that

$$q(Y(t)^-, t) = -k_V \frac{T(Y(t), t) - T_S}{Y(t) - 0} \ . \tag{13}$$

A heat balance (PROBLEM 16, also **§2.3.E**) shows that this flux must equal the conductive flux $-k_S T_x(Y(t), t)$, that is,

$$-k_V \frac{T(Y(t), t) - T_S}{Y(t)} = -k_S T_x(Y(t)^+, t), \qquad t > 0 \ . \tag{14}$$

It serves as an equation for the unknown temperature $T(Y(t), t)$ of the void-solid interface. Finally, across the solid-liquid interface, $x = X(t)$, the situation is similar to that in **§2.3.B**: enthalpy balance leads to the standard Stefan Condition, (15f), but the correct total energy balance (**§2.3.E**) produces the additional term $\frac{1}{2}\rho_L(1 - \frac{\rho_L^2}{\rho_S^2})X'^3$ on the left-hand side. Again, we will assume this term is *negligible* compared to $\rho_S L X'$, in order to retain the explicit solvability of the problem. Note that the neglected term is positive so its presence enhances the latent heat term, slowing down the freezing process. It follows that the explicit-solution interface will constitute an *upper bound* for the true interface.

Thus, we are led to the following [WILSON-SOLOMON]

MATHEMATICAL PROBLEM *(2-phase freezing for $\rho_L \le \rho_S$, with void at the wall):*

Find $T(x, t)$, $X(t)$, $Y(t)$ (and $v(t) = Y'(t)$) such that

$$T_t + v(t)\, T_x = \alpha_S\, T_{xx}, \quad Y(t) < x < X(t), \quad t > 0, \quad (solid) \tag{15a}$$

$$T_t = \alpha_L\, T_{xx}, \quad X(t) < x < \infty, \quad t > 0 \quad (liquid) \tag{15b}$$

$$Y(t) = (1 - \frac{\rho_L}{\rho_S}) X(t), \quad v(t) = Y'(t), \quad t > 0, \tag{15c}$$

$$\frac{k_V}{Y(t)} [\, T(Y(t), t) - T_S] = k_S\, T_x(Y(t)^+, t), \quad t > 0, \tag{15d}$$

$$T(X(t), t) = T_m, \quad t > 0, \tag{15e}$$

$$\rho_L L\, X'(t) = -k_L\, T_x(\, X(t)^+, t) + k_S\, T_x(\, X(t)^-, t), \quad t > 0 \quad (approximate) \tag{15f}$$

$$X(0) = Y(0) = 0, \quad T(x, 0) = T_L > T_m, \quad 0 < x < \infty\ . \tag{15g}$$

Despite the different-looking condition (15d) (which plays the role of (4g) here), this problem admits a similarity solution analogous to that in **§2.3.B**. Indeed, setting

$$\mu = \frac{\rho_L}{\rho_S}, \quad \nu = \sqrt{\frac{\alpha_L}{\alpha_S}}, \quad \boldsymbol{St}_L = \frac{c_L(T_L - T_m)}{L}, \quad \boldsymbol{St}_S = \frac{c_S(T_m - T_S)}{L}, \tag{16}$$

and remembering that we freeze here, we find (PROBLEM 17)

$$X(t) = 2\lambda\sqrt{\alpha_S t}, \quad Y(t) = 2\lambda(1 - \mu)\sqrt{\alpha_S t}, \quad t > 0\,, \tag{17a}$$

$$v(t) = \lambda(1 - \mu)\sqrt{\alpha_S / t} = \frac{\lambda(1 - \mu)}{\nu}\sqrt{\alpha_L / t}\,, \quad t > 0\,, \tag{17b}$$

$$T(x, t) = T_m - (T_m - T_S)\, \frac{\operatorname{erf}(\lambda\mu) - \operatorname{erf}\left(\dfrac{x}{2\sqrt{\alpha_S t}} - \lambda(1 - \mu)\right)}{\operatorname{erf}(\lambda\mu) + \dfrac{2\lambda(1 - \mu)}{\sqrt{\pi}} \dfrac{k_S}{k_V}}, \quad Y(t) \le x \le X(t), \quad t > 0\,, \quad (solid) \tag{17c}$$

$$T(x, t) = T_L - (T_L - T_m)\, \frac{\operatorname{erfc}\left(\dfrac{x}{2\sqrt{\alpha_L t}}\right)}{\operatorname{erfc}(\lambda / \nu)}, \quad X(t) \le x, \quad t > 0, \quad (liquid) \tag{17d}$$

where λ is the root of the transcendental equation

$$\frac{\boldsymbol{St}_S}{\lambda\mu e^{(\lambda\mu)^2}\left[\operatorname{erf}(\lambda\mu) + \dfrac{2\lambda(1 - \mu)}{\sqrt{\pi}} \dfrac{k_S}{k_V}\right]} - \frac{\boldsymbol{St}_L}{\dfrac{\lambda}{\nu}\, e^{(\lambda/\nu)^2} \operatorname{erfc}(\dfrac{\lambda}{\nu})} = \sqrt{\pi}\ . \tag{17e}$$

It posesses unique solution $\lambda > 0$ for *any* positive values of the parameters (defined in (16)) (PROBLEM 18), and therefore Problem (15) admits unique similarity solution.

Note that for $\rho_L = \rho_S$, we have $\mu = 1$ and we recover the Neumann solution to the 2-phase Stefan Problem for freezing (see PROBLEM 7 §2.2).

The temperature at the void-solid interface is found, from (17c), to be a constant:

$$T(Y(t), t) = T_m - (T_m - T_S) \frac{\operatorname{erf}(\lambda\mu)}{\operatorname{erf}(\lambda\mu) + \dfrac{2\lambda(1-\mu)}{\sqrt{\pi}} \dfrac{k_S}{k_V}}, \tag{18}$$

implying that this interface is also an isotherm.

What is the effect of the void? From (17e) we see that it affects λ *only* through its conductivity k_V, which is typically much smaller than k_S (e.g., for ice/water-vapour, $k_S/k_V \approx 130$). Hence, the left-hand side of (17e) is smaller compared with the $\rho_L = \rho_S$ case, resulting in a root considerably smaller than that of the standard Stefan Problem. It follows that the presence of the void reduces the rate of freezing considerably.

The insulating effect of the void may also be seen in (18). The larger k_S/k_V, the closer that temperature will be to T_m, hence the smaller the temperature gradient in the solid and therefore the slower the freezing.

EXAMPLE: Lithium Fluoride (LiF) is a candidate material for thermal energy storage in space. While its melting point and latent heat are high enough to meet the requirements of a power system for a space station, it has the very undesirable property of a 25% density-change at its melt point. Let us examine the effect this will have on the freezing of a long slab of this material, initially at its melt temperature, when exposed to a $T_S = 1000\,K$. The thermophysical properties of LiF are:

$\rho_L = 1.79,\quad \rho_S = 2.33\quad (g/cm^3),\quad c_L = 2.45,\quad c_S = 2.45\quad (J/g\,K),$
$k_L = 0.037,\quad k_S = 0.06,\quad k_v = 0.00037\quad (W/cm\,K),\quad T_m = 1121\,K,$
$L = 1037.1\ J/g.$

With $T_L = T_m$ and $T_S = 1000\,K$, the parameters in (16) have the values: $\mu = 0.768$, $\nu = 0.896$, $St_L = 0$, $St_S = 0.286$, and solving (17e) we find $\lambda = 0.06955$. If ρ_L were equal to ρ_S (no void) then the root would have been about 0.36. The reduction of λ by a factor of about 5 is due to the insulating effect of the void!

The interfaces $Y(t) = 3.25 \times 10^{-3}\sqrt{t}$ and $X(t) = 14 \times 10^{-3}\sqrt{t}$ are plotted in **Figure 2.3.5** for the first 6 hours of freezing. From (18), the void-solid interface temperature is found to be about 1117 K, so that the temperature drop across the solid is only about 4 K. Hence, the *effective* Stefan number in the solid is only 0.009 and we expect nearly linear temperature profiles in the solid; indeed, the temperature from (17c) is linear to four significant digits.

The example illustrates the fact that the appearance of a void affects the phase-change process drastically.

The case of shrinkage on melting ($\rho_L \geq \rho_S$) with a void forming at the wall may be studied similarly (PROBLEMS 21, 22).

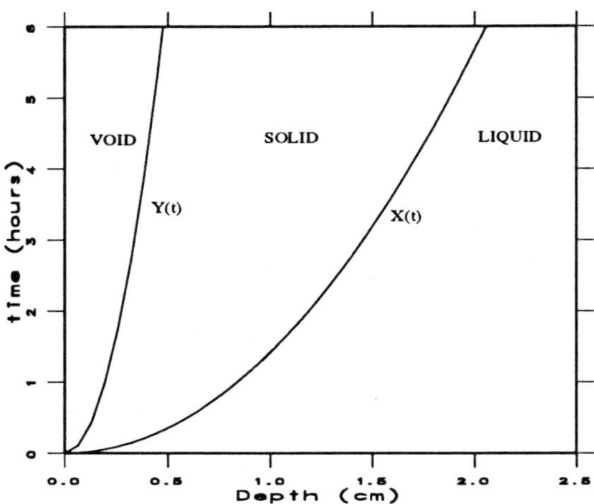

Figure 2.3.5. The void-solid and solid-liquid fronts.

2.3.E Conservation laws and interface conditions

Any density change whatsoever induces movement of material which adds *mechanical* effects to the thermal ones. Indeed, the **total** energy ε (per gram) is the *sum* of the **internal** energy u, the **kinetic** energy $\frac{1}{2}|\vec{v}|^2$ and the **potential** energy [BIRD et al]. The internal energy includes both heat and work and it is related to the enthalpy, e, by

$$u = e - \frac{P}{\rho}, \tag{19}$$

where P is the pressure and ρ the density. As long as we were operating under constant pressure with constant density and zero velocity, we were justified in considering the enthalpy as the total energy. When there is a density change however, this is no longer true.

Let us consider phase-change processes for a pure material under the following assumptions:

(i) *constant thermophysical properties* $\rho_L \neq \rho_S$, c_L, c_S, k_L, k_S, T_m, L, and a *sharp interface*;

(ii) *no viscous dissipation* (so that the stress is simply the pressure);

(iii) *no changes in potential energy* (so that total energy = internal + kinetic).

We restrict our attention to *1-dimensional* processes, for simplicity. Thus, each phase is *incompressible* and the volume change, due to the change of density upon change of phase, results in bulk displacement of one of the phases with *uniform* speed (but time-dependent in general) as we shall see shortly.

Conservation of thermal energy alone is no longer sufficient to determine the evolution of the system. Now *mass, momentum* and *total energy* must be conserved. Under assumption (ii) above, the general conservation laws [BIRD et al], take the following form (in 1-dimension):

mass conservation: $$\rho_t + (\rho \mathrm{v})_x = 0, \tag{20}$$

momentum conservation: $$(\rho \mathrm{v})_t + (\rho \mathrm{v}\mathrm{v} + P)_x = 0\,, \tag{21}$$

energy conservation: $$(\rho \varepsilon)_t + (\rho \varepsilon \mathrm{v} + q + P\mathrm{v})_x = 0\ . \tag{22}$$

Here ρ, $\rho\mathrm{v}$, $\rho\varepsilon$ are the mass, momentum and total energy per unit volume, P is the (thermodynamic) pressure and q is the *conductive* heat flux ($q = -kT_x$). The laws may be written in many other forms [BIRD et al] but (20-22) have the great advantage of being in "divergence form", that is of the form

$$A_t + B_x = 0\,, \tag{23}$$

with B the *flux* of the quantity A. Such a law implies that for any smooth space-time curve $x = X(t)$, the following **jump relation** (Rankine-Hugoniot or shock condition) must hold:

$$[\![A]\!]_-^+ X'(t) = [\![B]\!]_-^+ \quad \text{on} \quad x = X(t)\,, \tag{24}$$

expressing *conservation* of the quantity A across the curve (see e.g. [SEGEL], [TRUESDELL-TOUPIN], also **§1.2** and **§4.4**). Hence, we can immediately write down the jump conditions for mass, momentum and energy across an interface:

$$[\![\rho]\!]_-^+ X'(t) = [\![\rho\mathrm{v}]\!]_-^+ \qquad \text{on} \quad x = X(t)\,, \tag{25}$$

$$[\![\rho\mathrm{v}]\!]_-^+ X'(t) = [\![\rho\mathrm{v}^2 + P]\!]_-^+ \qquad \text{on} \quad x = X(t)\,, \tag{26}$$

$$[\![\rho\varepsilon]\!]_-^+ X'(t) = [\![\rho\varepsilon\mathrm{v} + P\mathrm{v} + q]\!]_-^+ \qquad \text{on} \quad x = X(t)\,. \tag{27}$$

We shall use them later to derive the correct interface conditions for the various situations considered in **§2.3.B, D**.

Inside the liquid or the solid, where $\rho \equiv$ constant, the equations may be simplified considerably. Indeed, the continuity equation, (20), implies that the velocity must be *uniform* in each phase, so v_L and v_S may only depend on time. Moreover, the energy equation, (22), re-written in terms of the *internal* energy, u, takes the form (PROBLEM 23)

$$(\rho u)_t + (\rho u \mathrm{v} + q)_x + P\mathrm{v}_x = 0\,, \tag{28a}$$

and therefore, *inside the liquid or solid* it simplifies to

$$\rho u_t + \rho u_x \mathrm{v} + q_x = 0\,, \tag{28b}$$

since $\mathrm{v}_x = 0$ there. Also u may be expressed simply in terms of temperature in each phase, as follows: ρ = *constant* implies $du = Tds - Pd(\frac{1}{\rho}) = Tds$;

considering the entropy s as a function of T and P, we have $Tds = T(\frac{\partial s}{\partial T})_P dT + (\frac{\partial s}{\partial P})_T dP$ but $T(\frac{\partial s}{\partial T})_P = c =$ *heat capacity under constant pressure*, and by the Maxwell relation [CALLEN]: $(\frac{\partial s}{\partial P})_T = -\left(\frac{\partial \frac{1}{\rho}}{\partial T}\right)_P$ equals zero, which leaves us with $du = c\,dT$. Therefore, $u_t = c\,T_t$, $u_x = c\,T_x$ and (28b) reduces to the heat conduction equations:

$$\rho_L c_L[T_t + \mathrm{v}_L T_x] = k_L T_{xx} \quad \text{in the } liquid \tag{29a}$$

and

$$\rho_S c_S[T_t + \mathrm{v}_S T_x] = k_S T_{xx} \quad \text{in the } solid\ . \tag{29b}$$

Let us emphasize that these equations express energy conservation only *inside* each phase (where $\rho = constant$). Across the interface, the density jumps and the correct energy balance is expressed by (27). To see what this really says, we note that, by (19),

$$\rho\varepsilon = \rho[u + \frac{1}{2}\mathrm{v}^2] = \rho e - P + \frac{1}{2}\rho\mathrm{v}^2\ . \tag{30}$$

Choosing solid at temperature T_m and pressure P_{ref} as the reference state, we can express the energy at any state (ρ, T, P) as (PROBLEM 24)

$$\rho\varepsilon = \begin{cases} \rho_L u_{ref} + \rho_L L - \left[1 - \frac{\rho_L}{\rho_S}\right] P_{ref} + \rho_L c_L[T - T_m] + \frac{1}{2}\rho_L \mathrm{v}_L^2 & \text{in the } liquid \\ \rho_S u_{ref} \qquad\qquad + \rho_S c_S[T - T_m] + \frac{1}{2}\rho_S \mathrm{v}_S^2 & \text{in the } solid\ , \end{cases} \tag{31}$$

where $u_{ref} := u(\rho_S, T_m, P_{ref})$ is the reference value of internal energy, and $L = \Delta e(T_m, P_{ref})$, so P_{ref} has to be the pressure at which the heat of fusion is measured. Then, (27) becomes

$$\begin{aligned} &\left\{(\rho_L - \rho_S)u_{ref} + \rho_L L - \left[1 - \frac{\rho_L}{\rho_S}\right]P_{ref} + \frac{1}{2}(\rho_L \mathrm{v}_L^2 - \rho_S \mathrm{v}_S^2)\right\}X' \\ &= (\rho_L \mathrm{v}_L - \rho_S \mathrm{v}_S)u_{ref} + \rho_L L \mathrm{v}_L - \left[1 - \frac{\rho_L}{\rho_S}\right]P_{ref}\mathrm{v}_L + P_0^L \mathrm{v}_L - P_0^S \mathrm{v}_S \\ &\quad + \frac{1}{2}(\rho_L \mathrm{v}_L^3 - \rho_S \mathrm{v}_S^3) + [\![q]\!]_S^L \end{aligned} \tag{32}$$

on $x = X(t)$. The pressure jump is found from (26) to be

$$P_0^L - P_0^S = (\rho_L \mathrm{v}_L - \rho_S \mathrm{v}_S)X' - (\rho_L \mathrm{v}_L^2 - \rho_S \mathrm{v}_S^2) \quad \text{on } x = X(t) \tag{33}$$

and (25) yields

$$(\rho_L - \rho_S)X' = \rho_L \text{v}_L - \rho_S \text{v}_S \qquad \text{on } x = X(t) \; . \tag{34}$$

Thanks to (34), the u_{ref} terms cancel out and (32) becomes

$$\begin{aligned} \rho_L L[X' - \text{v}_L] - \left[1 - \frac{\rho_L}{\rho_S}\right] P_{ref}[X' - \text{v}_L] + \frac{1}{2}\rho_L \text{v}_L^2 [X' - \text{v}_L] \\ - \frac{1}{2}\rho_S \text{v}_S^2 [X' - \text{v}_S] - [P_0^L \text{v}_L - P_0^S \text{v}_S] = [\![q]\!]_S^L \end{aligned} \tag{35}$$

Now, let us apply these to the situations examined in **§2.3.B** and **§2.3.D**, in order to derive the correct energy balance across the interface in each case.

Expansion on melting without void

In the situation of **§2.3.B**, the face $x = 0$ was assumed fixed, and therefore the uniform v_L must be identically zero. Then, the momentum equation in the liquid with $\text{v}_L \equiv 0$ implies $P_x^L \equiv 0$, whence $P^L(x, t) \equiv constant = P_0^L$, and we may choose $P_0^L = P_{ref}$. Also, with $\text{v}_L \equiv 0$, (34) yields

$$\text{v}_S(t) = \left(1 - \frac{\rho_L}{\rho_S}\right) X'(t) > 0 \, , \tag{36}$$

so the solid *must* move to the right with non-constant speed, and the momentum equation in the solid tells us that $P^S(x, t)$ will be linear in x. Its interfacial value is found from (33) to be

$$P_0^S - P_{ref} = P_0^L - P_{ref} + \rho_S \text{v}_S [X' - \text{v}_S] = P_0^L - P_{ref} + \rho_L \left(1 - \frac{\rho_L}{\rho_S}\right) X'^2 \, . \tag{37}$$

Using $\text{v}_L = 0$, (36), (37), in (35), we find

$$\rho_L L X' + \left(1 - \frac{\rho_L}{\rho_S}\right)[P_0^L - P_{ref}]X' + \frac{1}{2}\rho_L \left(1 - \frac{\rho_L}{\rho_S}\right)^2 X'^3 = [\![-kT_x]\!]_S^L, \quad x = X(t), \; t > 0 \, . \tag{38}$$

With the choice $P_0^L = P_{ref}$, this finally becomes

$$\rho_L L X' + \frac{1}{2}\rho_L \left(1 - \frac{\rho_L}{\rho_S}\right)^2 X'^3 = [\![-kT_x]\!]_S^L, \qquad x = X(t), \qquad t > 0 \, . \tag{39}$$

Equations (29a, b) are the same as (4a, b), (35) is (4c) and therefore the correct model is given by (4) but with (4e) replaced by (39), which does not allow a similarity solution. By dropping the (positive) X'^3-term we speed up the process, so we expect that the explicit solution will *overestimate* the true interface, especially in early times. The significance of the dropped term is examined in [ALEXIADES-DRAKE].

Similarly, one may derive the correct interface conditions for the following cases (PROBLEM 25):

– *expansion on freezing* ($\rho_S \le \rho_L$, $v_S = 0$, $v_L > 0$, PROBLEM 10):

$$\rho_S L X' - \frac{1}{2}\rho_S\left[1 - \frac{\rho_S}{\rho_L}\right]^2 X'^3 = [\![-kT_x]\!]_S^L \, ; \tag{40}$$

– *shrinkage on melting without void* ($\rho_L \ge \rho_S$, $v_L = 0$, $v_S < 0$, PROBLEM 11):

$$\rho_L L X' - \frac{1}{2}\rho_L\left[\frac{\rho_L}{\rho_S} - 1\right]^2 X'^3 = [\![-kT_x]\!]_S^L \, ; \tag{41}$$

– *shrinkage on freezing without void* ($\rho_S \ge \rho_L$, $v_S = 0$, $v_L < 0$, PROBL. 12):

$$\rho_S L X' + \frac{1}{2}\rho_S\left[\frac{\rho_S}{\rho_L} - 1\right]^2 X'^3 = [\![-kT_x]\!]_S^L \, . \tag{42}$$

Note that in (40) and (41), the X'^3 - term is *negative*, so we expect the explicit solutions (resulting by dropping the term) to *underestimate* the true interfaces in those cases.

Shrinkage on freezing with void at the wall

In the situation of **§2.3.D**, $\rho_S \ge \rho_L$ and the liquid was assumed stationary, implying $v_L = 0$. Then (34) yields

$$v_S(t) = (1 - \frac{\rho_L}{\rho_S})X'(t) > 0 \, , \tag{43}$$

so the solid *must* move to the right with non-constant speed. Applying (25-27) to the void-solid interface $x = Y(t)$, we find (since $\rho_{void} = 0$)

$$Y' = v_S, \qquad p^{void} = p^S, \qquad 0 = [\![q]\!]_{void}^{s} \, , \tag{44}$$

respectively, and (35) gives (PROBLEM 26)

$$\rho_L L X' + \frac{1}{2}\rho_L\left[1 - \frac{\rho_L}{\rho_S}\right]^2 X'^3 = [\![-kT_x]\!]_S^L \quad \text{on} \quad x = X(t), \quad t > 0 \, . \tag{45}$$

Therefore the complete model consist of (15) but with (45) replacing (15f). The X'^3-term is again positive, so the interface of the approximate model, (15), will *overestimate* the true interface.

Similarly, one may consider *shrinkage on melting with void at the wall* ($\rho_L \ge \rho_S$, $v_S = 0$, $v_L > 0$, PROBLEM 21), in which case

$$v_L(t) = (1 - \frac{\rho_S}{\rho_L})X'(t), \qquad \rho_S L X' - \frac{1}{2}\rho_S\left[1 - \frac{\rho_S}{\rho_L}\right]^2 X'^3 = [\![-kT_x]\!]_S^L \, . \tag{46}$$

Here we expect an *underestimate* by dropping the X'^3-term.

PROBLEMS

PROBLEM 1. Show that the interface condition (4e) remains unchanged if we choose a different reference temperature, say $T_{ref} < T_m$. [Hint:

$$E(t) = \int_0^X \{\rho_L c_S[T_m - T_{ref}] + \rho_L c_L[T - T_m] + \rho_L L\}dx + \int_X^l \rho_S c_S[T - T_{ref}]dx\,,$$

proceed as in §**1.2.E**].

PROBLEM 2. Derive the similarity solution (6), and verify that it satisfies (4).

PROBLEM 3. Show that the transcendental equation (6e) has unique solution for any $St_L > 0$, $St_S \geq 0$, $\mu > 0$, $\nu > 0$. [see PROBLEM 5 §**2.2**].

PROBLEM 4. Show that the root λ of (6e) increases as μ increases [see PROBLEM 4 §**2.2**].

PROBLEM 5. Consider the 1-phase analogue of (4), i.e. $T_S = T_m$.

(a) Formulate the melting problem when $\rho_L \leq \rho_S$.

(b) Find its similarity solution. Compare with (6), and with the case $\rho_L = \rho_S$ (§**2.1**).

(c) Show that the root λ is identical to that of the $\rho_L = \rho_S$ case. Will the solid still move? Why?

PROBLEM 6. Using any convenient method, verify the values of the root λ shown in **Table 1**.

PROBLEM 7. Under what conditions may one expect the root λ of (6e) to be small? Under such conditions derive an approximate expression for λ.

PROBLEM 8. Consider the melting of ice, initially at $T_S = -10°C$, via $T_L = 25°C$ at $x = 0$, with $c_L = 4.2$, $c_S = 2.05$, $k_L = 0.58 \times 10^{-3}$, $k_S = 2.28 \times 10^{-3}$, $T_m = 0°C$, $L = 333.4$ (in SI units, see (7)).

(a) Take $\rho_S = \rho_L = 1000\ kg/m^3$ and find the root λ for the standard Stefan Problem. Next, take $\rho_L = \rho_S = 917\ kg/m^3$ and find the root λ. Explain why they are the same.

(b) Take $\rho_L = 1000$, $\rho_S = 917\ kg/m^3$ and find the root λ of (6e).

(c) Plot the interface in each case and compare.

(d) Find the melt-time of $x = 1\ cm$ in each case and compare.

PROBLEM 9. For the problem of §**2.3.B**, calculate the latent and total sensible heats (see §**2.2.E**). Is the ratio of sensible to latent heat independent of time as it was in the case $\rho_L = \rho_S$? Find the ratio.

PROBLEM 10. Formulate a problem, similar to (4), modeling the *freezing* of an initially liquid slab from the left, when $\rho_L > \rho_S$ (water/ice), assuming the liquid moves uniformly (to the right) to accommodate the *expansion* (but see (37)).

PROBLEM 11. Formulate a problem, similar to (4), for *melting* an initially solid slab from the left, when $\rho_L > \rho_S$ (like water/ice), assuming the solid moves uniformly (to the left) to accommodate the *shrinkage* (but see (38)).

PROBLEM 12. Formulate a problem, similar to (4), modeling the *freezing* of an initially liquid slab from the left, when $\rho_L < \rho_S$, assuming the liquid moves uniformly (to the left) to accommodate the *shrinkage* (but see (39)).

PROBLEM 13. Find the similarity solution for the model of PROBLEM 10. Verify that it coincides with (6) if the subscripts "L" and "S" are interchanged.

PROBLEM 14. Derive (9) and (10).

PROBLEM 15. Answer the two questions in **§2.3.C** using the minimum and also the maximum values of the parameters in **Table 2**.

PROBLEM 16. In the situation of **§2.3.D**, fix a material point $\xi(t)$ in the solid (hence $\xi'(t) = \text{v}$) and set up mass and heat balances in $[0, \xi(t)]$ to derive the interface condition (14) across the void-solid interface $x = Y(t)$.

PROBLEM 17. Seek a similarity solution of (15) in the form

$$X(t) = 2\lambda\sqrt{\alpha_S t}, \qquad Y(t) = (1-\mu)X(t), \qquad T(x,t) = F(\xi)\,, \qquad \xi = x/\sqrt{t}\ .$$

Derive (17) and verify that it is a solution.

PROBLEM 18. Let $f(\lambda)$ denote the left-hand side of (17e) multiplied by λ, so that the equation becomes $f(x) = \lambda\sqrt{\pi}$. Show that $f(\lambda)$ is a decreasing function (see PROB. 4 **§2.2**), $f(\lambda) \to +\infty$ as $\lambda \downarrow 0$ and $f(\lambda) \to 0$ as $\lambda \uparrow +\infty$. Conclude that (17e) has unique solution $\lambda > 0$ for any $\mu > 0$, $\nu > 0$, $\boldsymbol{St}_S > 0$, $\boldsymbol{St}_L \geq 0$.

PROBLEM 19. Formulate the 1-phase problem with void at the wall, and derive its similarity solution. Verify that (15) and (17) reduce to the 1-phase case when $T_L = T_m$. Also check that $\rho_L = \rho_S$ leads to the 1-phase Stefan Problem **§2.1**).

PROBLEM 20. Is the ratio of sensible to latent heat constant (in time) for (15-17)?

PROBLEM 21. Formulate and solve a *melting* problem analogous to (15) for a material with $\rho_L \geq \rho_S$.

PROBLEM 22. Determine the effect of the void growing in a long (vertical) tube initially filled with ice and being melted at the top face. [see PROBLEM 21; for parameter values see PROBLEM 8].

PROBLEM 23. Using the continuity and momentum equations, (20-21), show that the energy equation, (22), becomes (28a) when re-written in terms of the internal energy.

PROBLEM 24. To derive the energy expression, (31), one may proceed as follows:

(a) In *each* phase, we can write $d[\varepsilon - \frac{1}{2}v^2] = du = cdT$ since $\rho \equiv const.$ Integrate separately in each phase, referring to interfacial values (subscripted by 0) to obtain

$$\varepsilon^L = u_0^L + c_L[T - T_m] + \tfrac{1}{2}v_L^2, \qquad \varepsilon^S = u_0^S + c_S[T - T_m] + \tfrac{1}{2}v_S^2 \ . \tag{47}$$

where $u_0^L = u(\rho_L, T_m, P_0^L)$ and $u_0^S = u(\rho_S, T_m, P_0^S)$, with P_0^L, P_0^S the pressures at the interface from the liquid and solid sides respectively.

(b) As common reference state, choose *solid* at T_m, P_{ref}, whence $u(\rho_S, T_m, P_{ref}) =: u_{ref}$. Show that

$$u_0^S \ = \ u_{ref} + [u_0^S - u_{ref}] \ = \ u_{ref} \ . \tag{48a}$$

Next, write

$$u_0^L = u_{ref} + [u(\rho_L, T_m, P_{ref}) - u(\rho_S, T_m, P_{ref})] + [u(\rho_L, T_m, P_0^L) - u(\rho_L, T_m, P_{ref})],$$

i.e. change phase at (T_m, P_{ref}). The last term is zero, as in (b). Using $\Delta u = \Delta e - \Delta(P/\rho)$ for the middle term, show that

$$u_0^L \ = \ u_{ref} + L - \left[\frac{P_{ref}}{\rho_L} - \frac{P_{ref}}{\rho_S}\right] . \tag{48b}$$

(c) From (47, 48) derive (31).

PROBLEM 25. Derive the interface conditions (39-42).

PROBLEM 26. Derive the interface conditions (44-46).

2.4. SOLIDIFICATION OF A SUPERCOOLED MELT

2.4.A How supercooling arises

The equilibrium melt temperature, T_m, of a pure material is that temperature at which its liquid and solid phases can coexist in thermodynamic equilibrium (under ambient pressure conditions). At temperatures higher than T_m, the free energy of liquid (**Figure 2.4.1**) is lower than that of solid, so according to the Second Law of Thermodynamics, liquid is the stable phase. The free energy curves cross each other at T_m, and below T_m the stable phase is solid. Nevertheless, many materials can be cooled to temperatures substantially below T_m before solid actually forms. For example, under very precise control, it is possible to cool water down to $-40°C$ before it will freeze spontaneously. Other materials, like silicates and polymers, easily supercool by hundreds of degrees turning into glass instead of crystalizing !

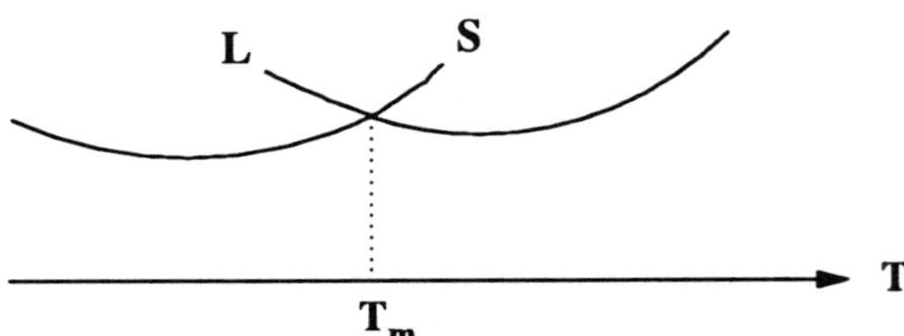

Figure 2.4.1. Free energy curves.

Liquid at a temperature $T < T_m$ is said to be **supercooled** and $\Delta T := T_m - T$ is the degree of supercooling. The supercooled state is thermodynamically **metastable,** meaning that it has free energy *higher* than that of the true stable state, the solid; thus it is a *local* minimum free energy state, not a global one. There is an energy barrier preventing the system from attaining the true stable state, and this barrier is attributed to nucleation difficulties (activation energy barrier for homogeneous nucleation, [KURZ-FISHER], [PORTER-EASTERLING]) arising from interfacial energy considerations. Container walls and any foreign particles in the melt constitute nucleation sites for heterogeneous nucleation, for which the activation barrier is considerably lower. As the supercooling increases, the nucleation barrier decreases and a large enough fluctuation in the system (from whatever source, thermal, vibrational, even random) may create a nucleus large enough to be stable (larger than a critical size), [TILLER], [WALDRAM], [PORTER-EASTERLING], [KURZ-FISHER], [ALEXIADES-SOLOMON-WILSON, 1988], [ALEXIADES-SOLOMON, 1989].

Once that happens, the nucleus grows very rapidly. The latent heat released upon freezing at the interface raises the temperature locally to the freezing temperature, at which more liquid freezes, and so on. At this point the *morphology* of the interface itself plays a role. If the interface is curved, with mean curvature $\frac{1}{2}\kappa$, then the *local freezing temperature*, T_f, is given by the

Gibbs – Thomson relation: $$T_f = T_m - \Gamma\kappa, \tag{1}$$

where $\Gamma > 0$ depends on the ratio of surface-tension coefficient to latent heat, see **§2.4.F**, also [PORTER-EASTERLING], [KURZ-FISHER], [TRIVEDI]. The mean curvature of a surface is the average of its two principal curvatures: $\frac{1}{2}\kappa = \frac{1}{2}(\frac{1}{R_1} + \frac{1}{R_2})$, with R_1, R_2 the principal radii of curvature (in our notation κ denotes the sum of the principal curvatures). For a planar (flat) interface, we have $\kappa = 0$, so $T_f = T_m$, and the supercooled liquid must be warmed up to T_m in order to freeze. On the other hand, at a protrusion of curvature $\kappa > 0$, the freezing point is lowered by $\Gamma\kappa$, so supercooled liquid of temperature $T_m - \Gamma\kappa$ will freeze on it making it grow. In fact, the more pointed the tip the more it is favored for growth! This is the mechanism by which columnar and dendritic growth takes place, creating a fascinating variety of morphologies, (**Figure 1.1.2**. Note that *the Gibbs-Thomson effect is strictly multidimensional,* since in one-dimension the interface is just a point, having no curvature, so the effect cannot be felt there.

In the presense of supercooling, the (multidimensional) Stefan Problem becomes considerably more complicated. The interface temperature, T_f, now depends on the local curvature instead of being a given constant, and the Stefan Condition (for constant c_L, c_S) takes the form:

$$\rho\Big[\, L - (c_L - c_S)\,[\,T_m - T_f\,]\,\Big] V_n \;=\; -k_L \frac{\partial T}{\partial \vec{n}} + k_S \frac{\partial T}{\partial \vec{n}}, \tag{2}$$

with V_n the velocity in the direction $\vec{n}$ normal to the interface. The modification to the latent heat term arises from the fact that the energy jump across the interface occurs at temperature $T_f \neq T_m$, see **§2.4.F**. The Stefan Problem with supercooling, consisting of the heat equation in the liquid and in the solid, the interface conditions (1) and (2), and appropriate initial and boundary conditions, is very difficult to analyze mathematically (see [CHADAM, 1991] for some theoretical and numerical approaches). The Mullins-Sekerka morphological stability analysis, based on approximate solutions to this problem (**§2.4.G**) shows that supercooling destabilizes a planar interface while surface tension acts as a stabilizing force. It is the competition between these two that creates the great variety of observed morphologies.

It should be remarked that Γ is very small typically and therefore the freezing point depression is appreciable only for very large curvatures. For example, in the case of water/ice $\Gamma \approx 18 \times 10^{-9}\, m\,K$, so in order for the freezing point to be depressed by one degree, the protrusion should have radius of curvature no larger than $2\Gamma \approx 36$ nanometers ! Hence, the Gibbs-Thomson effect, fundamental as it is for the morphology of the interface, is typically insignificant for the overall heat transfer process.

A typical experimental cooling curve exhibiting supercooling is seen in **Figure 2.4.2.** In this experiment, [DEAL-SOLOMON] a box was filled with a hydrated salt ($T_m \approx 28°C$) and 5 thermocouples were placed at various distances from a heat-exchange side (all other sides were kept insulated). The Figure shows the temperature recorded by two of the thermocouples during the 350 minutes of the experiment (the one marked "wall" was attached to the heat-exchange side of the box). Cooling began at $t = 60$ min. Observe the other curve. It drops steadily to several degrees below $T_m = 28°C$ until some disturbance set off nucleation (at $t \approx 250$ min). Then it rises rapidly to T_m, where freezing took place, before falling again as a solid.

Supercooling may also arise from causes other than nucleation difficulties. For example, doping of crystaline silicon results in an amorphous silicon layer near the surface, (due to the destruction of the lattice structure) and laser annealing is applied in an attempt to restore the crystaline structure, trapping the dopant. The melt temperature of amorphous Si is much lower that than of crystaline Si. When the laser energy is absorbed by the amorphous layer, liquid silicon appears but at a temperature far below its normal melting point. Thus, highly supercooled liquid silicon is formed, which subsequently solidifies.

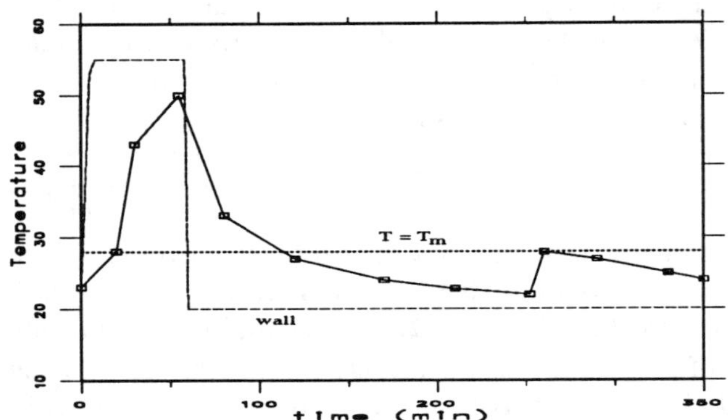

Figure 2.4.2. Experimental cooling curve exhibiting supercooling.

In solidification of alloys there is yet another source of supercooling, referred to as "constitutional supercooling". It will be discussed in §2.5.

Here we will examine the solidification of supercooled liquid, after nucleation has taken place, in a setting simple enough for the problem to be explicitly solvable. Thus, we will modify the Classical one-dimensional Stefan Problem to allow initial supercooling and will examine the effect on the solution. Clearly, this idealized setting leaves out all the interesting (microscopic) aspects of nucleation and interfacial morphology, subjects which continue to be under intense scientific study. The thermodynamics of phase equilibrium and a derivation of the Gibbs-Thomson relation will be discussed in §2.4.F. Finally, the classical approach to interfacial stability, known as the Mullins-Sekerka stability analysis, will be presented in §2.4.G.

2.4.B One-phase supercooled solidification

We begin with the simplest *one-dimensional* situation, in which a one-phase process arises. Recall that the interface *must* be assumed planar for the process to be one-dimensional.

PHYSICAL PROBLEM: A semi-infinite slab, $x \geq 0$, is initially at uniform temperature $T_{init} < T_m$ but *liquid*, hence supercooled. At time $t = 0$, the face $x = 0$ is brought to the melt temperature T_m and a (planar) solidification front begins propagating from $x = 0$ into the slab. We assume constant thermophysical properties, $\rho_S = \rho_L$, and that the freezing (liberation of the latent heat) occurs at the normal melt temperature T_m, consistent with the Gibbs-Thomson relation, (1).

Thus, the temperature at the front $x = X(t)$ is equal to T_m, ahead of it there is *supercooled liquid* at temperature below T_m, and behind it there is *solid* at temperature T_m.

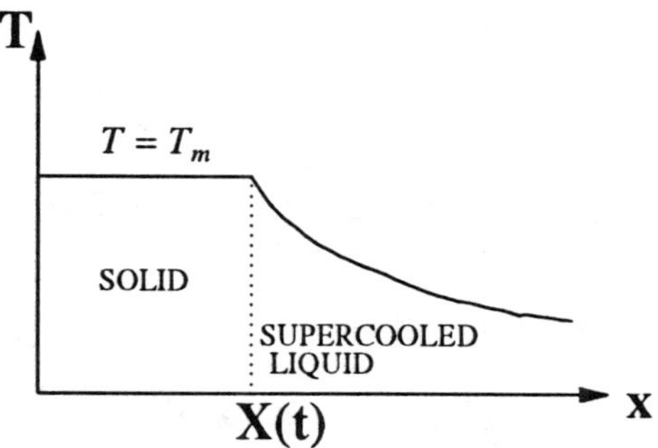

Figure 2.4.3. Typical temperature profile in 1-phase solidification of supercooled liquid.

Notice that exactly the same situation would arise if the face $x = 0$ were *insulated* and we assumed that nucleation started there at time $t = 0$ (from whatever cause).

In either case, we have a one-phase Stefan problem but, unlike the one-phase problem of **§2.1**, here the heat flows *from* the interface *to* the supercooled liquid ahead of it as depicted in **Figure 2.4.3**. The latent heat released upon freezing at the interface warms up the supercooled liquid ahead of it to the freezing temperature T_m enabling it to freeze. Clearly, this requires that the liquid *not* be hypercooled (**§1.1**), as we shall see below.

Our unknowns are the interface location $x = X(t)$ and the temperature $T(x, t)$ in the supercooled liquid, since $T \equiv T_m$ in the solid ($0 \le x \le X(t)$, $t > 0$). The interface conditions are the standard ones since $T_f = T_m$ here (PROBLEM 1). Thus we are led to the

MATHEMATICAL PROBLEM: (one-phase, supercooled semi-infinite slab freezing from the left):

Find $T(x, t)$ and $X(t)$ satisfying

$$T(x,t) = T_m, \qquad 0 < x < X(t), \qquad t > 0 \tag{3a}$$

$$T_t = \alpha_L T_{xx}, \qquad X(t) < x < \infty, \qquad t > 0 \tag{3b}$$

$$T(X(t),t) = T_m, \qquad t > 0 \tag{4a}$$

$$\rho L X'(t) = -k_L T_x(X(t)^+, t), \qquad t > 0 \tag{4b}$$

$$T(x,0) = T_{init} < T_m, \qquad 0 < x < \infty, \qquad \text{(but liquid)} \tag{5a}$$

$$X(0) = 0 \tag{5b}$$

$$T(0,t) = T_m \quad \textbf{or} \quad -k_s T_x(0,t) = 0, \qquad t > 0\ . \tag{5c}$$

The problem admits a similarity solution as before. Setting $\xi = x/\sqrt{t}$, $X(t) = 2\lambda\sqrt{\alpha_L t}$, $T(x,t) = F(\xi)$, we find (PROBLEM 2),

$$X(t) = 2\lambda\sqrt{\alpha_L t}, \quad t > 0 \tag{6a}$$

$$T(x, t) = T_m\,, \qquad 0 \le x \le X(t), \quad t > 0\,, \tag{6b}$$

$$T(x, t) = T_{init} + (T_m - T_{init})\,\frac{\operatorname{erfc}(\frac{x}{2\sqrt{\alpha_L t}})}{\operatorname{erfc}(\lambda)}\,, \quad X(t) \le x, \quad t > 0\,, \tag{6c}$$

where λ is a root of the transcendental equation

$$\lambda\sqrt{\pi}\; e^{\lambda^2}\operatorname{erfc}(\lambda) = \boldsymbol{St}_{init}, \qquad \boldsymbol{St}_{init} = \frac{c_L(T_m - T_{init})}{L}\,. \tag{6d}$$

Let us examine the solvability of (6d). The function $f(\lambda) := \lambda\sqrt{\pi}e^{\lambda^2}\operatorname{erfc}(\lambda)$ is increasing (PROBLEM 4 §2.2) for $\lambda > 0$, and satisfies $f(0) = 0$, $f(\infty) = 1$; hence $0 < f(\lambda) < 1$ for all $\lambda > 0$. It follows that the equation $f(\lambda) = \boldsymbol{St}_{init}$, i.e. (6d), has *unique* solution if and only if $0 \le \boldsymbol{St}_{init} < 1$, that is, if and only if

$$T_m - \frac{L}{c_L} < T_{init} \le T_m\,. \tag{7}$$

It says that the liquid *should not be hypercooled* initially (**§1.1**), as one would expect on physical grounds (see **§2.4.C** for further discussion). The qualitative appearance of temperature profiles and freezing curves is examined in PROBLEMS 3 and 4. A typical temperature profile is shown in **Figure 2.4.3**.

2.4.C Two-phase supercooled solidification

Now we consider the more general two-phase case, but still one-dimensional.

PHYSICAL PROBLEM: Supercooled liquid at $T_{init} < T_m$ occupies a semi-infinite slab $0 \le x < \infty$. At time $t = 0$, a temperature $T_s \le T_m$ is imposed at $x = 0$ and a (planar) solidification front begins propagating from $x = 0$ into the slab. We assume constant thermophysical properties, $\rho_S = \rho_L := \rho$ and freezing at the normal T_m.

Thus, the temperature at the front $x = X(t)$ is equal to T_m, ahead of it there is supercooled liquid at temperature below T_m, and behind it there is solid also at temperature below T_m. Heat flows *from* the interface in both directions !

The mathematical formulation of the problem (PROBLEM 5) appears to be the same as that of the standard 2-phase Stefan Problem for freezing from the left ((1)**§2.2** with subscripts "L" and "S" interchanged and latent heat L replaced by $-L$), except that the *initial condition* is

$$T(x,0) = T_{init} < T_m, \quad x > 0, \qquad X(0) = 0 \tag{8a}$$

and the boundary conditions are

$$T(0, t) = T_S < T_m, \qquad \lim_{x\to\infty} T(x, t) = T_{init} < T_m, \quad t > 0\,. \tag{8b}$$

Its similarity solution is

$$X(t) = 2\lambda\sqrt{\alpha_S t}, \quad t > 0 \tag{9a}$$

$$T(x,t) = T_S + (T_m - T_S)\,\frac{\operatorname{erf}\left(\frac{x}{2\sqrt{\alpha_S t}}\right)}{\operatorname{erf}(\lambda)}, \quad 0 \le x \le X(t), \quad t > 0 \quad \text{(solid)} \tag{9b}$$

$$T(x,t) = T_{init} + (T_m - T_{init})\,\frac{\operatorname{erfc}\left(\frac{x}{2\sqrt{\alpha_L t}}\right)}{\operatorname{erfc}(\nu\lambda)}, \quad X(t) \le x < \infty, \; t > 0, \; \text{(liquid)} \tag{9c}$$

where λ is a root of the transcendental equation

$$\frac{\boldsymbol{St}_S}{\lambda\sqrt{\pi}e^{\lambda^2}\operatorname{erf}(\lambda)} + \frac{\boldsymbol{St}_{init}}{(\nu\lambda)\sqrt{\pi}e^{(\nu\lambda)^2}\operatorname{erfc}(\nu\lambda)} = 1\,, \tag{9d}$$

with

$$\boldsymbol{St}_S = \frac{c_S(T_m - T_S)}{L}, \qquad \boldsymbol{St}_{init} = \frac{c_L(T_m - T_{init})}{L}, \qquad \nu = \sqrt{\frac{\alpha_S}{\alpha_L}}\,. \tag{10}$$

A typical temperature profile is shown in **Figure 2.4.4**.

Let us examine the solvability of (9d), thinking of it as $f(\lambda) = 1$; its left-hand side, $f(\lambda)$, is a strictly decreasing function of $\lambda > 0$, being the sum of two decreasing functions (see PROBLEM 4 §2.2) and we have $f(0) = +\infty$, $f(\infty) = \boldsymbol{St}_{init}$. Therefore, the equation $f(\lambda) = 1$ has unique positive solution *if and only if* $0 \le \boldsymbol{St}_{init} < 1$. We conclude that the 2-phase supercooled solidification problem has unique similarity solution if and only if the liquid is *not hypercooled* initially, namely if

$$T_m - \frac{L}{c_L} \;<\; T_{init} \;\le\; T_m\,. \tag{11}$$

The physical significance of this condition is clear. For freezing to occur the temperature must be raised to T_m, which requires the sensible heat amount of $c_L(T_m - T_{init})$ per gram. Since the only source of heat here is the latent heat, it is necessary (and sufficient) to have $c_L(T_m - T_{init}) < L$.

When $\boldsymbol{St}_{init} \ge 1$, the liquid is said to be **hypercooled**. Then, (11) is violated and our problem has *no* similarity solution. Moreover, the physical argument above shows that there can be no solution at all. Clearly, hypercooled liquid can only solidify at a freezing temperature *lower than* T_m, if at all. Many materials become glass instead of solidifying.

Note that for $T_S = T_m$, (9) reduces to the 1-phase solution (6), and for $T_{init} > T_m$ we recover the Neumann solution of the Stefan Problem for freezing (**§2.2**).

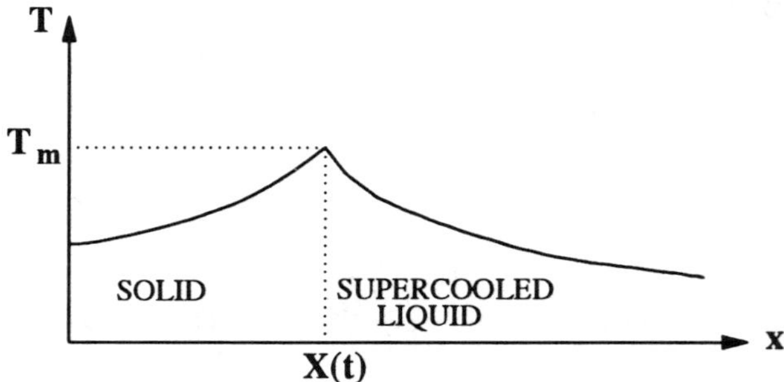

Figure 2.4.4. Typical temperature profile in 2-phase solidification of supercooled liquid.

EXAMPLE: *Solidification of supercooled water.* Consider a semi-infinite slab of supercooled water, initially at $T_{init} = -10°C$, solidifying under $T_S = -20°C$ imposed at $x = 0$ for $t \geq 0$. Using as parameter values $c_L = 4.2$, $c_S = 2.02 (kJ/kg°C)$, $k_L = 5.64 \times 10^{-4}$, $k_S = 2.24 \times 10^{-3} (kJ/m\, s°C)$, $\rho = \rho_L = \rho_S = 960\ kg/m^3$, $L = 333.4\ kJ/kg$, $T_m = 0°C$, we have $\boldsymbol{St}_S = 0.121$, $\boldsymbol{St}_{init} = 0.126$ and $\nu = 2.87$. Solving (9d) numerically, we find $\lambda \approx 0.266$ and we can now compute the interface location and temperature from (9). **Figure 2.4.5** shows the temperature profiles at times 2000 and 4000 seconds (observe that they are essentially linear in the solid, as expected due to the smallness of $\boldsymbol{St}_S$). The highest temperature occurs at the interface. **Figure 2.4.6** shows the temperature history at location $x^* = 2.5\, cm$ which solidified at time $t^* \approx 1900\, s$.

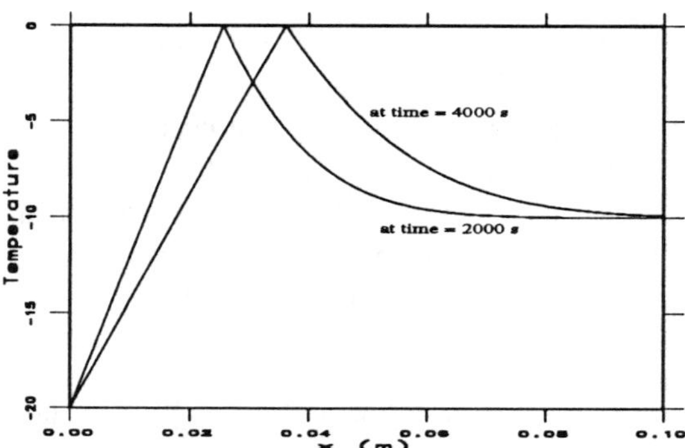

Figure 2.4.5. Temperature profiles in solidifying supercooled water.

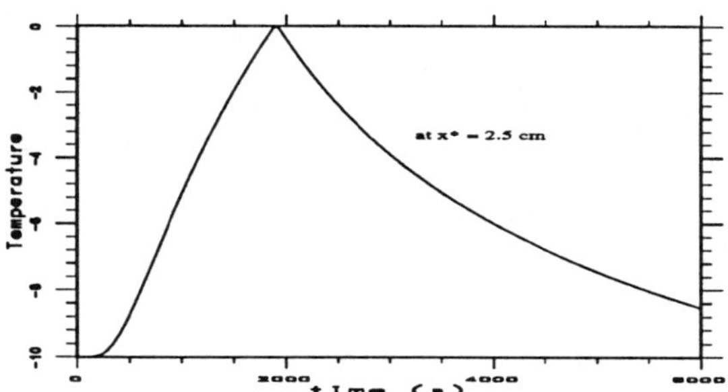

Figure 2.4.6. Temperature history at $x^* = 2.5\,cm$.

2.4.D Supercooled solidification with density-change

The model can be easily generalized to include density-change effects, as in §2.3. For example, let's consider the situation of §**2.4.C** but with $\rho_S \geq \rho_L$. There will be volume reduction upon freezing and let's assume that a void will be created at the face $x = 0$ to accommodate the shrinkage. The forming solid will then be moving to the right with uniform speed $\mathrm{v}(t)$, and we have the situation of §**2.3.D** except that now the liquid is supercooled initially.

The mathematical formulation of the problem will be as in (15)§**2.3.D** (with exact Stefan Condition (45)§**2.3.E**), but with the initial condition replaced by

$$T(x,0) = T_{init} < T_m, \qquad 0 < x < \infty \,. \tag{12}$$

As a result, $\boldsymbol{St}_L$ is replaced by $-\boldsymbol{St}_{init}$, and the similarity solution (for the *approximate* Stefan Condition (15f)§**2.3.D**) will be given by (17)§**2.3.D** (PROBLEM 7). Unique root exists provided $0 \leq \boldsymbol{St}_{init} < 1$, as above.

2.4.E Steady-state of a finite slab

Phase-change problems in a finite slab do *not* admit explicit solutions. On the other hand, steady-state (time independent) solutions are always of interest since they describe the long time behavior of the system.

PHYSICAL PROBLEM: Consider a finite slab, $0 \leq x \leq l$, insulated at both ends, and occupied by supercooled liquid at temperature $T_{init} < T_m$ initially. We assume that nucleation starts at $x = 0$, that $\rho_S = \rho_L$, and wish to find its steady-state.

Since there are no time variations, the interface will be a constant, say X_∞, the temperature a function of x only, and the heat equation reduces to $T_{xx} = 0$.

MATHEMATICAL PROBLEM (*steady-state of insulated supercooled finite slab*):

Find $T(x)$ and X_∞ such that

$$T_{xx} = 0, \quad 0 < x < X_\infty \quad \text{(solid)} \tag{13a}$$

$$T_{xx} = 0, \quad X_\infty < x < l \quad \text{(liquid)} \tag{13b}$$

$$T(X_\infty) = T_m \tag{14a}$$

$$0 = -k_L T_x(X_\infty^+) + k_S T_x(X_\infty^-) \tag{14b}$$

$$-k_S T_x(0) = 0, \quad -k_L T_x(l) = 0 \ . \tag{15}$$

This problem is trivial and leads to $T(x) \equiv T_m$, $0 \leq x \leq l$. Notice however that the initial condition has not been incorporated and X_∞ has not been found. To find it, we must ensure energy conservation between the initial and final states. Using the melt temperature as reference, initially the energy was

$$E = \rho L l + \rho c_L (T_{init} - T_m) l \ . \tag{16}$$

The final state has $T = T_m$ and $E = \rho L(l - X_\infty)$. Hence we must have

$$\frac{X_\infty}{l} = St_{init} := \frac{c_L(T_m - T_{init})}{L} \ . \tag{17}$$

If $0 \leq St_{init} < 1$, then $X_\infty = l \, St_{init}$. Therefore, the unique steady-state for the initially supercooled but *not* hypercooled slab consists of solid in $0 \leq x < X_\infty$ and liquid in $X_\infty < x \leq l$, both at temperature T_m.

If $St_{init} = 1$, then $X_\infty = l$, so we find all solid at T_m.

Finally, if $St_{init} > 1$, then (17) is impossible. In fact, this is the case of *hypercooling* and we already saw earlier (§**2.4.B** ,C) that we cannot demand phase-change at T_m. Relaxing (14a) to $T(X_\infty) = T_\infty$, we find $T \equiv T_\infty < T_m$, but now we have two unknowns (X_∞ and T_∞) and only one relation (the energy balance) from which to determine them. Clearly, the steady-state of a hypercooled slab cannot be determined without additional information or assumptions. One possibility is that the steady-state will be all solid at $T \equiv T_\infty$, i.e. $X_\infty = l$. Then, equating (16) with $E = \rho c_S (T_\infty - T_m) l$ we find

$$T_\infty = T_m - \frac{L}{c_S}(St_{init} - 1) < T_m \ . \tag{18}$$

Density change may be incorporated in the above considerations, leading to (17) again. The amount of void (created when $\rho_S \geq \rho_L$) or elongation (when $\rho_S \leq \rho_L$) of the slab is easily found by mass conservation to be $(1 - \rho_L / \rho_S) \, X_\infty$, see PROBLEMS 8-10.

2.4.F Phase equilibrium and the Gibbs-Thomson effect

The lowering of the freezing point at a curved interface is known as the Gibbs-Thomson effect, and it is due to surface tension. We present its derivation from first principles in order to make clear that it is a direct consequence of the conditions for thermodynamic equilibrium at a curved interface.

Phase equilibrium - The Laplace-Young relation

Consider a multi-component system, of fixed mass M and total volume V, consisting of solid and liquid separated by an interface (idealized as a surface) of area A. The total internal energy U of the system may be written as ([LUPIS], [ADAMSON])

$$U = T^L S^L + T^S S^S - P^L V^L - P^S V^S + \sum \mu_i^L N_i^L + \sum \mu_i^S N_i^S + \sigma A , \quad (19)$$

where T^j, S^j, P^j, V^j are respectively the temperature, entropy, pressure, and volume of phase $j = L, S$, μ_i^j, N_i^j are the chemical potential and number of moles of component i in phase j, and σ is the surface tension of the interface. The fundamental differential of energy is

$$dU = T^L dS^L + T^S dS^S - P^L dV^L - P^S dV^S + \sum \mu_i^L dN_i^L + \sum \mu_i^S dN_i^S + \sigma dA. \quad (20)$$

According to the Second Law of Thermodynamics, at equilibrium, the entropy is maximum for a given energy, or equivalently, the energy is minimum for a given entropy. Thus the general condition for thermodynamic equilibrium may be expressed as

$$(\delta U)_{S,V,N_i} \geq 0, \quad (21)$$

for all possible variations keeping S, V, N_i fixed.

Keeping the volumes and interface fixed, let us consider (non-mechanical) variations in entropies and compositions. Since the total S and N_i's should remain fixed, the variations are constrained by $\delta S^L + \delta S^S = \delta S = 0$, and $\delta N_i^L + \delta N_i^S = \delta N_i = 0$. Hence (21) implies $(T^L - T^S)\,\delta S^L + \sum(\mu_i^L - \mu_i^S)\,\delta N_i^L \geq 0$ for arbitrary δS^L, δN_i^L (positive or negative), and therefore we must have

$$T^L = T^S \qquad \text{and} \qquad \mu_i^L = \mu_i^S , \qquad i = 1, \ldots, m. \quad (22)$$

These are the conditions for thermal and chemical equilibrium between the two phases. Hence,

$$dU = T\,dS + \sum \mu_i\, dN_i - P^L\, dV^L - P^S\, dV^S + \sigma\, dA . \quad (23)$$

Next, keeping S and the N_i's fixed, let us consider mechanical variations δV^L, δV^S and δA. These are constrained by $\delta V^L + \delta V^S = \delta V = 0$, and by the geometric consistency relation : $\delta A = \kappa\, \delta V^S$. Then (21) implies $(P^L - P^S)\,\delta V^S + \sigma \kappa\, \delta V^S \geq 0$ for all δV^S, so we obtain the

$$\textbf{Laplace – Young relation:} \qquad P^S - P^L = \sigma \kappa , \quad (24)$$

expressing mechanical equilibrium at a *curved* interface. It follows, in particular, that for a flat (planar) interface ($\kappa = 0$) the pressures must be *equal*.

The **Gibbs free-energy** is defined as

$$G := U - TS + PV \equiv H - TS \equiv \sum \mu_i N_i , \tag{25}$$

where $H = U + PV$ is the enthalpy. The Gibbs free-energy is a thermodynamic potential with natural variables T, P, N_i and its fundamental differential is

$$dG = -S\,dT + V\,dP + \sum \mu_i\,dN_i . \tag{26}$$

It follows that $\mu_i = \left(\frac{\partial G}{\partial N_i}\right)_{T,P}$, so the chemical potential of component i is the *partial molar Gibbs free-energy of* i. For a one-component system (pure material), the chemical potential coincides with the *molar Gibbs free-energy* G/N.

The Clausius-Clapeyron relation

From now on we consider a pure material of fixed mass M, and for convenience, after division by M we deal with **specific** quantities (per gram), which we shall denote by lower case letters. In particular, we have the relations

$$g = u - Ts + Pv = h - Ts, \qquad dg = -s\,dT + v\,dP, \tag{27}$$

with $v = 1/\rho$ the specific volume, and the conditions for thermodynamic equilibrium are : equality of temperatures, equality of (specific) Gibbs free-energies, and the Laplace-Young relation. Denoting interfacial values by a subscript f, we must have

$$T_f^L = T_f^S =: T_f, \qquad P_f^S - P_f^L = \sigma\kappa, \qquad g^L(T_f, P_f^L) = g^S(T_f, P_f^S). \tag{28}$$

These constrain the interfacial values of the temperature and pressures so that only one of them remains independent, say T_f, in addition to κ (the set of phase-coexistence states has two degrees of freedom). Note that $g^L = g^S$ implies $h_f^L - T_f s_f^L = h_f^S - T_f s_f^S$, so that the enthalpy and entropy of fusion at T_f are related by

$$\Delta h(T_f) = T_f\,\Delta s(T_f). \tag{29}$$

As T varies, the pressures will also vary constrained by (28). Hence, at any coexistence state (T_f, P_f^L, P_f^S) we must have

$$dg^L \equiv -s^L\,dT_f + v^L\,dP^L = -s^S\,dT_f + v^S\,dP^S \equiv dg^S,$$

which yields the **Clapeyron relation**:

$$v^L\,dP^L - v^S\,dP^S = \Delta s(T_f)\,dT_f. \tag{30}$$

Phase coexistence for flat interface

For a flat interface ($\kappa = 0$), the pressures are equal :

$$P_f^S = P_f^L =: P_{flat}(T_f). \tag{31}$$

The *normal phase-change temperature* T_m is determined as the temperature at which the free-energy curves for liquid and solid intersect *at atmospheric pressure* ($P_{flat} = P_{atm}$). Then $\Delta h(T_m) = L$, the latent heat at (T_m, P_{atm}), and from (29) we have

$$\Delta s(T_m) = \frac{L}{T_m}. \tag{32}$$

At any other pressure, from (30) we get the

Clausius – Clapeyron relation: $$dP_{flat} = \frac{\Delta s(T_f)}{\Delta v(T_f)}\, dT_f \tag{33}$$

for the slope of the phase-coexistence curve $P_f = P_{flat}(T_f)$, where $\Delta v := v^L - v^S = \frac{1}{\rho^L} - \frac{1}{\rho^S}$ is the difference of specific volumes. In order to find the phase-coexistence curve itself, we need the temperature-dependence of the entropy of fusion. It changes as follows (PROBLEM 14) along this curve:

$$\begin{aligned}\frac{d}{dT}\Delta s &= \left(\frac{\partial \Delta s}{\partial \Delta T}\right)_P + \left(\frac{\partial \Delta s}{\partial \Delta P}\right)_T \frac{dP}{dT} = \frac{\Delta c}{T} - \left(\frac{\partial \Delta v}{\partial T}\right)_P \frac{\Delta s}{\Delta v} \\ &= \frac{\Delta c}{T} - \left[\frac{d}{dT}\Delta v - \left(\frac{\partial \Delta v}{\partial P}\right)_T \frac{\Delta s}{\Delta v}\right]\frac{\Delta s}{\Delta v},\end{aligned} \tag{34}$$

where we have used (33), the definition of specific heat: $c = T(\frac{\partial s}{\partial T})_P$, and the Maxwell relation: $\left(\frac{\partial s}{\partial P}\right)_T = -\left(\frac{\partial v}{\partial T}\right)_P$.

Now, the incompressibility of liquid and solid implies $\left(\frac{\partial \Delta v}{\partial P}\right)_T = 0$, and (34) becomes $\frac{d}{dT}(\Delta v\, \Delta s) = \frac{\Delta c}{T}\Delta v$. Integrating from the reference state (T_m, P_{atm}), we find the desired temperature-dependence of the entropy of fusion to be

$$\Delta s(T) = \frac{\Delta v(T_m)}{\Delta v(T)}\Delta s(T_m) + \frac{1}{\Delta v(T)}\int_{T_m}^{T} \Delta v(\tau)\frac{\Delta c(\tau)}{\tau}\, d\tau. \tag{35}$$

Then, integration of (33) yields the *pressure-temperature phase-coexistence curve for a flat interface*:

$$P_{flat}(T_f) = P_{atm} + \int_{T_m}^{T_f} \frac{1}{\Delta v(\tau)^2}\left[\Delta v(T_m)\frac{L}{T_m} + \int_{T_m}^{\tau} \Delta v(\bar\tau)\frac{\Delta c(\bar\tau)}{\bar\tau}\, d\bar\tau\right] d\tau. \tag{36}$$

In particular, if the densities ρ^L, ρ^S and specific heats c^L, c^S are *constants*, then (36) becomes

$$P_{flat}(T_f) = P_{atm} + \frac{1}{\Delta v}\left[L\left(\frac{T_f}{T_m} - 1\right) + \Delta c\left(T_f \ln\frac{T_f}{T_m} + T_m - T_f \right)\right]. \quad (37)$$

If the densities are *equal,* (37) is not valid, but (33) rewritten as $\Delta s(T_f)\dfrac{dT_f}{dP_f} = \Delta v(T_f) = 0$ implies $dT_f = 0$, so the *only* possible coexistence state is (T_m, P_{atm}) in that case.

Phase coexistence for curved interface

For a curved interface, the Laplace-Young relation, (24), implies $dP^L = dP^S + d(\sigma\kappa)$, so (30) becomes $dP^L = \dfrac{\Delta s}{\Delta v}dT_f + \dfrac{v^S}{\Delta v}d(\sigma\kappa)$, or, using (33),

$$d\left[P^L - P_{flat} \right] = \frac{v^S}{\Delta v} d(\sigma\kappa). \quad (38)$$

Integrating from (T_f, $\kappa = 0$) to (T_f, κ), we find

$$P^L - P_{flat} = \int_0^{\kappa} \frac{v^S}{\Delta v}\left[\sigma + \kappa\frac{\partial\sigma}{\partial\kappa} \right] d\kappa. \quad (39)$$

Assuming σ, v^S, v^L to be independent of κ, we get

$$P^L = P_{flat} + \frac{v^S}{\Delta v}\sigma\kappa, \quad (40)$$

and similarly

$$P^S = P_{flat} + \frac{v^L}{\Delta v}\sigma\kappa, \quad (41)$$

with P_{flat} given by (36).

If P^L is specified, then (40) constitutes an equation for T_f, determining the effect of curvature on the phase-change temperature, and (41) determines the pressure on the other side of the interface. Clearly, the roles of P^L and P^S may be interchanged. The classical Gibbs-Thomson relation, (1), is a particular case of these as we shall see.

The Gibbs-Thomson relation

If the densities and specific heats are *constants*, from (37),(40) we get

$$(\frac{1}{\rho^L} - \frac{1}{\rho^S})[P^L - P_{atm}] = L(\frac{T_f}{T_m} - 1) + \Delta c[T_f \ln\frac{T_f}{T_m} + T_m - T_f] + \frac{1}{\rho^S}\sigma\kappa. \quad (42)$$

In particular, for $\Delta c \approx 0$ and $\rho^L = \rho^S =: \rho$ (or $P = P_{atm}$), this reduces to the classical Gibbs-Thomson relation

$$T_f = T_m(1 - \frac{\sigma}{\rho L}\kappa) = T_m - \Gamma\kappa, \tag{43}$$

with *capillary constant* $\Gamma := \sigma T_m / \rho L$. Note that this may also be derived *directly* from the Clapeyron and Laplace-Young relations, (30),(24), under the assumptions: $\rho^L = \rho^S =: \rho$, $c_L = c_S$ (PROBLEM 15).

In the more general case of $\rho^L \neq \rho^S$, $\Delta c \neq 0$, for *small* dimensionless supercooling $\delta := (T_m - T_f)/T_m > 0$, we may approximate (42) to order δ^2 by

$$\frac{1}{2}\frac{T_m \Delta c}{L}\delta^2 - \delta + \bar{\Gamma} = 0, \qquad \text{where } \bar{\Gamma} := \frac{\sigma\kappa}{\rho^S L} - (\frac{1}{\rho^L} - \frac{1}{\rho^S})\frac{P^L - P_{atm}}{L},$$

from which we get (PROBLEM 16)

$$\delta = \bar{\Gamma}(1 + T_m \Delta c \bar{\Gamma}). \tag{44}$$

Therefore, the effect of $\Delta c \neq 0$ is only of second order in $\bar{\Gamma}$ (which is small), and (43) is usually an adequate approximation to (44) (PROBLEM 17). On the other hand, molecular attachment kinetics considerations introduce an additional term in (44) (or (43)) proportional to the interfacial velocity, see [SCHAEFFER-GLICKSMAN], [CORIELL-SEKERKA], [GURTIN], [TILLER].

The Stefan Condition in the presense of supercooling

The standard Stefan Condition is only valid for a *flat* interface at T_m and constant density. The effect of density change was discussed in **§2.3.E** . Here we assume constant density $\rho^L = \rho^S =: \rho$, but a *curved* interface at temperature $T_f = T_m - \Gamma\kappa$ and derive the correct Stefan Condition.

Energy conservation across an interface (for *constant* density) demands

$$[\![\rho\varepsilon]\!]_S^L V_n = [\![\vec{q}\cdot\vec{n}]\!]_S^L, \tag{45}$$

where ε is the total energy per gram, V_n is the interfacial velocity in the direction $\vec{n}$ normal to the interface, and $\vec{q} = -k\nabla T$ is the heat flux (see **§1.2, §2.3.E**). Note that $\rho = constant$ implies $\varepsilon \equiv u = h - P/\rho$, so the energy jump is

$$[\![\rho\varepsilon]\!]_S^L = \rho\Delta h(T_f) - [P_f^L - P_f^S], \tag{46}$$

where, by (29), $\Delta h(T_f) = T_f \Delta s(T_f)$ is the enthalpy of fusion at T_f and, by (24), $-[P_f^L - P_f^S] = \sigma\kappa$. Using (35) for $\Delta c = constant$, and (32), the jump in (46) becomes (PROBLEM 18)

$$[\![\rho\varepsilon]\!]_S^L = \rho\frac{T_f}{T_m}\left(L + T_m \Delta c \ln\frac{T_f}{T_m}\right) + \sigma\kappa. \tag{47}$$

But, (42) for $\rho^L = \rho^S =: \rho$ gives

$$\sigma\kappa = \rho L(1 - \frac{T_f}{T_m}) - \rho\Delta c\left(T_f \ln\frac{T_f}{T_m} + T_m - T_f \right),$$

so that (47) finally reduces to

$$[\![\rho\varepsilon]\!]_S^L = \rho\left(L - \Delta c\,[\,T_m - T_f\,] \right).$$

We conclude that for $\rho^L = \rho^S =: \rho \equiv constant$ and $\Delta c \equiv constant$, the Stefan Condition reads as mentioned in (2), namely

$$\rho\left(L - \Delta c\,[\,T_m - T_f\,] \right)V_n = -k_L \frac{\partial T}{\partial \vec{n}} + k_S \frac{\partial T}{\partial \vec{n}}, \tag{48}$$

2.4.G Mullins–Sekerka morphological stability analysis

The now classical approach to morphological stability was introduced by [MULLINS-SEKERKA,1963,1964] in the context of *directional solidification* and has been extended and generalized to more complicated problems and geometries, e.g. [SEKERKA,1968,1984a,1984b], [NASH–GLICKSMAN], [OCKENDON], [CHADAM–ORTOLEVA], [TURLAND–PECKOVER], [WOLKIND–NOTESTINE], [CORIELL–SEKERKA], [McFADDEN–CORIELL], and references therein. The ideas are easiest to see in the simple setting of (two-dimensional) steady-state directional solidification which we now describe.

Consider a thin layer of a pure material with thermophysical properties T_m, L, ρ, c_L, c_S, k_L, k_S (assumed constant). The layer has infinite extend in the x-direction and it will be solidified in the y-direction. We place two heaters at distance $2\cdot\Delta y$ apart, the lower one held at constant temperature $T_S < T_m$, the upper one held at constant temperature $T_L > T_m$, and we move this setup at a constant speed V_0 along the y-axis. We want the interface to remain *stationary* and flat at $y = 0$ with respect to this moving frame, i.e. we want to have

$$T_{xx} + T_{yy} = 0 \qquad \text{for } -\Delta y < y < 0 \text{ and for } 0 < y < +\Delta y, \tag{49a}$$

$$T(x,0) = T_m, \tag{49b}$$

$$\rho L V_0 = -k_L T_y + k_S T_y, \qquad on\ y = 0, \tag{49c}$$

$$T(x,-\Delta y) = T_S < T_m, \qquad T(x,\Delta y) = T_L > T_m. \tag{49d}$$

Clearly, T is linear in y, and the lines

$$T = T_m + g_S y, \quad -\Delta y < y < 0, \qquad \text{and} \qquad T = T_m + g_L y, \quad 0 < y < \Delta y$$

will satisfy this problem provided the (constant) gradients $g_S = \dfrac{T_m - T_S}{\Delta y}$ and $g_L = \dfrac{T_L - T_m}{\Delta y}$ satisfy

$$\rho L V_0 = -k_S g_S + k_L g_L , \tag{50}$$

which can be achieved by appropriate choice of T_S , T_L and Δy. Note that if the liquid is supercooled then $g_L < 0$.

We now study the stability of this stationary, flat-interface solution under small perturbations to the interface. Let

$$y = Y(x,t) := \delta(t) \sin \omega x \tag{51}$$

be the perturbed interface, with $\delta(t)$ small, so that to within a perturbation of order $O(\delta)$ the interface $y = Y(x,t)$ is close to $y = 0$ and flat (its curvature is $\kappa = -Y_{xx}(1+|Y_x|^2)^{-3/2} = \delta(t)\, \omega^2 \sin \omega x + O(\delta^2)$). We wish to find a solution $T(x, y, t)$ of the quasistationary problem

$$T_{xx} + T_{yy} = 0 \quad \text{for} \quad -\Delta y < y < 0 \quad \text{and for} \quad 0 < y < +\Delta y , \tag{52a}$$

$$T(x, Y(x,t), t) = T_m - \Gamma \kappa , \tag{52b}$$

$$\rho[L + \Delta c\, (T(x,Y,t) - T_m)]\,[V_0 + Y_t] = -k_L \frac{\partial T}{\partial \vec{n}} + k_S \frac{\partial T}{\partial \vec{n}} \quad \text{at} \quad y = Y(x,t), \tag{52c}$$

$$T(x, -\Delta y, t) = T_S < T_m , \qquad T(x, \Delta y, t) = T_L > T_m , \tag{52d}$$

where $\Delta c = c_L - c_S$ arises as in (48), and $\vec{n}$ denotes the unit normal to the interface towards the liquid. We seek an approximate solution in the form

$$T(x, y, t) = T_m + g_S + A_S(t)\, e^{\omega y} \sin \omega x , \quad y < 0 \quad (\text{ in the solid }), \tag{53a}$$

$$T(x, y, t) = T_m + g_L + A_L(t)\, e^{-\omega y} \sin \omega x , \quad y > 0 \quad (\text{ in the liquid }), \tag{53b}$$

with $A(t)$ and $\delta(t)$ to be found. These already satisfy Laplace's equation (52a), and (52b) demands

$$g_S Y + A_S(t)\, e^{\omega Y} \sin \omega x = \Gamma\, Y_{xx}\, (1+|Y_x|^2)^{-3/2} + O(\delta^2) = g_L Y + A_L(t)\, e^{-\omega Y} \sin \omega x .$$

Matching terms to order δ, we obtain (PROBLEM 19)

$$A_S(t) = -[\Gamma \omega^2 + g_S]\, \delta(t) , \qquad A_L(t) = -[\Gamma \omega^2 + g_L]\, \delta(t) . \tag{54}$$

Now we compute the fluxes at the interface $y = Y(x,t)$ to terms of order δ. The unit normal to the interface is

$$\vec{n} = \frac{(-Y_x , 1)}{(1+|Y_x|^2)^{1/2}} .$$

so the flux from the liquid is given by (PROBLEM 20)

$$-k_L \frac{\partial T}{\partial \vec{n}} = -k_L (T_x , T_y) \cdot \vec{n} = \frac{-T_x Y_x + T_y}{(1+|Y_x|^2)^{1/2}}$$

$$= -k_L \frac{-(A_L e^{-\omega Y}\, \omega \cos \omega x)(\delta \omega \cos \omega x) + g_L - A_L\, \omega\, e^{-\omega Y} \sin \omega x}{(1+\delta^2 \omega^2 \cos^2 \omega x)^{1/2}}$$

$$= -k_L g_L - k_L [\Gamma\omega^2 + g_L] \delta \omega \sin \omega x + O(\delta^2), \quad (55a)$$

and similarly

$$-k_S \frac{\partial T}{\partial \vec{n}} = -k_S g_S + k_S [\Gamma\omega^2 + g_S] \delta \omega \sin \omega x + O(\delta^2). \quad (55b)$$

Substituting into (52c) and matching terms, we obtain (50) and (PROBLEM 21)

$$\delta'(t) = -\Lambda(\omega) \delta(t), \quad (56)$$

$$\text{with } \Lambda(\omega) = \omega [\Gamma\omega (k_L + k_S + V_0 \rho \Delta c) + (k_L g_L + k_S g_S)] / \rho L.$$

It follows that the amplitude $\delta(t)$ of the perturbation *grows exponentially if* $\Lambda(\omega) < 0$, and decays exponentially if $\Lambda(\omega) > 0$. Note that Γ, ω, k_L, k_S, g_S, Δc are all positive, so *only* $g_L < 0$ *can create instability.*

We conclude that

- If $g_L > 0$, i.e. if the liquid is *not* supercooled, then $\Lambda(\omega) > 0$, so the flat interface is *stable.*
- If $g_L < 0$, i.e. if the liquid is supercooled, but $-g_L < \frac{k_S}{k_L} g_S$, then again $\Lambda(\omega) > 0$ and the flat interface is stable.
- If $g_L < 0$, and $-g_L > \frac{k_S}{k_L} g_S$ (large supercooling), then

— *without* surface tension ($\Gamma = 0$), the flat interface is *unstable,*

— with surface tension ($\Gamma > 0$), the interface will still be stable for large frequences ($\omega^2 > \frac{-(k_L g_L + k_S g_S)}{\Gamma(k_L + k_S + V_0 \rho \Delta c)}$),

— but *unstable* for small frequences.

This shows clearly the destabilizing effect of supercooling as well as the stabilizing effect of surface tension.

PROBLEMS

PROBLEM 1. Derive the interface condition for the 1-phase problem of **§2.4.B**.

PROBLEM 2. Derive the similarity solution (6) of the 1-phase problem (2-5).

PROBLEM 3. Show that at each $t > 0$ the temperature profile (6c) is decreasing and convex, as shown in **Figure 2.4.4**.

PROBLEM 4. Examine the convexity of the temperature - time curve $T = T(x^*, t)$ at a fixed location x^*, given by (6c).

PROBLEM 5. State precisely the mathematical problem modeling the 2-phase process of **§2.4.C** and derive its solution (9).

PROBLEM 6. Formulate and solve the problem arising when $T_S > T_m$ in the situation of **§2.4.C**.

PROBLEM 7. State precisely the mathematical model for the situation of **§2.4.D**. Verify that (17)**§2.3.E**, appropriately adjusted, is its solution.

PROBLEM 8. Find the steady-state of an insulated finite slab, initially supercooled but not hypercooled, when $\rho_S \geq \rho_L$ (assume a void between wall and solid).

PROBLEM 9. Find the steady-state of an insulated finite slab, initially supercooled but not hypercooled, when $\rho_S \leq \rho_L$ (assume expansion to the left). In particular, find the final length of the slab.

PROBLEM 10. Apply the results of PROBLEM 9 to water/ice to find the steady-state of an insulated slab of water, supercooled to −10°C, of thickness 1 *meter*. Use $\rho_L = 1000$, $\rho_S = 920\ kg/m^3$, $c_L = 4.2\ kJ/kg\,°C$, $L = 333\ kJ/kg$.

PROBLEM 11. A manufacturer of heat storage systems is selling a sealed container of a particular hydrated salt. The advertising brochure claims that the material "soaks up latent heat like a sponge" and it never supercools. The manufacturer has agreed to give you a unit for testing, provided you do not damage it in any way. How can you test their claim that their material never supercools? Assume you are able to measure surface temperatures and fluxes.

PROBLEM 12. An ice sculpture in the shape of a "dumbbell" (two spheres joined by a thin cylinder) is immersed totally in a large pool of water at 0°C. On the basis of the Gibbs-Thomson relation, (1), describe what you expect to happen.

PROBLEM 13. A cup of water being heated in a microwave oven often begins to boil with a small explosion spilling the water out of the cup (especially if the cup has smooth clean inside surface). To prevent this, one is advised to place a wooden stir stick (or any non-metallic object with rough surface) into the cup. Perform the experiment to see if this works. Explain why. Would you expect coffee, or any liquid containing small particles, to spill in the microwave oven ? why ?

PROBLEM 14. Justify (34) and then derive (35).

PROBLEM 15. Assuming $\rho^L = \rho^S \equiv \rho =$ constant and $\Delta c \equiv 0$, derive the Gibbs-Thomson relation (43) directly from (30), (24).

PROBLEM 16. For small dimensionless supercooling $\delta = (T_m - T_f)/T_m$, derive the more general Gibbs-Thomson relation (44).

PROBLEM 17. Compare the values of $\delta = (T_m - T_f)/T_m$ from (43) and (44) for a nucleus of ice in supercooled water at atmospheric pressure. The parameter values are : $T_m = 273\ K,\ L = 333\ J/g,\ c_L = 4.2, c_S = 2\ J/g\,K,\ \rho^L = 1,$ $\rho^S = 0.92\ g/cm^3$, and $\sigma = 22\ dyn/cm = 22 \times 10^{-7}\ J/cm^2$.

PROBLEM 18. Derive (47).

PROBLEM 19. Derive (54).

PROBLEM 20. Check the computations indicated in (55a,b).

PROBLEM 21. Derive (56).

2.5. CHANGE OF PHASE IN A BINARY ALLOY

2.5.A Introduction

Up to now we have considered only pure materials (one-component systems) or systems which behave as if they consist of only one component (such as hydrated salts or eutectics).

A binary alloy is a **two-component** system, *A-B* in which the molecules of *A* and *B* form *physical bonds* creating a new compound, but not a new chemical substance (they do not react chemically). Thus they are metallic or intermetallic solutions or mixtures, e.g. Al-Si, Cu-Ni, Ge-Si, (HgTe)-(CdTe). Component *A* is called the **solvent** and *B* the **solute**. Alloys are a primary source of new exotic materials and thus very important technologically. Their usable crystalline form is commonly prepared by casting from the melt, i.e. by solidification. This is what makes melting and solidification processes important in metallurgy and materials science.

The new feature here is that in addition to heat transfer we also have to deal with the mass transfer which arises from component segregation (see **§2.5.B**). Consider a mass, m grams, of a binary alloy *A-B*. It consists of m_A grams of *solvent A* and m_B grams of *solute B*, $m_A + m_B = m$. The mass ratios

$$C_A := \frac{m_A}{m}, \qquad C_B := \frac{m_B}{m}$$

measure the proportion of *A* and *B* respectively, in the total mass, and are called the **concentrations** of *A* and *B*. Since $C_A + C_B = 1$, only one of them is necessary to determine the composition of the alloy. We shall use the concentration of solute, $C := C_B$ as composition variable. In the chemical and physical literature, composition is expressed in terms of the **mole fraction** of solute,

$$x := x_B := N_B \,/\, (N_A + N_B)$$

for N_i the number of moles of component $i = A\,,\,B$. It is related to the mass fraction, C by

$$C = \frac{M_B x}{[M_A(1-x) + M_B x]}$$

where $M_i =$ molecular weight of $i = A, B$. [BIRD et al], [LUPIS].

The phenomena involved in alloy solidification occur at two distinct scales, naturally giving rise to separate modeling at each scale. *Macroscopically*, at the observer's scale, one is concerned with the global heat and mass transfer effects, namely the evolution of the phases and of the temperature and concentration fields during the course of solidification, driven by initial and boundary conditions. At the *microscopic* scale it is the crystalline structure and morphology of the interface that are of interest. Of course, effective control of the microstructure requires understanding of the large-small scale interactions in a unified multi-scale model, a task for nonlinear science that is yet to be accomplished. Alloy solidification is a fascinating subject both from the theoretical "pattern formation" point of view, as well as from the practical "usefulness" one.

2.5.B The phase diagram

Phase-changes in an alloy are governed by its **phase diagram**. Such a diagram describes the thermodynamic states at which various phases can co-exist in thermodynamic equilibrium ([CALLEN], [REYNOLDS], [WALDRAM], [LUPIS], [PORTER-EASTERLING]. The simplest phase-diagram has the appearance of **Figure 2.5.1**. More complicated ones are shown in **Figure 2.5.2**. T_A and T_B denote the melt temperatures of pure components A and B, respectively.

Let us consider a binary alloy A-B with phase diagram as in **Figure 2.5.1**. Under ambient pressure of one atmosphere, the thermodynamic state of the alloy is determined by the two variables: temperature, T, and concentration, C, $0 < C < 1$

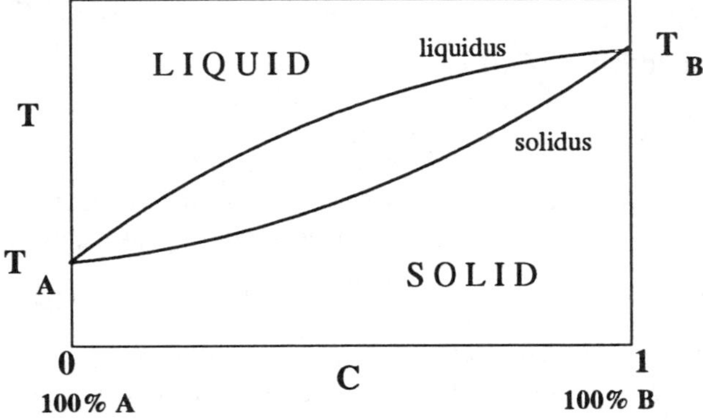

Figure 2.5.1. Phase diagram of a binary alloy.

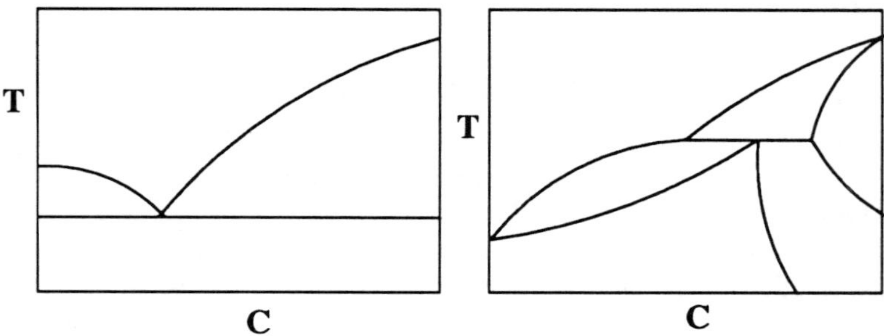

Figure 2.5.2. Other typical types of phase diagrams.

(see below). The curves marked liquidus and solidus demarcate the possible phases as follows: If the state, (C, T), lies *above* the

liquidus curve : $$T = T^L(C) \quad \text{or} \quad C = C^L(T) \tag{1}$$

then the alloy is **liquid**. If the state, (C, T), lies below the

solidus curve : $$T = T^S(C) \quad \text{or} \quad C = C^S(T)\,, \tag{2}$$

then the alloy is **solid**.

If the state, (C, T), lies in-between the liquidus and solidus, referred to as the **mushy region**, then the alloy consists of *liquid* at concentration $C^L(T)$ *coexisting* with *solid* at concentration $C^S(T)$, **Figure 2.5.1**. Thus, at any temperature T, $T_A < T < T_B$, solid and liquid can coexist in equilibrium only at these distinct concentrations. The **miscibility gap**, $[C^L(T)\,,\, C^S(T)]$, arises from thermodynamic considerations (the equality of chemical potentials of coexisting phases for each component) as required for thermodynamic equilibrium [CALLEN], [LUPIS]. Denoting by λ the percent of liquid present per unit mass, or **liquid fraction**, $0 < \lambda < 1$, the mean concentration, C, may be expressed via the

lever rule: $$C = \lambda\, C^L(T) + (1 - \lambda)\, C^S(T)\ . \tag{3}$$

It follows that, for given (C, T),

$$\lambda(C,T) = \frac{C^S(T) - C}{C^S(T) - C^L(T)}\ . \tag{4}$$

In general, the number of variables, f (equal to the degrees of freedom), needed to specify the state of a multicomponent, multi- phase alloy is given by the

Gibbs phase rule : $$f = 2 + N_{components} - N_{phases} \tag{5}$$

(see [CALLEN], [LUPIS]). Thus, for a pure material, $N_{components} = 1$, so in the liquid or solid: $f = 2 + 1 - 1 = 2$, namely the temperature and pressure are needed, whereas at the interface, where both solid and liquid coexist, $f = 2 + 1 - 2 = 1$, so only one of T or P is sufficient to determine the state. For a binary alloy, $N_{components} = 2$, so in the liquid or solid $f = 2 + 2 - 1 = 3$, namely T, P, C are needed to specify the state; in the mushy region, $N_{phases} = 2$, hence $f = 2$, i.e. only two variables suffice. So, actually, the phases change in the 3-dimensional space (T, P, C). Since, however, most phase-change processes take place under constant pressure, we fix the pressure at 1 atm, leaving only two variables; the phase diagram is the projection on the (C, T) plane of the intersection of the full three-dimensional equilibrium surfaces with the plane $P = 1$ atm.

Let us consider the solidification of a binary alloy A-B with phase diagram that of **Figure 2.5.3**. Let us fix attention on a certain amount of alloy, initially liquid at temperature T_0 and (uniform) concentration C_0, $T_0 > T^L(C_0)$. As it is slowly cooled, the temperature drops until $T_1 := T^L(C_0)$, the liquidus temperature at C_0, is reached, while the concentration remains at C_0 due to the absence of concentration gradients. At temperature $T^L(C_0)$, the stability of the liquid breaks down (under ideal equilibrium conditions) and a thin layer of solid will begin to form (in the absence of nucleation difficulties). However, at this temperature, solid and liquid can co-exist only at the distinct concentrations indicated on the phase diagram. Hence, the first solid must have concentration $C^S(T_1)$, which is higher than $C^L(T_1) \equiv C_0$, so solid will contain more B than the liquid with which it can coexist. This requires redistribution (segregation) of components A, B in order to enrich the solid (and deplete the liquid) in B. As solid forms, latent heat is released, which must be conducted away before the temperature can further reduce. As the temperature drops, the remaining liquid is being depleted of B (its state goes down the liquidus), while the forming solid contains progressively less B (its state goes down the solidus). The mean concentration remains about C_0, while the phase-change temperature continues to fall. When the temperature drops to $T_2 := T^S(C_0)$, all of the liquid has frozen and the process is complete. We see

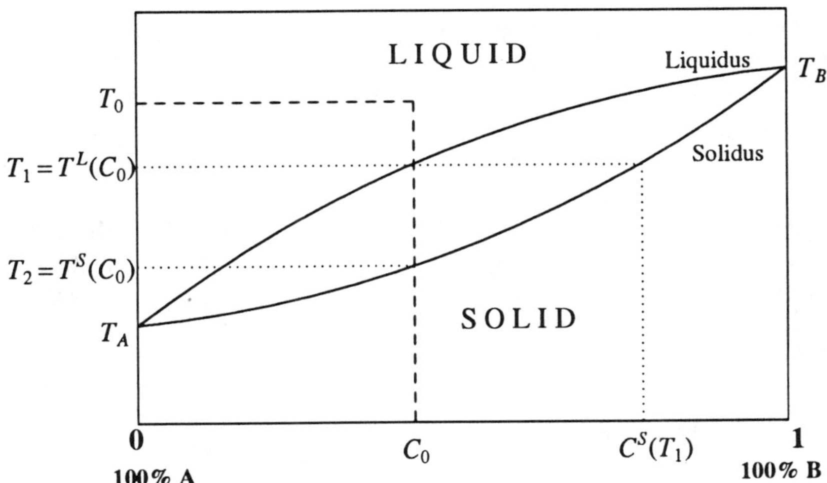

Figure 2.5.3. Solidification from initial state (C_0, T_0).

that alloy solidification requires both the removal of latent heat and the redistribution of solute, resulting in a coupled heat and mass transfer process.

Observe that while $T^S(C_0) < T < T^L(C_0)$, the liquid, being colder than the liquidus temperature $T^L(C_0)$ appears supercooled and the solid appears superheated. We say the alloy is **constitutionally supercooled** [TURNBULL], [CHALMERS], [PORTER-EASTERLING]. The effect on temperature is that the freezing takes place not at a fixed freezing point but rather over an extended range, from $T^L(C_0)$ down to $T^S(C_0)$. Thus, the liquid gives off its latent heat gradually over this extended range of temperatures.

Constitutional supercooling may have a dramatic effect on the structure of the interface and the resulting solid. In fact the interface is likely to be a spatially extended zone, a two-phase region, commonly referred to as a **mushy zone**, in which solid and liquid co-exist in some spatial arrangement (micro-structure). The possibilities are shown schematically in **Figure 1.1.2** (see [PORTER-EASTERLING], [KURZ-FISHER], [CHALMERS], [FLEMINGS]). When the diffusivities for heat and mass transfer in the liquid are of the same order, a *planar* interface may be possible (**Figure 1.1.2a**), see **§2.5.F**. Under very slow cooling, homogeneous nucleation in the supercooled melt may occur, leading to an *amorphous* mushy region (**Figure 1.1.2d**) and equiaxed growth; the resulting solid will consist of packed grains with radially non-uniform concentration. At faster cooling, *columnar* growth may arise (**Figure 1.1.2b**), and at still faster cooling rates side-branching may occur resulting in *dendritic* morphology (**Figure 1.1.2c**). Of course, various combinations of these are possible and likely to occur. Theories of interfacial instabilities play a prominent role in Materials Science and Metallurgy [PORTER-EASTERLING], [KURZ-FISHER]. Such microstructures create irregular concentration distributions in the resulting solid with, usually, detrimental effects on its physical characteristics such as strength or photoelectric properties. In addition to heat and mass transfer, the microstructure is controlled by capilarity (surface tension) effects and by atomic attachment kinetics (see [TILLER] for a detailed account). Clearly, alloy solidification is an extremely complicated process. Even though the physical phenomena involved are individually well-understood (at least qualitatively), the highly coupled overall process is far from having yielded to satisfactory *quantitative* modeling.

Solute can be redistributed by *inter-diffusion* and by *convection*. The former is always present, it is by far the simplest of the two, and it is often the limiting (controlling) factor for the growth and morphology of the inter-phase. Convection in the melt is usually ignored or accounted for approximately (by enhancing the diffusivities).

2.5.C Interdiffusion

The relative movement of molecules of one substance into another is called **interdiffusion**.

Consider a binary alloy A-B. Let ρ be the local density (mass per unit volume) of the alloy and $\rho^A(x,t)$, $\rho^B(x,t)$ the densities of A and B, so that

$$\rho = \rho^A + \rho^B \ . \tag{6}$$

Denote by J^A, J^B the mass fluxes of A and B (mass crossing a unit area per unit time). The total mass flux is zero, because the alloy itself does not move, with only molecules of A and B interchanging locations. Hence

$$J^A + J^B = 0 \ . \tag{7}$$

Consider an elementary volume of unit cross-sectional area and length Δx centered at x. The mass of component A in this volume (per unit area) at time t is $\rho^A(x,t)\cdot\Delta x$ and at time $t+\Delta t$, $\rho^A(x,t+\Delta t)\cdot\Delta x$. Hence, during the time interval Δt

$$\textbf{mass gain} = [\rho^A(x,t+\Delta t) - \rho^A(x,t)]\cdot\Delta x \ , \tag{8a}$$

which must equal the net amount of mass that came into the volume through the boundaries, namely

$$\Delta t \cdot [J^A(x - \frac{\Delta x}{2}, t) - J^A(x + \frac{\Delta x}{2}, t)]. \tag{8b}$$

Dividing this mass balance by Δx and Δt and letting them tend to zero, we obtain the (one-dimensional)

$$\textbf{mass conservation law:} \qquad \rho_t^A + J_x^A = 0 \tag{9}$$

for substance A. Similarly for B,

$$\rho_t^B + J_x^B = 0 \ . \tag{10}$$

The simplest constitutive law relating flux and density is

$$\textbf{Fick's law:} \qquad J^A = -D^{AB}\rho_x^A \ , \qquad J^B = -D^{BA}\rho_x^B \ , \tag{11}$$

where D^{AB}, D^{BA}(m^2/sec) are the diffusivities of A into B and of B into A respectively. Note that since the alloy density remains constant we have $\rho_x^A + \rho_x^B = \rho_x \equiv 0$, and then from (7), (11) we find

$$0 = J^A + J^B = -[D^{AB}\rho_x^A + D^{BA}\rho_x^B] = -[D^{AB} - D^{BA}]\rho_x^A \ . \tag{12}$$

Therefore, $D^{AB} = D^{BA}$, and the common value D is called the **interdiffusion coefficient** of components A and B [BIRD-STEWART-LIGHTFOOT]. We conclude that the mass conservation laws are expressed by the

$$\textbf{diffusion equations:} \qquad \rho_t^A = (D\rho_x^A)_x \ , \qquad \rho_t^B = (D\rho_x^B)_x \tag{13}$$

or, more generally (in three dimensions),

$$\rho_t^i = div(D\nabla\rho^i) \ , \qquad i = A\ , B \ . \tag{14}$$

Dividing by the constant total density, ρ, equation (14) may be expressed in terms of the concentration

$$C_i = \frac{\rho^i}{\rho} = \frac{\rho^i}{\rho^A + \rho^B}, \qquad i = A, B,$$

as

$$\frac{\partial}{\partial t} C_i = div(D \nabla C_i), \qquad i = A, B. \tag{15}$$

Recalling that $\rho^A + \rho^B = \rho$, implying that $C_A + C_B = 1$, only one of the two diffusion equations is needed and, following standard practice, we consider the diffusion of solute B into the solvent A, in terms of $C := C_B$:

$$C_t = (DC_x)_x, \qquad C = C_B = 1 - C_A. \tag{16}$$

In general, the inter-diffusion coefficient is a function of the thermodynamic state, $D = D(C, T)$, and very difficult to measure [HULTGREN et al]. It is of the order of 10^{-5}cm^2/sec for liquids and 10^{-10}cm^2/sec for solids, so a common approximation is $D_S \approx 0$.

Attention should be drawn to the similarity between Fick's law of diffusion, (11), and Fourier's law of conduction, (10)§**1.2**. As a result, the diffusion equation (15) and the heat conduction equation (17)§**1.2** are of the same type. This mathematical analogy allows us to think of heat as "diffusing".

For convenience, let us define the "concentration flux" $j = -DC_x = J^B/\rho$, so that (16) may be written in the standard conservation-law form as

$$C_t + j_x = 0, \qquad j = -DC_x. \tag{17}$$

Across a solid-liquid interface, $x = X(t)$, both the concentration C and its flux j experience jumps:

$$[\![C]\!]_S^L \equiv C(X(t)^+, t) - C(X(t)^-, t) = C^L(T_f) - C^S(T_f) \tag{18a}$$

where $C^L(T_f)$, $C^S(T_f)$ are the liquidus and solidus concentrations (**Figure 2.5.3**) corresponding to the current interface temperature T_f, and

$$[\![j]\!]_S^L = -D_L C_x + D_S C_x \qquad \text{on} \quad x = X(t), \tag{18b}$$

where D_L, D_S are the inter-diffusion coefficients for liquid and solid. Then, conservation across the interface dictates

$$[\![C]\!]_S^L X' = [\![j]\!]_S^L \qquad \text{on} \quad x = X(t), \tag{19}$$

which may be derived from (17) exactly as the Stefan Condition was derived from the heat equation in §**1.2.D**.

2.5.D A simple binary alloy solidification model

We consider the simplest possible *coupled* model of alloy solidification which admits a similarity solution [RUBINSTEIN] and then expose its shortcomings.

PHYSICAL PROBLEM: Consider a semi-infinite slab of an alloy A-B in which the components are soluble in all proportions (e.g. Cu-Ni) so that the phase diagram has the form of **Figure 2.5.3**, with $T = T^L(C)$ and $T = T^S(C)$ the liquidus and solidus temperature curves (**§2.5.B**). Initially the alloy is liquid at uniform concentration C_0 and temperature T_0: $T_0 > T^L(C_0)$. The face $x = 0$ is impermeable ($j(0,t) \equiv 0$) and a cold temperature $T_{cold} < T_A$ is imposed there starting at time $t = 0$. As a result, a solidification front will begin propagating from $x = 0$ into the slab. We assume the following:

(a) the phase-change front is planar and sharp (no mushy zone forming); this excludes constitutional supercooling effects, see **§2.5.G**;
(b) the density of the alloy is a constant ρ;
(c) the latent heat of fusion L is a constant, independent of concentration (presumably the value at C_0);
(d) the thermophysical parameters c_L, c_S, k_L, k_S, D_L, D_S are constants;
(e) the liquidus and solidus curves are smooth (differentiable), strictly increasing and non-intersecting (except, of course, at $C = 0$ and $C = 1$);
(f) heat is transferred exclusively by conduction, and solute is transferred exclusively by diffusion. In particular, the heat carried by the diffusing molecules is ignored or assumed negligible (see [ALEXIADES-WILSON-SOLOMON, 1985], [BENNON-INCROPERA] for such and other terms).

Under these assumptions, in each phase the heat conduction and mass diffusion equations are valid, and solid faces liquid across a planar front, $x = X(t)$, which, at each time $t > 0$, is an isotherm at the current (unknown) phase change temperature $T_f(t)$:

$$T(X(t),t) = T_f(t). \tag{20}$$

Conservation of heat and mass across the interface are expressed by the Stefan conditions. Thus, we arrive at the following generalization of the Stefan Problem.

MATHEMATICAL PROBLEM: (*semi-infinite alloy freezing from the left*):

Find $X(t)$, $T(x,t)$, $C(x,t)$, $T_f(t)$ such that

$$T_t = \alpha_S T_{xx}, \quad C_t = D_S C_{xx} \quad \text{in } 0 < x < X(t),\ t > 0 \ \text{(solid)} \tag{21a}$$

$$T_t = \alpha_L T_{xx}, \quad C_t = D_L C_{xx} \quad \text{in } X(t) < x < \infty,\ t > 0 \ \text{(liquid)} \tag{21b}$$

$$T(X(t),t) = T_f(t), \quad t > 0, \tag{22a}$$

$$C(X(t)^+,t) = C^L(T_f(t)), \quad C(X(t)^-,t) = C^S(T_f(t)), \quad t > 0, \tag{22b}$$

$$\rho L X'(t) = -k_L T_x + k_S T_x \quad \text{on } x = X(t),\ t > 0, \tag{22c}$$

$$\left[C^L(T_f(t)) - C^S(T_f(t))\right] X'(t) = -D_L C_x + D_S C_x \quad \text{on } x = X(t),\ t > 0, \tag{22d}$$

$$X(0) = 0, \tag{23a}$$

$$T(x,0) = T_0\,, \quad C(x,0) = C_0 \quad \text{with } C_0 < C^L(T_0) \text{ i.e. liquid}\,, \tag{23b}$$

$$T(0,t) = T_{cold} < T_A\,, \quad C_x(0,t) = 0\,, \quad t > 0\,. \tag{24}$$

The only coupling between heat and mass transfer here is at the interface. We may view (22c) as determining $X(t)$ as before, and the additional interface condition (22d) as determining the additional unknown $T_f(t)$.

This basic model, but for a *finite* slab, is discussed in [TAYLER]; power series expansions have been obtained by [BOLEY], [TAO] and numerical methods have been proposed by [FIX], [CROWLEY-OCKENDON], [MEYER], [BERMUDEZ-SAGUEZ], [WILSON-SOLOMON-ALEXIADES, 1984], [WHITE].

2.5.E The Rubinstein similarity solution

It is remarkable that problem (21)-(24) admits a similarity solution, given in [RUBINSTEIN]. It is a rather curious solution: the concentration in the solid is a constant,

$$C(x,t) \equiv C_S^*\ .$$

Then (22b) implies that T_f is a constant, $T_f(t) \equiv T_f^*$, and therefore also $C^L(T_f^*) =: C_L^*$ is a constant. These three constants must be related by (**Figure 2.5.4**),

$$T_f^* = T^S(C_S^*) = T^L(C_L^*)\ . \tag{25}$$

The similarity solution is

$$X(t) = 2\lambda\sqrt{\alpha_S t}\,, \quad t > 0 \tag{26}$$

$$C(x,t) = C_S^*\,, \tag{27a}$$

$$0 < x < X(t)\,,\ t > 0 \ (\textit{solid})$$

$$T(x,t) = T_{cold} + (T_f^* - T_{cold})\frac{\operatorname{erf}(\frac{x}{2\sqrt{\alpha_S t}})}{\operatorname{erf}\lambda}\,, \tag{27b}$$

$$C(x,t) = C_0 + (C_L^* - C_0)\frac{\operatorname{erfc}(\frac{x}{2\sqrt{D_L t}})}{\operatorname{erfc}(\lambda\sqrt{\frac{\alpha_S}{D_L}})}\,, \tag{28a}$$

$$X(t) < x\,,\ t > 0 \ (\textit{liquid})$$

$$T(x,t) = T_0 + (T_f^* - T_0)\frac{\operatorname{erfc}(\frac{x}{2\sqrt{\alpha_L t}})}{\operatorname{erfc}(\lambda\sqrt{\frac{\alpha_S}{\alpha_L}})}\,, \tag{28b}$$

where the constants λ, T_f^*, C_S^*, C_L^* satisfy

$$C_S^* = C^S(T_f^*)\,, \quad C_L^* = C^L(T_f^*)\,, \tag{29}$$

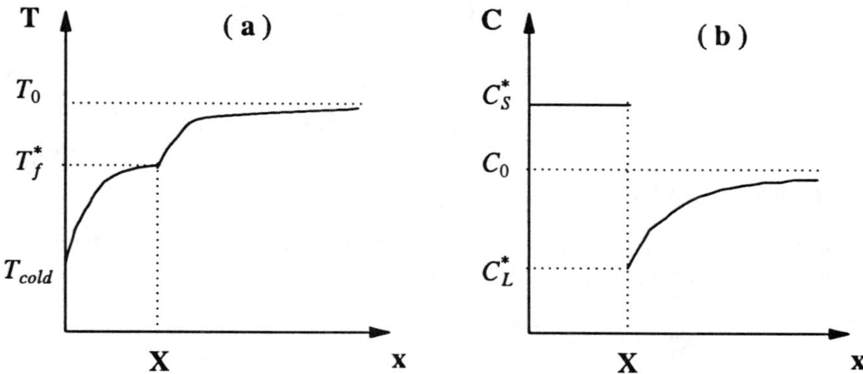

Figure 2.5.4. Temperature and Concentration profiles expected in the Rubinstein solution.

$$\frac{C_0 - C_L^*}{C_S^* - C_L^*} = H_2(\lambda\sqrt{\frac{\alpha_S}{D_L}}) , \tag{30a}$$

$$T_f^* = \frac{\rho L + T_{cold}\ A_1(\lambda) + T_0\ A_2(\lambda)}{A_1(\lambda) + A_2(\lambda)} , \tag{30b}$$

where

$$A_1(\lambda) = \frac{k_S}{\alpha_S\ H_1(\lambda)} , \qquad A_2(\lambda) = \frac{k_L}{\alpha_L\ H_2(\lambda\sqrt{\frac{\alpha_S}{\alpha_L}})} , \tag{31}$$

and

$$H_1(z) = \sqrt{\pi}\, z\, e^{z^2} \mathrm{erf}\, z , \qquad H_2(z) = \sqrt{\pi}\, z\, e^{z^2} \mathrm{erfc}\, z . \tag{32}$$

With C_S^*, C_L^* being functions of T_f^* via (29), the two equations (30a, b) constitute a system of transcendental equations for λ and T_f^*. The temperature and concentration profiles expected here are shown schematically in **Figure 2.5.4.**

It may be shown (PROBLEMS 1, 2) that this system is solvable for any values of the parameters (see also [RUBINSTEIN, p. 55], [WILSON- SOLOMON-ALEXIADES 1982]), but uniqueness of the solution may only be shown under restrictive assumptions on the phase-diagram and initial concentration ([RUBINSTEIN, p. 57]).

2.5.F Shortcomings of the simple model

Even though it is not stated explicitly in the formulation of the problem, it is implicitly assumed that the material on the "solid side" of the interface ($x < X(t)$) is all solid, while on the "liquid side" ($x > X(t)$) it is all liquid! More precisely, this means (see (1)-(2)) that

$$T(x, t) < T^S(C(x, t)) \qquad \text{for } x < X(t)\,, \tag{33a}$$

and

$$T(x, t) \geq T^L(C(x, t)) \qquad \text{for } x > X(t)\ . \tag{33b}$$

The similarity solution of **§2.5.E** clearly satisfies (33a) since $T(x, t) < T_f^* = T^S(C_S^*) \equiv T^S(C(x, t))$, but it may violate (33b) [WILSON-SOLOMON-ALEXIADES, 1982], namely, we

Claim: *If D_L / α_L is small enough, then there is mushy material ahead of the interface.* Indeed, (33b) will be violated if

$$\frac{\partial}{\partial x}\,[\,T(x, t) - T^L(C(x, t))\,] < 0 \qquad \text{at } x = X(t)^+\,, \tag{34}$$

since on the interface it is (by (25))

$$T(X, t) - T^L(C(X, t)) \;=\; T_f^* - T^L(C_L^*) = 0$$

Computing the derivatives C_x, T_x from (28) and using (31), condition (34) takes the form

$$\frac{T_0 - T_f^*}{C_S^* - C_L^*} \;<\; T^{L\prime}(C_L^*) \cdot \frac{\alpha_L}{D_L} \cdot H_2(\frac{\lambda}{\nu})\,, \tag{35}$$

where $\nu = \sqrt{\alpha_L/\alpha_S}$, $T^{L\prime}(C_L^*)$ denotes the slope of the liquidus curve at $C = C_L^*$ and H_2 is defined by (32b). Now, the left-hand side of (35) is a finite non-negative number and $H_2(\frac{\lambda}{\nu})$ is bounded away from zero (PROBLEM 3). It follows that (35), and therefore (34), will hold if α_L/D_L is large enough, as claimed. Notice, moreover, that inequality (35) is independent of time, so if a mushy region forms at all it will be there for the duration of the process.

According to the claim, the formation of a mushy zone may be attributed to the incompatibility between heat and mass transfer in the liquid. Indeed, $\alpha_L \gg D_L$ implies that conduction is much faster than diffusion, so the latent heat is conducted away rapidly while the solute cannot redistribute itself fast enough to bring the state to the solidus where full freezing occurs. As a result, the material becomes constitutionally supercooled and the mushy region appears.

In fact, it is almost always the case that the material diffusivity D_L is much smaller than the thermal diffusivity α_L, typically by several orders of magnitude. Hence, in most cases, the similarity solution will produce mushy (constitutionally supercooled) material ahead of the interface, contradicting the assumptions of the model itself. It seems that even the solution tries to tell us that constitutional supercooling may not be ignored !

A general model incorporating constitutional supercooling was developed by [ALEXIADES-WILSON-SOLOMON, 1985], and it has been used to simulate axisymmetric casting of Mercury-Cadmium-Telluride (an infrared detector alloy) pseudo-binary $HgTe_{1-x}CdTe_x$ in microgravity, [ALEXIADES, 1991]. The model consists of strongly coupled mass and energy conservation laws accounting for diffusion, conduction, the heat carried by the diffusing species, and also for Soret and Dufour cross effects. These laws are posed globally (irrespectively of phase) in a weak sense (see §4.4), and they update the concentration and energy fields, from which the temperature is found via an equation of state derived from solution thermodynamics. To present it here would carry us too far afield.

2.5.G Uncoupled models of alloy solidification

In the heat and mass transfer literature there are many papers proposing simplified models of alloy solidification. The two common approaches are to ignore either diffusion or conduction. Thus, alloy solidification is viewed either as conduction-limited (a uniform concentration is assumed to prevail) or as diffusion-limited (a uniform temperature prevails). The latter, isothermal, approach is usually used in studies of interface morphology and stability [PORTER-EASTERLING], [KURZ-FISHER].

The former, isoconcentration, approach leads to considering solidification over an extended freezing range [CARSLAW-JAEGER], [TIEN-GEIGER], [CHO-SUNDERLAND], [CLYNE], [ALEXIADES-CANNON]). Since the concentration is assumed constant, the liquidus and solidus temperatures, T_0^L and T_0^S are known numbers, between which the freezing temperature T_f may vary, creating three phases: liquid, solid, and, between them, a mushy (two-phase) region according to $T > T_S^L$, $T < T_0^S$ and $T_0^S < T < T_0^L$, respectively. In the liquid and solid we have plain heat conduction. In the mushy region, the release of the latent heat is accounted for by a heat source term equal to $\frac{\partial}{\partial t}[\rho L f_S]$, where f_S denotes the solid fraction present (solidification of liquid of mass ρf_S releases latent heat $\rho L f_S$). Hence, heat conduction in the mushy zone is described by

$$\rho c_m T_t = k_m T_{xx} + \rho L \frac{\partial f_S}{\partial t}, \tag{36}$$

with c_m, k_m some average values of specific heat and conductivity of mushy material. The solid fraction, f_S, is considered as a known function of temperature and/or of location; several expressions have been proposed for it, obtained from various considerations (lever rule, Scheil rule, intermediates,)which amount to assumptions about the micro-structure of the mushy zone. The simplest one is

$$f_S(T) = \frac{T_0^L - T}{T_0^L - T_0^S}, \tag{37}$$

in which case (36) becomes

$$\rho[c_m + \frac{L}{T_0^L - T_0^S}]\, T_t = k_m T_{xx} . \tag{38}$$

Hence, the release of the latent heat is simply accounted for by enhancing the specific heat inside the mushy region. Another simple choice, appropriate for a eutectic, is [TIEN-GEIGER]

$$f_S(x) = f_0 \frac{X_L - x}{X_L - X_S} \tag{39}$$

where $x = X_L$, $x = X_S$ are the liquidus and solidus interfaces and f_0 is the solid fraction, $0 < f_0 < 1$, at the solidus (where the remaining liquid completely freezes also; f_0 may be found from the Scheil rule, (see (63) below).

On the liquidus, $x = X_L(t)$ and solidus $x = X_S(t)$ interfaces (where $T = T_0^L$ and $T = T_0^S$ respectively), jump conditions analogous to the Stefan condition must hold (PROBLEM 4), namely

$$\rho L f_S X_L'(t) = -k_L T_x(X_L^+, t) + k_m T_x(X_L^-, t) \quad (on\ liquidus) \tag{40a}$$

$$\rho L(1 - f_S) X_S'(t) = -k_m T_x(X_S^+, t) + k_S T_x(X_S^-, t) \quad (on\ solidus) \tag{40b}$$

Note that under (37), f_S equals zero at the liquidus and one at the solidus, so the left-hand sides of both (40a) and (40b) are zero (the latent heat is gradually and totally released within the mushy zone). Under (39) on the other hand, $f_S = 0$ at the liquidus and $f_0 < 1$ at the solidus.

Clearly, various other ad hoc expressions may be considered for f_S within the above framework, (e.g. [SEKHAR et al], [CLYNE]) and in some cases explicit similarity solutions may be found (see **§2.5.H**, [CARSLAW-JAEGER, p. 290], [CHO-SUNDERLAND], [SOLOMON-WILSON-ALEXIADES, 1982]). The solvability of transcendental equations arising in multi-phase problems is shown in [WILSON, 1978]. The basic 3-phase model described above has been generalized by [ALEXIADES-CANNON] to the nonlinear heat conduction equation in any number of dimensions, with the solid fraction being any given function of temperature, and its well-posedness was established.

2.5.H The Tien-Geiger model for freezing over an extended range

As an example of an uncoupled model (**§2.5.G**) with explicit solution we present the following simple model for the "solidification of a binary eutectic system" proposed in [TIEN-GEIGER].

PHYSICAL PROBLEM: Consider a semi-infinite slab of a binary alloy of uniform concentration C_0 at its *liquidus* temperature $T_0^L = T^L(C_0)$. A constant temperature $T_{cold} < T_0^S := T^S(C_0)$ is imposed at $x = 0$, and we assume conditions of "normal non-equilibrium freezing", meaning that $D_S \approx 0$ and $D_L \approx \infty$, $\alpha_L \approx \infty$ (effected by continuous stirring for example, see **§2.5.I**). Since there is an infinite supply of solute here (due to infinite volume !) *no* segregation will occur in this case. The concentration will remain at C_0, and the liquid will remain at temperature T_0^L. However, the latent heat is gradually released over the range $T_0^S < T < T_0^L$, creating a mushy zone between the solidus interface $x = X_S(t)$ and the liquidus interface $x = X_L(t)$. The solid fraction in (36) is assumed to be given by (39), and $\rho_S = \rho_L =: \rho$, c_S, c_m, k_S, k_m, f_0 are given constants.

This is an uncoupled (isoconcentration) model with two active phases (solid and mushy) and two interfaces. The Stefan condition on the liquidus, (40a), reduces to $k_m\, T_x(X_L(t)^-, t) = 0, \quad t > 0$, and we are led to the following

MATHEMATICAL PROBLEM: Find $T(x, t)$ and $X_L(T)$, $X_S(t)$ such that

$$T_t = \alpha_S T_{xx}, \qquad 0 < x < X_S(t), \quad t > 0 \tag{41a}$$

$$T_t = \alpha_m T_{xx} + \frac{L}{c_m}\frac{\partial f_S}{\partial t}, \quad X_S(t) < x < X_L(t), \ t > 0 \ \text{with} \ f_S = f_0 \frac{X_L - x}{X_L - X_S} \tag{41b}$$

$$T(X_S(t), t) = T_0^S, \qquad T(X_L(t), t) = T_0^L, \quad t > 0 \tag{42a}$$

$$\rho L(1 - f_0) X_S'(t) = -k_m T_x(X_S(t)^+, t) + k_s T_x(X_S(t)^-, t), \quad t > 0 \tag{42b}$$

$$0 = k_m T_x(X_L(t)^-, t), \quad t > 0 \tag{42c}$$

$$X_S(0) = X_L(0) = 0 \tag{43}$$

$$T(0, t) = T_{cold} < T_0^S, \quad t > 0. \tag{44}$$

The similarity solution (PROBLEM 5) is given by

$$X_S(t) = 2\lambda_S\sqrt{\alpha_S t}, \qquad X_L(t) = 2\lambda_L\sqrt{\alpha_S t}, \qquad t > 0 \tag{45}$$

$$T(x, t) = T_{cold} + (T_0^S - T_{cold}) \frac{\operatorname{erf}\left(\frac{x}{2\sqrt{\alpha_S t}}\right)}{\operatorname{erf}(\lambda_S)}, \quad 0 \le x \le X_S(t), \ t > 0 \tag{46a}$$

$$T(x, t) = T_0^L + \frac{f_0 L}{c_m(\lambda_L - \lambda_S)}\left[\lambda_L - \frac{x}{2\sqrt{\alpha_S t}}\right]$$
$$-\left[T_0^L - T_0^S + \frac{f_0 L}{c_m}\right] \frac{\operatorname{erf}\left(\frac{x}{2\sqrt{\alpha_m t}}\right) - \operatorname{erf}(\nu\lambda_L)}{\operatorname{erf}(\nu\lambda_S) - \operatorname{erf}(\nu\lambda_L)}, \quad X_S(t) \le x \le X_L(t), \ t > 0, \tag{46b}$$

where λ_L, λ_S are roots of the system

$$f_0\sqrt{\pi}\, e^{(\nu\lambda_L)^2}\left[\operatorname{erf}(\nu\lambda_L) - \operatorname{erf}(\nu\lambda_S)\right] = \nu(\lambda_L - \lambda_S)[\boldsymbol{St}_0 + f_0], \tag{47a}$$

$$\frac{\boldsymbol{St}_{cold}}{\sqrt{\pi}\,\lambda_S e^{\lambda_S^2}\operatorname{erf}(\lambda_S)} = 1 - f_0 + f_0\frac{e^{\nu^2(\lambda_L^2 - \lambda_S^2)} - 1/2}{\nu^2\lambda_S(\lambda_L - \lambda_S)}, \tag{47b}$$

with

$$\boldsymbol{St}_0 = \frac{c_m(T_0^L - T_0^S)}{L}, \qquad \boldsymbol{St}_{cold} = \frac{c_S(T_0^S - T_{cold})}{L}, \qquad \nu = \sqrt{\frac{\alpha_S}{\alpha_m}}. \tag{48}$$

It can be shown that the Jacobian of the system (47) is non-zero and that the system is solvable (PROBLEM 7).

Clearly, this problem is a generalization of the 1-phase Stefan Problem to which it reduces when $T_0^L = T_0^S$ and $f_0 = 0$ (PROBLEM 6).

It should be pointed out that it is the assumption of *semi-infinite* slab that leads to no-segregation. Indeed, "normal non-equilibrium freezing" of a *finite* slab leads to *maximum* segregation, [CHALMERS], but then there is no explicit solution. We shall discuss such a model in the following subsection.

2.5.I Rapid freezing of a finite slab with stirring of melt

The case of *complete mixing* of the liquid and *no diffusion* in the solid arises in several technologically important applications, such as zone-refining [PFANN], [HERINGTON], and "dunking" of cold solid into the liquid for temperature control [ALEXIADES, 1983]. Complete mixing, which amounts to $D_L \approx \infty$ and $\alpha_L \approx \infty$, may be achieved by mechanical or electromagnetic stirring of the melt. The condition of no solid diffusion, i.e. $D_S = 0$, is easily met in any fairly rapid freezing process since D_S is always very small. The combination is often referred to as "normal non-equilibrium freezing" ([KURZ-FISHER], [FLEMINGS]), a rather inappropriate term since the liquid is actually assumed to remain in thermal and material equilibrium with the interface; see [CHALMERS] for a qualitative description. For a *finite* slab, we expect maximum segregation since the liquid is depleted of solute as the forming solid traps more and more of it. The mixing of the melt inhibits constitutional supercooling, resulting in a single (planar) interface across which the concentration jumps as dictated by the phase diagram.

PHYSICAL PROBLEM: Consider a finite slab, $0 \le x \le l$, of binary alloy of concentration C_0 initially at its liquidus temperature $T_0 = T^L(C_0)$. A low temperature $T_{cold} < T^S(C_0)$ is imposed at $x = 0$, while the back face $x = l$ is insulated (and both ends are impermeable). We assume:

(i) heat conduction but no material diffusion in the solid (rapid freezing);
(ii) liquid maintained in thermal and material equilibrium with the interface by, say, mixing;
(iii) monotonic phase diagram, as in **Figure 2.5.1**;
(iv) $\rho_S = \rho_L =: \rho$ and all thermophysical properties are constants (independent of both temperature and concentration); moreover, $D_S = 0$, $D_L = \infty$ and $\alpha_L = \infty$.

Thus, the solidification front $x = X(t)$, where $T = T_f(t) =$ current freezing temperature, starts at $x = 0$ and advances into the melt which is being depleted of solute as more and more solid forms. Since $T_{cold} < T^S(C_0) < T^L(C_0)$, it is initutively clear (also see comments following the mathematical formulation below) that no remelting can occur and each point x freezes at a unique time $t_f(x) := X^{-1}(x)$, at temperature $T_f(t_f(x))$, and concentration $C^S(T_f(t_f(x))$ which remains constant thereafter. The deeper a point x is in the solid, the later it froze, at a lower T_f and lower C^S. Therefore the solid will be *segregated.* Heat conduction through the solid is described by the standard heat equation ((53a) below), but by assumption (ii), the melt has uniform temperature $T_f(t)$ and uniform concentration $C^L(T_f(t))$, which decrease in time as the freezing temperature decreases.

We see that both concentrations are determined by $T_f(t)$, so the only unknowns are $X(t)$, $T_f(t)$ and $T(x, t)$ of the solid. The necessary interface conditions are obtained from conservation of energy and mass. However, the time-dependence of the freezing temperature complicates matters and one must be careful in defining the "energy" of the system (see [ALEXIADES-SOLOMON-WILSON, 1981b]).

Let us measure the energy *drop* from a high reference temperature T_{ref} (say T_0). The sensible heat in the liquid is $\int_X^l \rho c_L[T_{ref} - T_f(t)]dx$. The sensible heat in the solid is that needed to lower the temperature to the freezing temperature in the liquid state plus that needed to lower it further to $T(x, t)$ in the solid state. Remembering that each point x froze at temperature $T_f(t_f(x))$, we find that the total heat released until time t is

$$\begin{aligned} E(t) = &\int_0^{X(t)} \rho c_L[\, T_{ref} - T_f(t_f(x))\,]\, dx + \int_0^{X(t)} \rho L\, dx \\ &+ \int_0^{X(t)} \rho c_S[\, T_f(t_f(x)) - T(x, t)\,]\, dx + \int_{X(t)}^{l} \rho c_L[\, T_{ref} - T_f(t)\,]\, dx\,. \end{aligned} \tag{49}$$

By standard heat balance (PROBLEM 8) we obtain the interface condition

$$\rho\, LX'(t) = +\, k_S T_x(X(t)^-, t) + [l - X(t)\,]\, \rho\, c_L \frac{dT_f(t)}{dt}\,, \tag{50}$$

which generalizes the (1-phase) Stefan Condition to the case of *variable* melt temperature. Clearly, it reduces to the standard one when $T_f \equiv$ constant.

Similarly, the total solute mass (fraction) is

$$M(t) = \int_0^{X(t)} C^S(T_f(t_f(x)))dx + \int_{X(t)}^{l} C^L(T_f(t))dx\,, \tag{51}$$

and mass balance (PROBLEM 9) yields the interface condition

$$[\, C^L(T_f(t)) - C^S(T_f(t))\,]\, X'(t) = [l - X(t)\,] \left.\frac{dC^L}{dT}\right|_{T = T_f(t)} \frac{dT_f(t)}{dt}\,. \tag{52}$$

Thus, we are led to the following

MATHEMATICAL PROBLEM (*freezing of finite alloy slab with mixing of melt*):

Find $T(x, t)$, $X(t)$ and $T_f(t)$ such that

$$T_t = \alpha_S T_{xx}, \quad 0 < x < X(t), \quad t > 0 \quad (solid) \tag{53a}$$

$$T(x, t) = T_f(t), \quad X(t) \le x \le l, \quad t > 0 \quad (liquid) \tag{53b}$$

$$T(X(t)^-,t) = T_f(t), \quad t > 0 \tag{54a}$$

$$\rho L X'(t) = k_S T_x(X^-,t) + [l - X]\rho c_L T_f'(t), \quad t > 0 \tag{54b}$$

$$[C^L(T_f) - C^S(T_f)]X'(t) = [l - X] \cdot C^{L\prime}(T_f) \cdot T_f'(t), \quad t > 0 \tag{54c}$$

$$X(0) = 0, \qquad T_f(0) = T_0, \qquad T(x,0) = T_0, \quad 0 < x < l, \tag{55}$$

$$T(0,t) = T_{cold}, \qquad -k_L T_x(l,t) = 0, \qquad t > 0 \tag{56}$$

Recall that the curves $C = C^L(T_f)$ and $C = C^S(T_f)$ are known from the phase diagram of the alloy (see (1,2)). Thus, (54c) relates $T_f(t)$ to $X(t)$, and the rest constitutes a 1-phase problem but with phase-change temperature a function of the interface location. The well-posedness of such general problems has been established in [FASANO-PRIMICERIO, 1977]; those results imply the existence of a smooth monotone interface for our problem, thus justifying the assumption of no-remelting we made in the formulation of the problem.

Once the thermal problem is solved, the concentrations are given by

$$C(x,t) = C^S(T_f(t_f(x))) \text{ in the solid and } C(x,t) = C^L(T_f(t)) \text{ in the liquid.} \tag{57}$$

Unfortunately, no explicit solution is possible due to the finiteness of the slab. Perturbation solutions (§3.3) are presented in [ALEXIADES, 1983] for the above problem (and also for one with a convective boundary condition at $x = 0$) for the case of a dilute alloy which we now consider in order to derive the "Scheil rule".

Dilute Alloy

For dilute alloys, the phase diagram curves may, very conveniently, be approximated by *straight lines*. Thus, the solidus and liquidus concentrations may be expressed as (**Figure 2.5.5**)

$$C^S(T_f) = \kappa_S(T_f - T_A), \qquad C^L(T_f) = \kappa_L(T_f - T_A), \tag{58a}$$

or equivalently,

$$T_f - T_A = \frac{1}{\kappa_S} C^S = \frac{1}{\kappa_L} C^L . \tag{58b}$$

The ratio of the slopes κ_S/κ_L is commonly known as the

$$\textbf{distribution coefficient}: \qquad k = \frac{\kappa_S}{\kappa_L} = \frac{C^S}{C^L} \tag{59}$$

of the dilute alloy [CHALMERS], [PFANN], [KURZ-FISHER], [PORTER-EASTERLING]. Note that k may be > 1 or < 1.

In this case, (54c) simplifies to

$$\frac{T_f'}{T_f - T_A} = (1-k)\frac{X'}{l - X} . \tag{60}$$

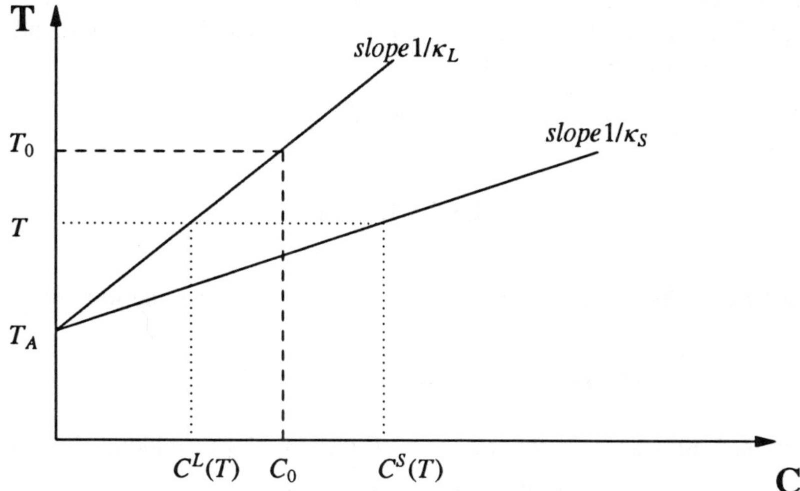

Figure 2.5.5. Liquidus and solidus curves for dilute alloy.

It can be integrated directly (using (55)) to give (PROBLEM 10)

$$\frac{T_f(t) - T_A}{T_0 - T_A} = \left(1 - \frac{X(t)}{l}\right)^{k-1}, \tag{61}$$

which expresses $T_f(t)$ in terms of $X(t)$.

This allows us to calculate the concentration distribution in the forming solid. Indeed, from (57), (58a), (61), the solid concentration is independent of time and given by

$$\begin{aligned} C(x, t) &= \kappa_S\,[\,T_f(t_f(x)) - T_A\,] = \kappa_S\,[\,T_0 - T_A\,]\left(1 - \frac{X(t_f(x))}{l}\right)^{k-1} \\ &= C^S(T_0)\left(1 - \frac{x}{l}\right)^{k-1}. \end{aligned} \tag{62}$$

Note that $C^S(T_0) = \kappa_S[T_0 - T_A] = k\,C_0$ is the initial *solidus* concentration, and $f_S = \frac{x}{l}$ represents the fraction frozen. Therefore (62) takes the convenient form known as the

Scheil rule : $$C_{solid} = k\,C_0\,(1 - f_S)^{k-1}, \tag{63a}$$

or *non-equilibrium lever rule* [PORTER-EASTERLING], [KURZ-FISHER]. It expresses the composition of solid in terms of the initial composition C_0 and the phase diagram for a dilute alloy solidifying directionally with no diffusion in the solid and perfect mixing in the liquid. Note that it is based on mass balance alone, (PROBLEM 10), hence independent of the heat transfer process, and it constitutes the fundamental equation for directional freezing and zone-refining [PFANN],

[HERINGTON]. The (uniform) composition of the remaining liquid, which changes with time, (57), may also be simply expressed in terms of the liquid fraction $f_L = 1 - f_S = 1 - X(t)/l$ as

$$C_{liquid} = C_0 \, f_L{}^{k-1} \, . \tag{63b}$$

PROBLEMS

PROBLEM 1. To prove that the system (30a, b) has at least one solution, λ, T_f^*, let

$$W(\lambda) = \frac{\rho L + T_{cold} \, A_1(\lambda) + T_0 \, A_2(\lambda)}{A_1(\lambda) + A_2(\lambda)} \, , \qquad \lambda > 0 \, ,$$

so that (30b) takes the form: $T_f^* = W(\lambda)$ (see (31) for definitions of A_1, A_2).

(a) Show that $W(\lambda)$ is differentiable and increasing for $\lambda > 0$.

(b) Show that $W(0) = T_{cold} < T_A$ and $W(\infty) > T_0 \geq T^L(C_0) > T_A$.

(c) Conclude that there exist values λ_1 and λ_2, $\lambda_1 < \lambda_2$, such that $W(\lambda_1) = T_A$ and $W(\lambda_2) = T^L(C_0)$.

PROBLEM 2. Continuing the previous problem, for $\lambda_1 < \lambda < \lambda_2$ define the function

$$Z(\lambda) := \frac{C_0 - C^L(W(\lambda))}{C^S(W(\lambda)) - C^L(W(\lambda))} \, ,$$

so that (30a) becomes $Z(\lambda) = H_2(\lambda\kappa)$, $\kappa = \sqrt{\alpha_S / D_L}$. Show that

(a) $Z(\lambda) > 0$ for $\lambda_1 < \lambda < \lambda_2$.

(b) $\lim_{\lambda \to \lambda_1^+} C^L(W(\lambda)) = C^L(T_A) < C_0$, and $\lim_{\lambda \to \lambda_1^+} C^S(W(\lambda)) = C^S(T_A) = C^L(T_A)$.

(c) Conclude that $\lim_{\lambda \to \lambda_1^+} Z(\lambda) = +\infty$, and show that $\lim_{\lambda \to \lambda_2^-} Z(\lambda) = 0$.

(d) Conclude that $Z(\lambda) = H_2(\lambda\kappa)$ has at least one root between λ_1 and λ_2.

PROBLEM 3. (a) Derive (35) from (34) using (28), (31).

(b) On the basis of PROBLEM 2(d-f), show that any root λ lies strictly between λ_1 and λ_2. Conclude that $T_f^* > T_A$ and $C_S^* - C_L^* > 0$. Therefore the left-hand side of (35) is always finite.

(c) Show that $H_2(\lambda/v) > H_2(\lambda_1/v) > 0$, so that $H_2(\lambda/v)$ is bounded away from zero.

PROBLEM 4. Derive the Stefan conditions (40).

PROBLEM 5. Derive the similarity solution (45-48) of the Tien-Geiger model (41-44).

PROBLEM 6. Verify that for a pure material ($T_0^L = T_0^S$, $f_0 = 0$) the Tien-Geiger model and its solution (§**2.5.H**) reduce to the 1-phase Stefan Problem.

PROBLEM 7. Show that the system of transcendental equations (47) is solvable in general.

PROBLEM 8. Derive the interface condition (50), expressing energy conservation across the interface in the problem of §**2.5.I**. [Hint: Proceed as in §**1.2.E**].

PROBLEM 9. Derive the interface condition (52) of mass conservation for the problem of §**2.5.I**.

PROBLEM 10. Consider a dilute alloy of concentration C_0 at its liquidus temperature T_0^L, freezing unidirectionally with no diffusion in the solid and perfect mixing in the liquid (§**2.5.I**). Set up the mass balance and derive the Scheil rule, equations (63a, b).

2.6. SIMILARITY SOLUTIONS IN CYLINDRICAL AND SPHERICAL GEOMETRIES

2.6.A Similarity solutions

In cylindrical-polar coordinates, the heat equation with axial symmetry has the form (§**1.2**)

$$T_t = \alpha\,(T_{rr} + \frac{1}{r}\,T_r) \equiv \frac{\alpha}{r}\frac{\partial}{\partial r}\left(r\,\frac{\partial T}{\partial r}\right). \tag{1}$$

The only similarity solution (in $\xi = r/\sqrt{t}$) is given by (PROBLEM 1)

$$T(r,t) = E_1(\frac{r^2}{4\,\alpha\,t}), \qquad r > 0, \quad t > 0, \tag{2}$$

where E_1 denotes [ABRAMOWITZ-STEGUN] the

exponential integral: $$E_1(x) = \int_x^\infty \frac{e^{-s}}{s}\,ds\,, \qquad x > 0\,. \tag{3}$$

The spherically symmetric heat equation has the form (§**1.2**)

$$T_t = \alpha\,(T_{rr} + \frac{2}{r}\,T_r) \equiv \frac{\alpha}{r^2}\frac{\partial}{\partial r}(r^2\,\frac{\partial T}{\partial r}), \tag{4}$$

and admits the similarity solution (PROBLEM 2)

$$T(r,t) = \frac{\sqrt{\alpha t}}{r}\,e^{-\frac{r^2}{4\alpha t}} - \frac{\sqrt{\pi}}{2}\,\mathrm{erfc}(\frac{r}{2\sqrt{\alpha t}}), \qquad r > 0, \quad t > 0\,. \tag{5}$$

As a result, the only phase-change problems that may be solved in terms of these ([CARSLAW-JAEGER]) are those for the entire space

- due to a line heat source, (resulting in axially-symmetric melting or freezing);
- due to a point source whose strength increases linearly with time (of no practical interest!)
- when initially supercooled and nucleation occurs along a line or at a point.

As examples, we present the line source melting and the cylindrical or spherical freezing of *supercooled* liquid, which are of some physical interest. All other physically important problems in cylindrical or spherical geometries must necessarily be treated by approximation methods (CHAPTER **3**) or numerical methods (CHAPTER **4**).

2.6.B Axially-symmetric melting due to a line source

Consider an infinite cylinder (the whole space), initially solid at $T_S < T_m$, which is melted by a *line* source of strength Q (watts/m) at $r = 0$. An interface $r = R(t)$ demarcates liquid $(0 < r < R(t))$ from solid $(r > R(t))$, and we assume constant $\rho_L = \rho_S$, c_L, c_S, k_L, k_S, L, T_m. Here, the heat flow rate (= flux × area) is prescribed at $r = 0$, which has to be expressed as a limit (see (6f)). With the Stefan condition as in (47)§**1.2**, we have the

MATHEMATICAL PROBLEM *(axially-symmetric melting due to a line source):*

Find $T(r, t)$ and $R(t)$ such that

$$T_t = \alpha_L[T_{rr} + \frac{1}{r} T_r], \quad 0 < r < R(t), \quad t > 0 \quad (liquid) \tag{6a}$$

$$T_t = \alpha_S[T_{rr} + \frac{1}{r} T_r], \quad R(t) < r < \infty, \quad t > 0 \quad (solid) \tag{6b}$$

$$T(R(t), t) = T_m, \quad t > 0, \tag{6c}$$

$$\rho L R'(t) = -k_L T_r(R(t),^- t) + k_S T_r(R(t),^+ t), \quad t > 0, \tag{6d}$$

$$R(0) = 0, \qquad T(r, 0) = T_S < T_m, \tag{6e}$$

$$\lim_{r\to 0}(-2\pi r k_L T_r) = Q > 0, \qquad \lim_{r\to\infty} T(r, t) = T_S. \tag{6f}$$

Its similarity solution [CARSLAW-JAEGER p. 294] is given by (PROBLEM 3)

$$R(t) = 2\lambda\sqrt{\alpha_L t}, \quad t > 0 \tag{7a}$$

$$T(r, t) = T_m + \frac{Q}{4\pi k_L}[E_1(\frac{r^2}{4\alpha_L t}) - E_1(\lambda^2)], \quad 0 < r \le R(t), \quad t > 0. \quad (liquid) \tag{7b}$$

$$T(r,t) = T_S + (T_m - T_S)\,\frac{E_1(\dfrac{r^2}{4\alpha_S t})}{E_1(\nu^2\lambda^2)}, \quad R(t) \le r,\ t > 0\ (solid) \tag{7c}$$

with λ the (unique) root of the transcendental equation

$$\frac{Q/(\rho L \alpha_L)}{4\pi\lambda^2 e^{\lambda^2}} + \frac{St_S}{(\nu\lambda)^2 e^{(\nu\lambda)^2} E_1(\nu^2\lambda^2)} = 1, \tag{7d}$$

and

$$\nu = \sqrt{\frac{\alpha_L}{\alpha_S}}, \qquad St_S = \frac{c_S(T_m - T_S)}{L}. \tag{7e}$$

2.6.C Freezing of supercooled liquid

Consider *supercooled liquid* (but not hypercooled) at temperature $T_{init} < T_m$ occupying the whole space (see §2.4). At time $t = 0$ nucleation occurs and we examine the following two cases:

Case I: **Nucleation along an infinite line.**

Then an axially-symmetric freeze-front, $r = R(t)$, will begin propagating from $r = 0$ into the supercooled liquid. We assume that the freezing-point depression due to curvature (Gibbs-Thomson effect) is negligible, so that freezing occurs at T_m. Hence the resulting solid will be at temperature T_m. The problem is easily formulated (PROBLEM 6) and its similarity solution is found to be [CARSLAW-JAEGER p. 295]

$$R(t) = 2\lambda\sqrt{\alpha_L t}, \qquad t > 0 \tag{8a}$$

$$T(r,t) = T_{init} - (T_m - T_{init})\,\frac{E_1(\dfrac{r^2}{4\alpha_L t})}{E_1(\lambda^2)}, \quad r > R(t),\ \ t > 0, \tag{8b}$$

with λ the root of

$$St_{init} - \lambda^2 e^{\lambda^2} E_1(\lambda^2) = 0, \qquad St_{ini} = \frac{c_L(T_m - T_{init})}{L}, \tag{8c}$$

provided $0 \le St_{init} < 1$.

Case II: **Nucleation at a single point**

Then a spherical front $r = R(t)$ will begin propagating from $r = 0$. Again, we ignore the Gibbs-Thomson effect by assuming freezing at temperature T_m. This problem has similarity solution [CARSLAW-JAEGER p. 295]

$$R(t) = 2\lambda\sqrt{\alpha_L t}, \qquad t > 0 \tag{9a}$$

$$T(r,t) = T_{init} + \frac{2\lambda(T_m - T_{init})}{e^{-\lambda^2} - \lambda\sqrt{\pi}\,\mathrm{erfc}(\lambda)} \left[\frac{\sqrt{\alpha_L t}}{r}\, e^{-\frac{r^2}{4\alpha_L t}} - \frac{\sqrt{\pi}}{2}\,\mathrm{erfc}\!\left(\frac{r}{2\sqrt{\alpha_L t}}\right) \right], \tag{9b}$$

$$r > R(t)\,, \quad t > 0\,,$$

with λ the root of the equation

$$2\lambda^2 e^{\lambda^2}\,[\,e^{-\lambda^2} - \lambda\sqrt{\pi}\,\mathrm{erfc}(\lambda)] = St_{init}, \qquad St_{init} = \frac{c_L(T_m - T_{init})}{L}, \tag{9c}$$

provided $0 \le St_{init} < 1$, (PROBLEM 7).

PROBLEMS

PROBLEM 1. Verify that (2) solves the axially symmetric heat equation (1).

PROBLEM 2. Verify that (5) solves the spherically symmetric heat equation (4).

PROBLEM 3. Verify that (7) solves problem (6).

PROBLEM 4. (a) Setting $T_S = T_m$ in (7), deduce the solution of the corresponding *1-phase* problem.

(b) Using the first two terms of the series expansion [ABRAMOWITZ-STEGUN]

$$E_1(x) = -\gamma - \ln x + \sum_{n=1}^{\infty} \frac{(-x)^n}{n\,n!}, \qquad \gamma = \text{Euler's constant},$$

show that for small $Q/(\rho L \alpha_L)$ the solution to the 1-phase problem is approximately given by

$$R \approx \sqrt{\frac{Q}{\pi\rho L}\,t}, \qquad T \approx T_m - \frac{Q}{2\pi k_L} \ln\frac{r}{R}, \qquad 0 < r < R, \qquad t > 0\,. \tag{10}$$

In **§3.2.A** we shall see that this is precisely the quasistationary approximation.

PROBLEM 5. Formulate and solve the axially-symmetric *freezing* problem due to a line *sink* (see **§2.6.B**).

PROBLEM 6. Formulate precisely the (1-phase) problem modeling the axially-symmetric freezing of supercooled liquid described in Case I of **§2.6.C**. Verify that (8) is its solution.

PROBLEM 7. Formulate precisely the (1-phase) problem modeling the spherically-symmetric freezing of supercooled liquid described in Case II of **§.2.6.C**. Verify that (9) is its solution.

2.7. BENCHMARK SOLUTIONS OF MULTIDIMENSIONAL PROBLEMS FOR SIMULATION VERIFICATION

2.7.A Necessity of benchmark solutions

In recent years computing power has grown to the point where it is now possible to simulate realistic two- and three- dimensional heat transfer and phase-change processes. The possibility of actually performing such calculations brings with it the very real problem of knowing if the mountain of numbers and pictures emerging from the simulation is actually correct. Errors can enter both in obvious and non-obvious ways. Simple programming errors can sometimes be easily caught; instability causes an almost immediate end to a run. Errors which are not so simple to discover are those leaving intact the *appearance* of correctness: drifts in results, realistic temperatures. Because one cannot hope to output all the results, those in obvious error may remain unseen. Thus in a building simulation performed by one of the authors (ADS), the structure's temperature seemed high but reasonable; it was only after extensive examination of detailed output that an error was discovered causing one small wall section to have a temperature of several thousand degrees !

The most obvious tool for checking on the correctness of a computer code is to compare its results with the exact solution of some *benchmark* problems. Several such problems in one space dimension have already been presented in this chapter. Yet, *for multi-dimensional problems no genuinely multi-dimensional solutions are known explicitly.* In this section we will present a simple tool for converting a one-dimensional explicit solution of a phase-change problem, into a benchmark solution to a multi-dimensional problem. The solution is obviously contrived, but can serve as a validation tool for codes. Furthermore, the solution presented is genuinely multi-dimensional, with time-varying boundary conditions. For the sake of simplicity we will show how to produce such a benchmark solution for a box-like region in three dimensions. Its application to other more general regions will be clear. The same method may be used to construct an explicit multi-dimensional solution out of any one of the similarity solutions discussed so far, and thus applicable to the multi-dimensional analogue of the physical situation to which it pertains.

2.7.B Phase-change in a box

Let us suppose that a computer code has been prepared, simulating heat transfer and phase-change processes in a box-like region in three dimensions. Let the box occupy the region

$$0 \le x \le A, \quad 0 \le y \le B, \quad 0 \le z \le C. \tag{1}$$

We denote the interior of the box by D and its bounding surface by Γ. Then the computer code will, hopefully, tell us both the temperature $T = T(x, y, z, t)$ and the evolution of the liquid and solid regions with increasing time; the code should be capable of accepting *non-uniform* and *time-dependent* initial and boundary data.

The idea for generating a benchmark solution is simple. Choose a plane Ξ not intersecting the box and regard it as the face of a semi-infinite slab of PCM filling the half-space containing the box, and having the same thermophysical properties (assumed constants) as the PCM inside the box. Assume that at time $t = 0$ the entire PCM is, say, solid at $T_S < T_m$ and a temperature $T_L > T_m$ is imposed on the plane Ξ. The resulting melting front will be a plane parallel to the "face" Ξ and the temperature into the slab will be *uniform* on planes Ξ' parallel to Ξ and will *only depend on the distance between* Ξ' *and* Ξ and time. This (one-dimensional) solution is given by the Neumann solution for a slab melting from $r = 0$ (§2.2). By evaluating this Neumann solution on the faces of the box, we obtain time-dependent boundary conditions for our code. Then we can compare the computed temperatures with those from the Neumann solution for points inside the box ! All we need is a formula for the distance r of a given point (x, y, z) from the "face" Ξ, which we now develop.

Pick any point $O'(x_0, y_0, z_0)$ *not* in D, and choose a plane Ξ through O', *not* intersecting D. Consider the translated coordinate system (x', y', z') with origin O', i.e.,

$$x' = x - x_0, \qquad y' = y - y_0, \qquad z' = z - z_0, \tag{2}$$

and set up spherical coordinates (r, θ, ϕ) there:

$$x' = r\cos\theta\sin\phi, \qquad y' = r\sin\theta\sin\phi, \qquad z' = r\cos\phi\ . \tag{3}$$

Thus, θ is the polar angle on the $x'y'$-plane, ϕ the azimouthal angle measured off the positive z' axis, and r the radial distance from O'. The orientation of the plane Ξ is specified by the direction of its normal, that is, by choosing particular values θ_0 and ϕ_0 for the two angles. Now we need a formula for the distance of any point $P(x, y, z)$ from the chosen plane Ξ. It is easy to see that this distance is given by (PROBLEM 1)

$$r_P = (x - x_0)\cos\theta_0\sin\phi_0 \ + \ (y - y_0)\sin\theta_0\sin\phi_0 \ + \ (z - z_0)\cos\phi_0\ . \tag{4}$$

Returning to the problem of melting the semi-infinite slab $0 \le r < \infty$ via an imposed temperature T_L at $r = 0$, we see that at time $t > 0$ the melting front will be a plane parallel to the face Ξ at distance

$$r = R(t) := 2\lambda\sqrt{\alpha_L t}, \quad t > 0\ . \tag{5}$$

Given any point $P(x, y, z)$ in D, we find its distance r_P from (4) and compare it with the value of $R(t) = 2\lambda\sqrt{\alpha_L t}$. If $r_P < R(t)$ then P lies in the *liquid* and has temperature

$$T(r_P, t) = T_L - (T_L - T_m)\ \frac{\operatorname{erf}\left(\dfrac{r_P}{2\sqrt{\alpha_L t}}\right)}{\operatorname{erf}(\lambda)}\ ; \tag{6a}$$

if, on the other hand, $r_P > R(t)$ then P is still *solid* with temperature

$$T(r_P, t) = T_S + (T_m - T_S) \frac{\operatorname{erf}\left(\dfrac{r_P}{2\sqrt{\alpha_S t}}\right)}{\operatorname{erfc}(\nu\lambda)}, \tag{6b}$$

where $\nu = \sqrt{\alpha_L/\alpha_S}$ and λ is the root of the transcendental equation (2d)§**2.2**. By evaluating the temperature on the faces of the box D, we obtain time-dependent boundary conditions for our code; for example, the temperature on the face $x = 0$, $0 \le y \le B$, $0 \le z \le C$, will be given by (6a,b) with $r_P = -x_0 \cos\theta_0 \sin\phi_0 + (y - y_0)\sin\theta_0 \sin\theta_0 + (z - z_0)\cos\phi_0$.

Notice that *any* boundary conditions can be handled. If, for example, we were testing a code with flux input through one (or more) of the faces, then we would evaluate the flux normal to that face by

$$q = -k\,\frac{\partial T}{\partial \vec{n}} = -k\,\frac{\partial T}{\partial r}\,\nabla r \cdot \vec{n}\,,$$

with $\vec{n}$ the unit normal to that face, $r = r_P$ from (4) for P on that face, and T from (6).

2.7.C Summary of the method

The method described above provides a recipe for developing 2- or 3- dimensional melting or freezing test problems with explicit solutions, based on explicitly solvable 1-dimensional models.

Choose two angles θ_0, ϕ_0 and a point $O'(x_0, y_0, z_0)$ not in the PCM, thus fixing a plane Ξ through O'. Assuming a, say, melting process begins at the plane Ξ at $t = 0$, relations (6) give us the temperature distribution throughout the half-space containing the PCM for any $t > 0$. For any point $P(x, y, z)$, we use (4) to evaluate r_P and comparison with the value of $R(t)$ from (5) tells us if the point P has melted yet, while (6) gives us its temperature. The boundary temperatures, or fluxes, serve as inputs to the code being tested. Then the computed interior temperatures and liquid/solid zones can be compared with the explicit solution values from (5, 6).

Note that even though the phase-change front is planar (parallel to the plane Ξ), the choice of the orientation of Ξ (via θ_0, ϕ_0) allow the front to be skew relative to the PCM (e.g. the box faces) and thus genuinely 3-dimensional. In addition, the time-dependence of the boundary data will provide a rather serious test of the code.

PROBLEMS

PROBLEM 1. Let (x_0, y_0, z_0) and θ_0, ϕ_0 be given. Show the following:

(a) The vector $\vec{n} = (cos\theta_0 \sin\phi_0, \sin\theta_0 \sin\phi_0, \cos\phi_0)$ is a unit vector.

(b) The equation of the plane Ξ through $O'(x_0, y_0, z_0)$ with normal $\vec{n}$ is $(x - x_0)\cos\theta_0 \sin\phi_0 + (y - y_0)\sin\theta_0 \sin\phi_0 + (z - z_0)\cos\theta_0 = 0$.

(c) The distance of a point $P(x, y, z)$ from the plane Ξ is given by (4).

PROBLEM 2. Describe the *two*-dimensional analogue of the method of constructing explicitly solvable benchmark phase-change problems.

PROBLEM 3. As a benchmark problem for two-dimensional melting of a rectangle $0 < x < A$, $0 < y < B$ of ice, choose: $O'(x_0, y_0) = O'(-A, 0)$, $\theta_0 = \frac{\pi}{4}$, $T_{init} = -10°C$, $T_L = 25°C$. Then the role of Ξ is played by the line $y = -x - A$. The Stefan numbers for the melting process are $\boldsymbol{St}_L = 0.314$, $\boldsymbol{St}_S = 0.06$ and the resulting value of λ is $\lambda = 0.09178$. Show that the lower-left corner $(0, 0)$ of the rectangle will melt at time $A^2/(8\lambda^2\alpha_L)$, and that the upper-right corner (A, B) will melt at time $(2A + B)^2/(8\lambda^2\alpha_L)$.

CHAPTER 3

ANALYTICAL APPROXIMATIONS

The most important feature of explicit solutions to phase change problems is that they furnish us with a complete picture of how the various parameters of the process under study interact with each other and influence those factors that are of interest to us. Thus with the aid of explicit solutions we know what the effect of doubling the conductivity or halving the latent heat will be on the melting time or surface temperature history of a body. It is only with the aid of explicit solutions that we have perfect knowledge of the process under study. Unfortunately, the collection of problems for which explicit solutions can be found is extremely small, and does not include processes of relevance to most realistic situations. As mentioned already in Chapter 2 explicit solutions exist only for *semi – infinite* problems with *parameters constant in each phase* and *constant initial* and *imposed* temperatures. This leaves out even problems with constant imposed flux! Very few explicit solutions are known in cylindrical and spherical geometries and absolutely none for finite domains and higher dimensions! Clearly then, for most realistic problems one is forced to seek approximate solutions. They come in two very distinct varieties: analytical approximations and numerical approximations. The latter will be the subject of CHAPTER 4.

In this chapter we present some of the standard analytical approximation methods for phase change problems, [OZISIK]. Our discussion is not exhaustive, but centers on four widely used analytical methods, commonly referred to as the *quasistationary approximation* method, the *Megerlin* method, the *heat balance integral* and *perturbation* methods. Let us emphasize, at the outset, the shortcomings of all such methods: their applicability depends on being able to simplify the problem to a form that fits the method. Such simplification is achieved by making various physical and/or mathematical (simplifying) assumptions on the underlying process and/or the mathematical problem modeling the process. The resulting solution may turn out to be quite accurate, especially for standard, simple, laboratory-type processes. The fundamental difficulty is that we have no way of knowing a priori how accurate the solution will be because, on the one hand, we have no way of checking the validity of the physical simplifications and, on the other hand, there are no error estimates for the mathematical approximations. The validity of an analytical approximation may be ascertained only by comparing its results with some other independently validated method and only for a very restricted class of entirely similar problems, e.g. see [DILLEY-LIOR] and the nice survey in [TOKSOY-ILKEN].

Consequently, the usefulness of analytical approximation methods is restricted to providing qualitative and "order-of-magnitude" information rather than precise quantitative information. Given the generality, versatility and relative ease of applying direct numerical (discretization) methods today (CHAPTER 4), only simple, easy to apply and easy to compute analytical solutions are really useful nowadays. In fact, such simple methods are necessary as both "system sizing" tools and as debugging tools for numerical codes. Driven by this philosophy and attitude, which grew out of our actual experience in the field, we stress the usefulness of the quasistationary approximation as a tool for estimating e.g. melt times and heat absorbed by simple "back-of-the-envelope" calculations.

The main usefulness of analytic approximations lies in the fact that, being analytic expressions, they reveal qualitative behavior (such as dependence of parameters) in ways that numerical solutions cannot. It is usually more expedient to work with dimensionless formulations and we present various undimensionalizations and reformulations of problems as the opportunity arises.

It should be remarked that often in the literature analytic approximations are advertised as "exact solutions" because an approximate problem is solved "exactly," sometimes by dictating the form of the solution (and imposing whatever conditions fail to be satisfied). We shall try to always make clear what explicit or implicit assumptions need to be made in order to obtain a solution.

3.1. THE QUASISTATIONARY APPROXIMATION

3.1.A Introduction

The simplest approximation, producing easily computable results for fast "back-of-the-envelope" estimates is the so called **quasistationary** (or, quasistatic) approximation. The method consists of replacing the heat conduction equation by the steady-state equation $T_{xx} = 0$, while allowing the phase-change front to vary in time. Consequently, it is impossible to meet initial conditions and the method can only be applied to 1-phase problems (see however EXAMPLE 5 below).

The basic physical assumption underlying the method is that the *sensible heat is negligible compared to the latent heat*. Since the Stefan number, $\boldsymbol{St} = c\Delta T / L$, is strongly related to the ratio of sensible to latent heat, (**§2.1**) this amounts to assuming $\boldsymbol{St} \approx 0$. In dimensionless variables (**§2.1**, also see (3) below), $\boldsymbol{St}$ multiplies the time-derivative of the dimensionless temperature, so $\boldsymbol{St} \to 0$ leads to the steady-state heat equation (see [SOLOMON-WILSON-ALEXIADES, 1984]).

Since the quasistationary approximation ignores the sensible heat, *all* of the heat must be used to drive the phase-change and thus we expect that this approximation will overestimate the actual interface location (see (24) below). This makes it useful for obtaining an upper bound on the melt depth but, unfortunately, *no precise error estimates are known.*

We shall apply the method to 1-phase problems with imposed temperature (**§3.1.B**), imposed flux (**§3.1.C**), convective flux (**§3.1.D**) and an internal heat source (**§3.1.E**) for a slab, and for cylinders and spheres (**§3.2**). For definiteness, we discuss melting processes and leave freezing for the Problems.

3.1.B One-phase Stefan Problem with imposed temperature

We consider a slab, $0 \le x \le l$, initially solid at its melt temperature T_m, and impose a *time-dependent* temperature $T_L(t) > T_m$ at $x = 0$. The mathematical model of the melting process is the 1-phase Stefan problem, which admits the Neumann solution only when T_L is constant, (**§2.1**).

MATHEMATICAL PROBLEM: Find $X(t)$, $T(x,t)$ such that

$$\rho c_L T_t = k_L T_{xx}, \quad 0 < x < X(t), \quad t > 0, \tag{1a}$$

$$T(X(t),t) = T_m, \quad t > 0, \tag{1b}$$

$$\rho L X'(t) = -k_L T_x(X(t),t), \quad t > 0, \tag{1c}$$

$$X(0) = 0 \tag{1d}$$

$$T(0,t) = T_L(t), \quad t > 0. \tag{1e}$$

It may be undimensionalized in various ways (PROBLEM 1) of which we prefer the following:

$$\xi = \frac{x}{\hat{x}}, \quad \tau = St_L Fo = St_L \frac{\alpha_L}{\hat{x}^2} t, \quad u(\xi,\tau) = \frac{T - T_m}{\Delta T_L}, \quad \Sigma(\tau) = \frac{X(t)}{\hat{x}}, \tag{2a}$$

where $\hat{x}$ is an arbitrary length (the problem has no natural length scale), and

$$St_L = \frac{c_L \Delta T_L}{L}, \quad \alpha_L = \frac{k_L}{\rho c_L}, \quad Fo = \frac{\alpha_L}{\hat{x}^2} t = \text{Fourier Number},$$

$$\Delta T_L = \max(T_L(t) - T_m). \tag{2b}$$

In terms of the dimensionless variables in (2) the problem reads

$$St_L u_\tau = u_{\xi\xi}, \quad 0 < \xi < \Sigma(\tau), \quad \tau > 0, \tag{3a}$$

$$u(\Sigma(\tau),\tau) = 0, \quad \tau > 0, \tag{3b}$$

$$\Sigma'(\tau) = -u_\xi(\Sigma(\tau),\tau), \quad \tau > 0, \tag{3c}$$

$$\Sigma(0) = 0, \tag{3d}$$

$$u(0,\tau) = U_L(\tau), \quad \tau > 0, \tag{3e}$$

where

$$U_L(\tau) = \frac{T_L(t) - T_m}{\Delta T_L}. \tag{4}$$

The quasistationary approximation replaces the heat equation (3a) simply by

$$u^{qs}_{\xi\xi} = 0, \quad 0 < \xi < \Sigma^{qs}(\tau), \quad \tau > 0, \tag{5}$$

which has as its general solution the linear profile $u^{qs} = A\xi + B$, with A, B independent of ξ. Imposing the boundary conditions (3b, e) leads to

$$u^{qs}(\xi,\tau) = U_L(\tau)\,[1 - \frac{\xi}{\Sigma^{qs}(\tau)}], \quad 0 < \xi < \Sigma^{qs}(\tau), \quad \tau > 0, \tag{6a}$$

whence (3c) becomes $\Sigma^{qs\prime}(\tau) = \dfrac{U_L(\tau)}{\Sigma(\tau)}$, or $\dfrac{(\Sigma^{qs}(\tau)^2)'}{2} = U_L(\tau)$. Integrating and using (3d) we obtain

$$\Sigma^{qs}(\tau) = \left[2 \int_0^\tau U_L(s)ds \right]^{1/2}, \qquad \tau \geq 0 \,. \tag{6b}$$

Thus the quasistationary solution of (3) is given by (6). In terms of the original physical variables the quasistationary solution becomes (PROBLEM 2)

$$X^{qs}(t) = \left[2 \frac{k_L}{\rho L} \int_0^t [T_L(s) - T_m]\, ds \right]^{1/2}, \qquad t \geq 0 \,, \tag{7a}$$

$$T^{qs}(x,t) = T_L(t) - [T_L(t) - T_m] \frac{x}{X^{qs}(t)}, \quad 0 \leq x \leq X(t)\,, \quad t \geq 0\,, \tag{7b}$$

and the heat flux is given by

$$q_L^{qs}(x,t) = -k_L T_x^{qs} = k_L \frac{T_L(t) - T_m}{X^{qs}(t)} \,. \tag{7c}$$

Thus at each time $t \geq 0$ the temperature profile $T^{qs}(x,t)$ is a straight line connecting the points ($x = 0$, $T = T_L(t)$) and ($x = X^{qs}(t)$, $T = T_m$) of the (x, T) plane.

The solution for *freezing* is entirely analogous and may be obtained formally by replacing every subscript "L" by "S" and the latent heat "L" by "$-L$" in (7), (PROBLEM 3).

EXAMPLE 1. Constant imposed temperature : $T_L(t) = T_L$.

In this case $U_L(\tau) \equiv 1$ and (6) give

$$\Sigma^{qs}(\tau) = \sqrt{2\tau}\,, \qquad u^{qs}(\xi,\tau) = 1 - \frac{\xi}{\sqrt{2\tau}}, \quad 0 < \xi < \Sigma^{qs}(\tau),\ \tau > 0\,, \tag{8}$$

or, from (7),

$$X^{qs}(t) = \sqrt{2\alpha_L \cdot St_L \cdot t} = 2\sqrt{St_L/2} \cdot \sqrt{\alpha_L t}\,, \qquad t > 0\,, \tag{9a}$$

$$T^{qs}(x,t) = T_L - [T_L - T_m] \frac{\dfrac{x}{2\sqrt{\alpha_L t}}}{\sqrt{St_L/2}}, \quad 0 < x < X^{qs}(t)\,, \quad t > 0\,. \tag{9b}$$

Comparison with **§2.1** reveals that this is precisely the expression for the Neumann solution when $St_L \approx 0$, as expected in view of the remarks in **§3.1.A**. In fact, one may easily show (PROBLEM 4b) that the transcendental root λ of the Neumann solution is *always smaller* than $\sqrt{St_L/2}$, which by (9) implies that the quasistationary front *overestimates* the actual (Neumann) front:

$$X^{qs}(t) > X(t)\,, \qquad t > 0\,; \tag{10}$$

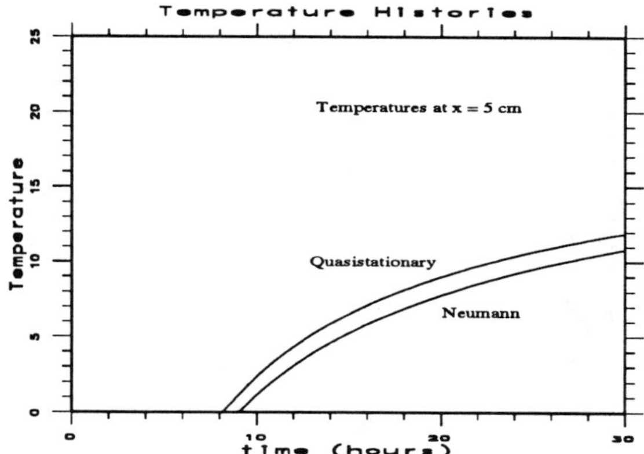

Figure 3.1.1. Comparison of temperature histories, EXAMPLE 1.

hence the linear temperature profile $T^{qs}(x,t)$ overestimates the actual (Neumann) temperature at each $t > 0$ (see **Figures 3.1.1,2,3** where we compare (9) with the case of melting of an ice slab of §**2.1.E** (**Figures 2.1.4,5,6**)). Unfortunately, no precise error estimate is known.

EXAMPLE 2. Linear growth of imposed temperature. Consider the case where

$$T_L(t) = T_m + (T_L - T_m)\frac{t}{t_0}, \qquad 0 \le t \le t_0, \tag{11}$$

i.e. $T_L(t)$ grows linearly from T_m to some $T_L > T_m$ during a time interval of

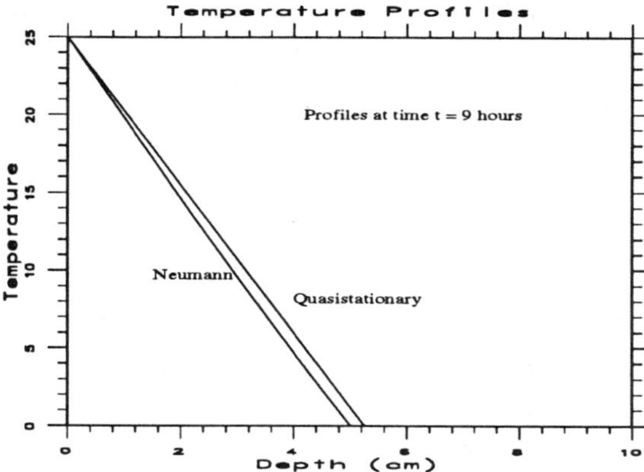

Figure 3.1.2. Comparison of temperature profiles, EXAMPLE 1.

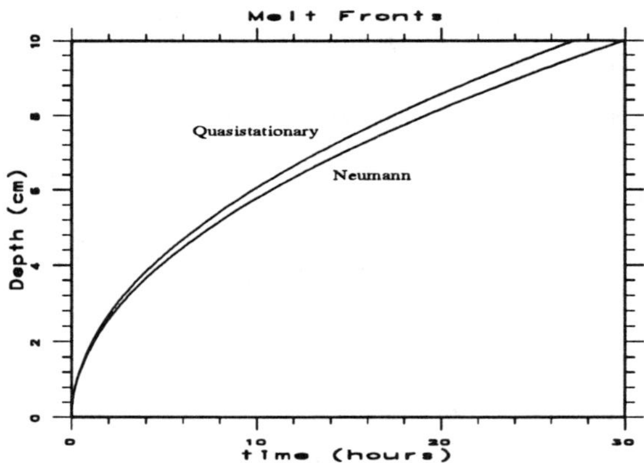

Figure 3.1.3. Comparison of melt fronts, EXAMPLE 1.

length t_0. Evaluating the integral in (7a), we find a linear front

$$X^{qs}(t) = t\,(\,\alpha_L \cdot St_L / t_0\,)^{1/2}\,, \tag{12a}$$

a temperature linear in space and time,

$$T^{qs}(x,t) = T_m + \frac{t}{t_0}\,\Delta T_L \left[1 - \frac{x}{X^{qs}(t)} \right], \quad 0 \le x \le X^{qs}(t)\,,\ 0 \le t \le t_0\,, \tag{12b}$$

and a constant flux throughout the liquid

$$q^{qs}(x,t) = (\,\rho L k_L \Delta T_L / t_0\,)^{1/2}\,, \quad 0 \le x \le X^{qs}(t)\,, \quad 0 \le t \le t_0\,. \tag{12c}$$

In **Figure 3.1.4** we see the melt front history for the case of melting ice with a face temperature rising from $0°C$ to $25°C$ over a $t_0 = 10$ hour period. **Figure 3.1.5** shows the quasistationary temperature history at the points $x=0$ and $x=1$ cm, and **Figure 3.1.6** shows the temperature distribution at time $t_0 = 10$ hours.

EXAMPLE 3. "Saw-tooth" imposed temperature. We now consider the case of the face temperature

$$T_L(t) = \begin{cases} T_m + \Delta T_L \dfrac{t}{t_0}\,, & 0 \le t \le t_0\,, \\ T_m + \Delta T_L \dfrac{t_1 - t}{t_1 - t_0}\,, & t_0 \le t \le t_1\,, \end{cases} \tag{13}$$

$\Delta T_L = T_L - T_m$. For $0 \le t \le t_0$, the solution was found in EXAMPLE 2. For $t_0 \le t \le t_1$, (7a, 13) yield

$$X^{qs}(t) = \sqrt{\frac{k_L \Delta T_L}{\rho L}} \sqrt{t_1 - \frac{(t_1 - t)^2}{t_1 - t_0}}\,, \qquad t_0 \le t \le t_1\,, \tag{14}$$

which is the upper part of an ellipse, with zero slope at $t = t_1$. Notice that the front location at time t_1, $X^{qs}(t_1) = \sqrt{k_L \Delta T_L t_1 / \rho L}$, is independent of the

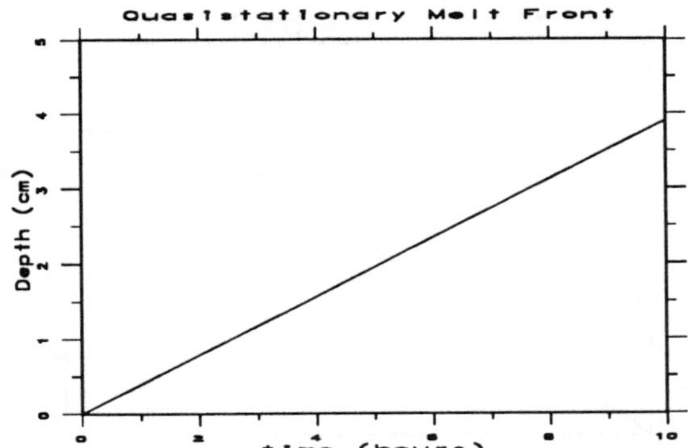

Figure 3.1.4. Quasistationary melt front history, EXAMPLE 2.

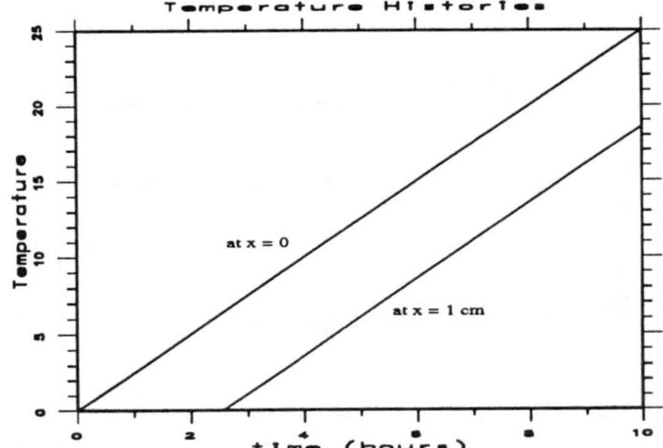

Figure 3.1.5. Quasistationary temperature histories, EXAMPLE 2.

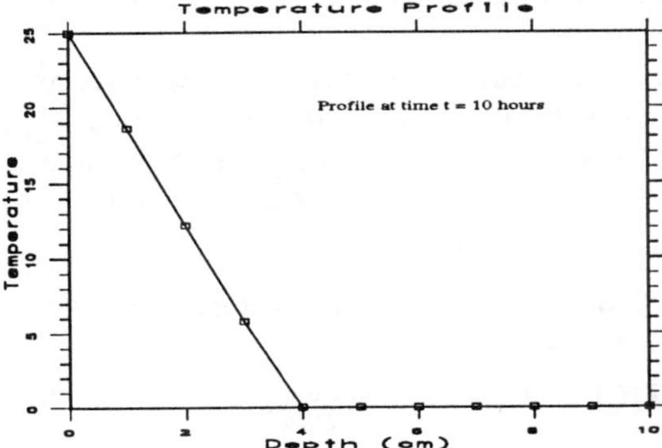

Figure 3.1.6. Quasistationary temperature profile, EXAMPLE 2.

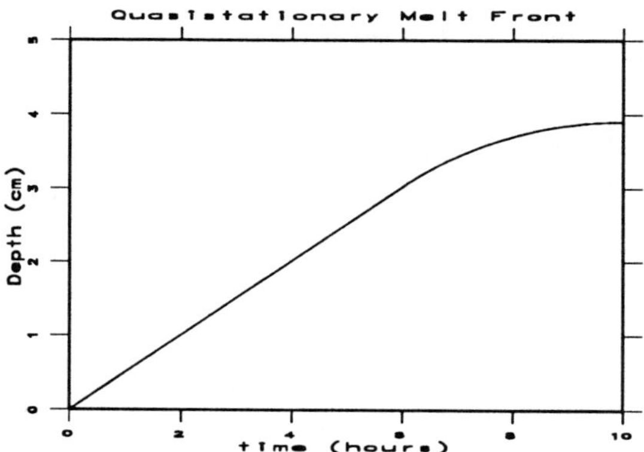

Figure 3.1.7. Quasistationary front history of EXAMPLE 3.

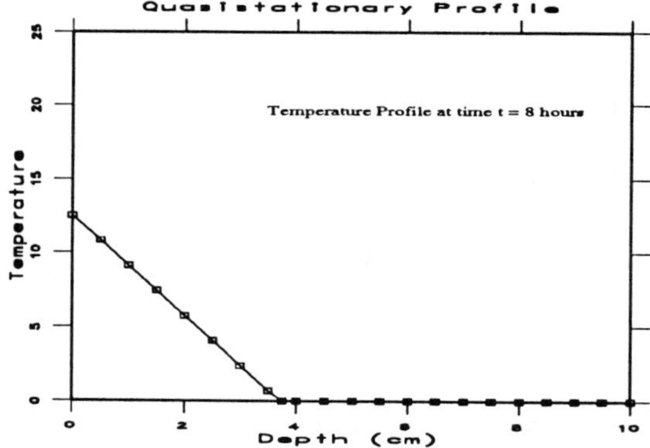

Figure 3.1.8. Quasistationary temperature profile of EXAMPLE 3.

time t_0 at which $T_L(t)$ peaks. In **Figures 3.1.7,8,9** we see again the phase front history, temperature distribution at a particular time (8 hours) and the temperature history at the face ($x = 0$) and at the point $x = 1$, for melting of ice with a face temperature rising from $0°C$ to $25°C$ over $t_0 = 6$ hours and then falling to $0°C$ in 4 hours ($t_1 = 10$ hours).

EXAMPLE 4. Sinusoidal imposed temperature. We now consider the case of the face temperature

$$T_L(t) = T_m + \Delta T_L \sin(\frac{\pi t}{2t_0}), \qquad 0 \leq t \leq 2t_0. \tag{15}$$

The quasistationary front turns out to also be sinusoidal (PROBLEM 5),

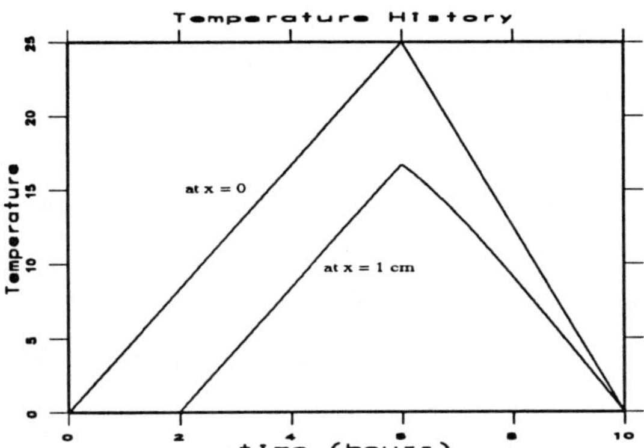

Figure 3.1.9. Quasistationary temperature histories of EXAMPLE 3.

$$X^{qs}(t) \;=\; \sqrt{\frac{8\alpha_L St_L t_0}{\pi}}\cdot \sin(\frac{\pi t}{4t_0})\,, \qquad 0 \le t \le 2t_0\,, \tag{16a}$$

$$T^{qs}(x,t) \;=\; T_m + \Delta T_L \sin(\frac{\pi t}{2t_0}) - \Delta T_L \cdot x \cdot \cos(\frac{\pi t}{4t_0})\sqrt{\frac{\pi\rho L}{2k_L t_0 \Delta T_L}}\,, \tag{16b}$$

$$q^{qs}(x,t) \;=\; \cos(\frac{\pi t}{4t_0})\sqrt{\frac{\pi\rho L k_L \Delta T_L}{2t_0}}\,, \qquad 0 \le x \le X^{qs}(t)\,, \quad t \ge 0. \tag{16c}$$

EXAMPLE 5. Melting of an initially cold slab. Strictly speaking, the quasistationary approximation cannot be applied to 2-phase problems (**§3.1.A**). However, if $\alpha_S \approx 0$ then the effect of a cold initial temperature may be accounted for as follows.

Consider a slab (finite or semi-infinite) initially at a uniform temperature $T_S < T_m$. A temperature $T_L > T_m$ is imposed at $x = 0$ to induce melting. If $\alpha_S \approx 0$, then $T_t \approx 0$ in the solid, hence $T(x,t) \approx T_S$ there. Thus, ignoring a boundary layer to the right of the front (**Figure 3.1.10**), we may assume that the limiting temperature at $x = X(t)$ from the solid side is essentially T_S, resulting in the Stefan condition

$$\rho[\,L + c_S(T_m - T_S)\,]\;X'(t) \;=\; -\,k_L T_x(\,X(t),t)\;,\; t > 0. \tag{17}$$

Then, the quasistationary approximation may be applied in the liquid as before. The only change will be the replacement of L by the enhanced $L + c_S(T_m - T_S)$.

We note that this approach exaggerates the influence of the initial subcooling, because it replaces the decreasing sensible heat by the constant $c_S(T_m - T_S)$, slowing down the melt front. Fortunately, this compensates somewhat for the over-advancing $X^{qs}(t)$ (instead of compounding the error).

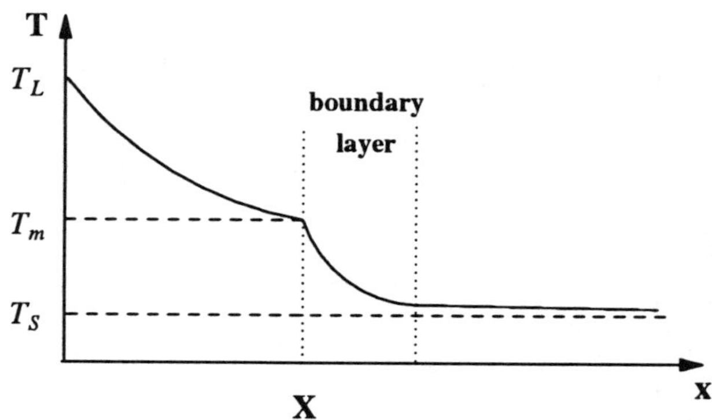

Figure 3.1.10. Profile expected in the 2-phase case, EXAMPLE 5.

Clearly, such considerations may only be used for first, rough estimates, since we have no way of knowing the error involved. In particular, the above reasoning is not applicable to multiple phases generated by a cycling hot-cold imposed temperature (see EXAMPLE 7, §**3.1.C**).

3.1.C One-phase Stefan Problem with imposed flux

We now turn to the case of a slab of solid at its melt temperature T_m, which is being melted by an imposed flux, $q_L(t) > 0$ at $x = 0$:

$$-k_L T_x(0,t) = q_L(t) > 0\,, \quad t > 0. \tag{18}$$

The mathematical problem is the 1-phase Stefan Problem (1),(18). Even for constant $q_L(t) = q_L > 0$ there is no explicit solution available, which is why this boundary condition was not considered in CHAPTER **2**.

Consider the case of constant $q_L(t) \equiv q_L$. The undimensionalization of this problem is interesting in that there are no parameters left whatsoever, i.e. *there exists a single, universal, dimensionless* 1-*phase Stefan Problem with imposed flux !* This is because there are *no* natural length, time or temperature scales in this problem, allowing us to choose these scales to make all coefficients equal to 1. Indeed, introducing the dimensionless variables

$$\xi = \frac{x}{\hat{x}}\,, \quad Fo = \frac{t}{\hat{t}}\,, \quad u(\xi\,, Fo) = \frac{T(x,t) - T_m}{\hat{T}}\,, \quad \Sigma(Fo) = \frac{X(t)}{\hat{x}}\,, \tag{19}$$

with $\hat{\xi}, \hat{t}, \hat{T}$ scales to be chosen, problem (1),(18) becomes

$$u_{Fo} = \frac{\alpha_L \hat{t}}{\hat{x}^2} u_{\xi\xi}\,, \quad 0 < \xi < \Sigma(Fo)\,, \quad Fo > 0\,, \tag{20a}$$

$$u(\Sigma(Fo)\,, Fo) = 0\,, \quad Fo > 0\,, \tag{20b}$$

$$\frac{\rho L \hat{x}^2}{\hat{t}\hat{T}k_L}\,\Sigma'(Fo) \;=\; -\,u_\xi(\Sigma(Fo)\,,\,Fo)\,, \qquad Fo > 0\,, \tag{20c}$$

$$\Sigma(0) \;=\; 0\,, \tag{20d}$$

$$-u_\xi(0,\,Fo) \;=\; \frac{\hat{x}}{k_L\hat{T}}\,q_L\,, \qquad Fo > 0\;. \tag{20e}$$

We observe three dimensionless groups appearing: $\dfrac{\alpha_L\hat{t}}{\hat{x}^2}\,,\;\dfrac{\rho L\hat{x}^2}{\hat{t}\hat{T}k_L}\,,\;\dfrac{\hat{x}q_L}{\hat{T}k_L}$; hence we may choose the three scales, $\hat{x}, \hat{t}, \hat{T}$, to make the coefficients equal to 1. This leads to the scales

$$\hat{x} = \frac{\rho L\,\alpha_L}{q_L}\,, \qquad \hat{t} = \frac{\hat{x}^2}{\alpha_L}\,, \qquad \hat{T} = \frac{\hat{x}q_L}{k_L} = \frac{L}{c_L}\,, \tag{21}$$

and the dimensionless, *parameter-free* 1-phase Stefan Problem with constant imposed flux is

$$u_{Fo} \;=\; u_{\xi\xi}\,, \qquad 0 < \xi < \Sigma(Fo)\,, \quad Fo > 0\,, \tag{22a}$$

$$u(\Sigma(Fo)\,,\,Fo) \;=\; 0\,, \qquad Fo > 0\,, \tag{22b}$$

$$\Sigma'(Fo) \;=\; -\,u_\xi(\Sigma(Fo)\,,\,Fo)\,, \qquad Fo > 0\,, \tag{22c}$$

$$\Sigma(0) \;=\; 0 \tag{22d}$$

$$u_\xi(0,\,Fo) \;=\; -\,1\,, \qquad Fo > 0\;. \tag{22e}$$

When $q_L(t)$ is not constant, we may choose any scale $\hat{q}_L$ (e.g. $\hat{q}_L = \max q_L(t)$) which will play the role of the constant q_L above, and set

$$Q(Fo) \;=\; \frac{q_L(t)}{\hat{q}_L}\,. \tag{23}$$

Then the boundary condition (20e) takes the form

$$u_\xi(0,\,Fo) \;=\; -\,Q(Fo)\,, \quad Fo > 0 \tag{24}$$

replacing (22e).

The quasistationary approximation to (22a-d), (24) assumes $u_{Fo} = 0$ in (22a), whence $u = A\xi + B$. From (24), $A = -\,Q(Fo)$, and from (22b), $B = Q\Sigma$, so then (22c) demands $\Sigma'(Fo) = Q(Fo)$. Therefore the quasistationary solution is

$$\Sigma^{qs}(Fo) \;=\; \int_0^{Fo} Q(s)ds\,, \qquad Fo \geq 0\,, \tag{25a}$$

$$u^{qs}(\xi, Fo) \;=\; Q(Fo)\,[\,\Sigma^{qs}(Fo) - \xi\,]\,, \qquad 0 \leq \xi \leq \Sigma^{qs}(Fo)\,, \quad Fo \geq 0\,. \tag{25b}$$

In particular, when $q_L(t) \equiv q_L$, this becomes very simple:

$$\Sigma^{qs}(Fo) \;=\; Fo\,, \qquad u^{qs}(\xi, Fo) \;=\; Fo - \xi. \tag{26}$$

Expressed in terms of the original physical variables, the quasistationary solution of (1), (18) takes the form (PROBLEM 6)

$$X^{qs}(t) = \frac{1}{\rho L}\int_0^t q_L(s)ds\,, \quad t \geq 0\,, \tag{26a}$$

$$T^{qs}(x,t) = T_m + \frac{q_L(t)}{k_L}\,[X^{qs}(t) - x]\,, \quad 0 \leq x \leq X^{qs}(t)\,, \quad t \geq 0. \tag{26b}$$

On physical grounds, we expect that $X^{qs}(t)$ overestimates the actual front as before:

$$X^{qs}(t) > X(t)\,, \qquad t > 0. \tag{27}$$

The quasistationary approximation to the *freezing* problem with imposed flux $q_S(t) < 0$ is found similarly (PROBLEM 7).

EXAMPLE 6. Constant imposed flux.

Consider the case $q_L(t) \equiv q_L > 0$. Then (26), yield

$$X^{qs}(t) = \frac{q_L}{\rho L}\,t\,, \qquad t \geq 0 \tag{28a}$$

$$T^{qs}(x,t) = T_m + \frac{q_L}{k_L}\,[\frac{q_L}{\rho L}\,t - x]\,, \quad 0 \leq x \leq X^{qs}(t)\,, \quad t \geq 0\,, \tag{28b}$$

which is linear in both x and t. In particular, the face temperature, $T^{qs}(0, t)$ grows linearly in time, and the melt-time of a location x is given by

$$t_{melt} = \frac{\rho L}{q_L}\,x\,. \tag{28c}$$

Further examples are suggested in PROBLEM 8.

EXAMPLE 7. Sign-switching flux. The flux

$$q_L(t) = \begin{cases} +q_0 > 0\,, & 0 \leq t \leq t_0\,, \\ -q_1 < 0\,, & t_0 \leq t \leq t_1\,, \end{cases} \tag{29}$$

models a charge-discharge cycle in a latent heat thermal energy storage system (see CHAPTER 5). Assuming the material is initially solid at its melt temperature T_m, the quasistationary approximation is applicable during $0 \leq t \leq t_0$ and its solution appears in EXAMPLE 6. After time t_0, however, heat is being **withdrawn** from the face $x = 0$, lowering its temperature until it reaches T_m at some later time, say $t = t^*$. If t^* happens to be less than t_1 then a second, **freezing**, front will emerge at $x = 0$ and will propagate into the material. During this transition time, $t_0 \leq t \leq t^*$, the quasistationarity approximation is **not** applicable because the existing liquid at time t_0 will **not** be at a uniform temperature (recall that the quasistationary solution cannot satisfy initial conditions). While the face temperature is falling, the melt front, X, continues to advance but progressively slower, until the temperature in the liquid near X reduces to T_m at some time t^{**}. At that moment the melt front becomes stationary, separating liquid at T_m from solid at T_m. If $t^{**} = t^*$ (which is unlikely), then at $t = t^*$ the liquid will be at T_m and the quasistationary approximation will again be applicable during $t^* \le t \le t_1$ for the emerging solid phase.

This example illustrates the limitations of the quasistationary approximation and also shows how a simple charge-discharge process can lead to a complicated problem. An analogous process will be examined in detail in CHAPTER **5**.

3.1.D The case of convective boundary condition

We now turn to the case of a slab of solid at its melt temperature T_m which is being heated convectively at the face $x = 0$,

$$-k_L T_x(0, t) = h\,[T_L(t) - T(0, t)]\,, \qquad t \ge 0\,. \tag{30}$$

This boundary condition replaces (1e), and its dimensionless form (in terms of the variables in (2)):

$$-u_\xi(0, \tau) = Bi \cdot [U_L(\tau) - u(0, \tau)] \tag{31}$$

replaces (3e), where

$$Bi := \frac{h\,\hat{x}}{k_L} = \textit{Biot Number}, \quad U_L(\tau) = \frac{T_L(t) - T_m}{\Delta T_L}, \quad \Delta T_L = \max(T_L(t) - T_m)\,. \tag{32}$$

Such a process is characterized by the Stefan Number $St_L = c_L \Delta T_L / L$ and the Biot Number for the "characteristic length" $\hat{x}$ (which may be any length of interest in the process).

The quasistationary approximation is easily found (PROBLEM 11) to be

$$\Sigma^{qs}(\tau) = -\frac{1}{Bi} + \left(\frac{1}{Bi^2} + 2\int_0^\tau U_L(s)ds \right)^{1/2}, \qquad \tau \ge 0\,, \tag{33a}$$

$$u^{qs}(\xi, \tau) = \frac{Bi\; U_L(\tau)}{1 + Bi\,\Sigma^{qs}(\tau)}\,[\Sigma^{qs}(\tau) - \xi], \quad 0 \le \xi \le \Sigma^{qs}(\tau), \quad \tau \ge 0\,, \tag{33b}$$

or, in physical variables,

$$X^{qs}(t) = -\frac{k_L}{h} + \left(\frac{k_L^2}{h^2} + 2\frac{k_L}{\rho L}\int_0^t [T_L(s) - T_m]ds \right)^{1/2}, \qquad t \ge 0\,, \tag{34a}$$

$$T^{qs}(x, t) = T_m + [T_L(t) - T_m]\,\frac{h[\,X^{qs}(t) - x]}{h\,X^{qs}(t) + k_L}\,, \quad 0 \le x \le X(t), \quad t > 0\,. \tag{34b}$$

Hence, the temperature and flux at the face $x = 0$ are

$$T^{qs}_{face}(t) = \frac{T_m + \dfrac{h\,X^{qs}(t)}{k_L}\,T_L(t)}{1 + \dfrac{h\,X^{qs}(t)}{k_L}}, \qquad t \geq 0, \tag{35a}$$

$$q^{qs}_{face}(t) = \frac{k_L}{X^{qs}(t)}\,[T^{qs}_{face}(t) - T_m], \qquad t \geq 0, \tag{35b}$$

and T^{qs} may be also written as

$$T^{qs} = T^{qs}_{face} - [T^{qs}_{face} - T_m]\,\frac{x}{X^{qs}}\ . \tag{35c}$$

When $T_L(t) \equiv T_L = constant$, solving (34a) for t yields the following expression for the **melt-time** of a location x_o:

$$t^{qs}_{melt} = \frac{x_o^2 \rho L}{2k_L \Delta T_L}\left[1 + \frac{2k_L}{h\,x_o}\right]. \tag{35d}$$

Note that $h \approx \infty$ corresponds to *imposing* T_L directly on the face $x = 0$, [SOLOMON-WILSON-ALEXIADES, 1984], and therefore (34) ought to reduce to (7). Indeed, taking $h \to \infty$, we have $T^{qs}_{face} \to T_L$ from (35a) and we see that (34) does in fact reduce to the imposed temperature solution, (7).

These approximations turn out to be very useful in practice for the following reasons:

a) Often we know **how** we want a particular system to behave, but the critical feature is to find the required value of h in order that it behave in that fashion. The quasistationary approximation offers a particularly simple way of figuring out if the design is feasible at all, and gives useful "sizing" estimates. Thus, if we know that a thermal cycle is to take a certain period of time to melt a layer that is X units thick, then relation (34a) can tell us if it is possible to find an h that will do this. While the accuracy of the approximation is *not* known, it does give us useful, first-cut , answers to such questions without the need for costly computations or more analytical power.

b) The estimate provided for the face temperature, (35a), is particularly important when constraints on how high or low this temperature may go must be taken into account.

Such questions will be examined in more detail in CHAPTER **5**, in the context of thermal energy storage problems.

A Word Of Caution.

Having praised the quasistationary approximation as a useful method for obtaining simple approximations to phase-change problems, we hasten to point out that there are problems in which this approximation produces physically unacceptable results! Indeed, consider melting via convective heat input, (30), with constant $T_L(t) \equiv T_L$ and h. Clearly, this process is *slower* than the melting process resulting from imposing T_L itself at $x = 0$, since the later corresponds to $h \approx \infty$. It follows that the melt time t_0^{qs} of a given point x_0, predicted by the quasistationary approximation (34), *should be greater* than the melt time $t_0^{\infty} = \dfrac{x_0^2}{4\lambda^2\alpha_L}$ predicted by the Neumann solution to the 1-phase Stefan Problem with T_L imposed at $x = 0$. Using (34a), one easily sees (PROBLEM 15) that

$$\frac{t_0^{qs}}{t_0^{\infty}} = \frac{2\lambda^2}{St_L}[1 + 2\frac{k_L}{h\,x_0}], \tag{36}$$

and therefore $t_0^{qs} > t_0^{\infty}$ if and only if

$$Bi = \frac{h\,x_0}{k_L} < \frac{2}{\dfrac{St_L}{2\lambda^2} - 1} =: Bi^*. \tag{37}$$

Note that Bi^* depends only on the Stefan Number $St_L = c_L(T_L - T_m)/L$ of the process (shown in **Figure 3.1.11**), and Bi is the Biot Number for the "characteristic length" x_0 and heat transfer coefficient h. We conclude that

*if the Biot Number of the process exceeds the Bi * corresponding to the Stefan Number of the process then the quasistationary approximation (34) is physically unacceptable.*

In such a case, a better approximation is obtained from the Neumann solution **§2.1.B** (by assuming $h \approx \infty$, i.e. imposing T_L directly on the face $x = 0$)! Clearly, as characteristic length x_0, one chooses some particular length that is of interest in the process under study.

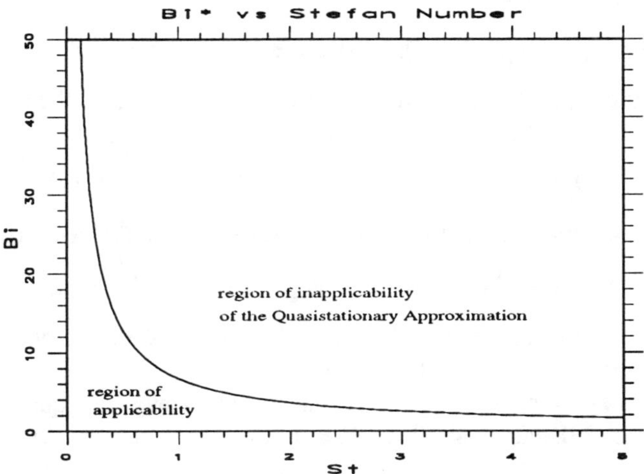

Figure 3.1.11. Bi^* of (37) vs Stefan Number.

3.1.E. Volumetric heating

Several mechanisms, such as radioactive decay, electrical resistance heating or irradiation (by light, microwaves, etc), provide energy directly to each point of the material, thus generating *internal* heating. We consider the following simple

PHYSICAL PROBLEM: A slab is subjected to internal heating by a volumetric heat source of strength Q (kW/m^3), assumed uniform for simplicity (see PROBLEM 17). Assume that at time $t = 0$ the slab is solid at T_m and a constant temperature $T_L > T_m$ is imposed at $x = 0$. Thus, a melt front starts at $x = 0$ and propagates to the right, while the whole slab is being heated internally by the source Q. All material properties are assumed constant, and so is Q.

Clearly, after a while, the heat source *alone* will deliver an amount of heat equal to the latent heat at each point, thus melting the *whole* slab at once. So, let us consider the situation at *early times*, before total melting occurs (i.e. before the enthalpy reaches the value L). At such times, the slab to the right of the melt front is still isothermal at T_m, but no longer solid. At each point there, the enthalpy satisfies $0 < e(x,t) < L$, so this region is "mushy", [LACEY-SHILLOR]. The energy conservation law ((15)§**1.2**): $\rho e_t + q_x = Q$, reduces to $\rho e_t = Q$, whence $\rho e = Q \cdot t$ as long as $\rho e \le \rho L$. Therefore we have

$$e(x,t) = \frac{Q}{\rho} t \quad \text{for} \quad 0 \le t < t^* := \frac{\rho L}{Q}, \quad x > X(t). \tag{38}$$

Up to the time t^*, the mushy region remains isothermal at T_m while the melt front advances into it. Therefore, during $0 < t < t^*$, we have a 1-phase problem for the liquid, with a heat source:

$$\rho c_L T_t = k_L T_{xx} + Q, \quad 0 < x < X(t), \quad 0 < t < t^*. \tag{39a}$$

The enthalpy jump $[\![\rho e]\!]_-^+$, across the melt-front, is $Qt - \rho L$, so the Stefan Condition: $[\![\rho e]\!]_-^+ X'(t) = [\![q]\!]_-^+$ now takes the form

$$[\rho L - Qt]X' = -k_L T_x \quad \text{on } x = X(t), \quad 0 < t < t^*. \tag{39b}$$

The remaining conditions are the standard ones for a 1-phase problem, namely,

$$T(X(t),t) = T_m, \quad X(0) = 0, \quad T(0,t) = T_L > T_m, \tag{39c}$$

for $0 < t < t^*$. Note that for $Q = 0$, (39) reduce to problem (1), but even with $T_L \equiv$ *constant*, there is *no* explicit solution to (39).

Assuming $\boldsymbol{St}_L = c_L(T_L - T_m)/L$ is small, we apply the quasistationary approximation to (39), and obtain (PROBLEM 16)

$$X^{qs}(t) = \left[\frac{2k_L(T_L - T_m)}{\rho L - Qt} t \right]^{1/2}, \quad 0 \le t < t^* \tag{40a}$$

$$T^{qs}(x,t) = T_L - (T_L - T_m)\frac{x}{X^{qs}(t)} + \frac{Q}{2k_L}\, x\,[X^{qs}(t) - x], \quad 0 \le x \le X^{qs}(t), \quad 0 \le t < t^* . \tag{40b}$$

Note that as $t \uparrow t^*$, $X^{qs}(t) \to \infty$, exactly as we expect, since the whole slab completely liquifies at that time!

It is apparent from this discussion that internal heating creates a situation quite different from the previous ones. The temperature alone can no longer describe the phases and we have to deal with the enthalpy itself, [LACEY-SHILLOR], [LACEY-TAYLER]. This approach will be explored fully in CHAPTER 4.

PROBLEMS

PROBLEM 1. Undimensionalizations of the 1-phase problem (1):

(a) Let $\xi = x / \hat{x}$, $Fo = t / \hat{t}$, $u = T - T_m / \hat{T}$, $\Sigma = X / \hat{x}$ in (1) and choose the scalings $\hat{x}, \hat{t}, \hat{T}$ so that (1) becomes

$$u_{Fo} = u_{\xi\xi}, \qquad U(\Sigma, Fo) = 0, \qquad \Sigma' = -u_\xi, \qquad u(0, Fo) = u_L(Fo).$$

(b) With the same choices as in (a), but with $\hat{T} = \Delta T_L := \max\,(T_L(t) - T_m)$, show that the Stefan number, $St_L = c_L \Delta T_L / L$, appears in the Stefan condition: $\Sigma'(Fo) = -St_L \cdot u_\xi(\Sigma, Fo)$.

(c) Show that (2) lead to the form (3), in which the Stefan number appears in the heat equation.

PROBLEM 2. Verify the quasistationary solution (7) by

(a) changing (6) to physical variables via (2);

(b) applying the quasistationary approximation directly to (1).

PROBLEM 3. Derive the quasistationary solution for a 1-phase *freezing* problem with imposed temperature $T_S(t) < T_m$.

PROBLEM 4. Proof of (10):

(a) Show that for any $\lambda > 0$, $\frac{\sqrt{\pi}}{2} \cdot \lambda \cdot e^{\lambda^2} \cdot \operatorname{erf} \lambda > \lambda^2$ [use series expansions of both e^{λ^2} and $\operatorname{erf} \lambda$].

(b) Show that the root of $\sqrt{\pi} \cdot \lambda \cdot e^{\lambda^2} \operatorname{erf} \lambda = St_L$ is smaller than the root of $\lambda^2 = St_L/2$ for any $St_L > 0$, and deduce (10).

(c) Show that $q(0,t) > q^{qs}(0,t)$, $t > 0$.

(d) Show that the quasistationary approximation to the total heat absorbed $Q^{qs}(t) = \int_0^t q^{qs}(0,s)\,ds$, **underestimates** the actual total heat, $Q(t) = \int_0^t q(0,s)\,ds$.

PROBLEM 5. Derive the quasistationary solution (16) for the sinusoidal imposed temperature given in (15) and apply it to the case of melting of the slab of ice (of EXAMPLES 2,3) for a sinusoidal temperature wave with $t_0 = 5$ hours. Graph the resulting temperature history at a depth of .01 meters, the interface curve and the temperature distribution at both 5 and 10 hours.

PROBLEM 6. Verify the quasistationary solution (26) by

(a) Changing (25) to physical variables via (19,21).

(b) Applying the quasistationary approximation directly to (1,18).

PROBLEM 7. Derive the quasistationary solution for a 1-phase freezing problem with imposed flux $q_S(t) < 0$. Verify that it coincides with (26) after replacing q_L by $-q_S$, L by $-L$ and k_L by k_S.

PROBLEM 8. Derive the quasistationary solution for a 1-phase melting problem with (a) linearly increasing flux, $q_L(t) = \gamma_0 t$, $\gamma_0 > 0$;

(b) "saw-tooth" flux,

$$q_L(t) = \begin{cases} \gamma_0 t , & 0 \le t \le t_0 , \\ \gamma_0 t_0 \dfrac{t_1 - t}{t_1 - t_0} , & t_0 \le t \le t_1 \end{cases}$$

(c) sinusoidal flux, $q_L(t) = q_{\max} \sin(\frac{\pi t}{2t_0})$, $0 \le t \le 2t_0$. Such a flux may be used to model solar beam radiation on a horizontal surface. Verify that at the completion time, $t = 2t_0$,

$$X^{qs}(2t_0) = \frac{4t_0 q_{\max}}{\rho L \pi} , \qquad T^{qs}(x, 2t_0) = T_m , \quad \text{(but liquid)}.$$

PROBLEM 9. A slab of ice 1 meter long, initially at 0 ° C is being melted by imposing a constant flux q_L at $x = 0$, the back face being insulated. Use the quasistationary approximation to fine (an estimate of) the value q_L required to melt the slab in 1 hour. For parameter values see **§2.1.E**.

PROBLEM 10. Examine the effectiveness of the relation (17) for the two-phase problem of melting ice of **§2.2**.

PROBLEM 11. Formulate the complete mathematical problem modeling the process of EXAMPLE 7, for $0 \le t \le t_1$.

PROBLEM 12. Find the quasistationary solution suggested in EXAMPLE 7 during $t^* \le t_1$, assuming $t^{**} = t^*$.

PROBLEM 13. Derive the dimensionless quasistationary solution (33) of the problem with convective flux (**§3.1.D**).

PROBLEM 14. Derive (34, 35) directly, and also by a change of variables from (33).

PROBLEM 15. Derive the relation (36) and explain the condition (37) in your own words.

PROBLEM 16. Apply the quasistationary approximation to problem (39) to derive (40).

PROBLEM 17. Carry out the analysis of §3.1.E for the case where

$$Q = Q_0 e^{-\gamma_0 x}$$

where Q_0 , γ_0 are positive constants.

PROBLEM 18. For each of the cases examined in this section derive an expression for the total energy in the system as a function of time.

PROBLEM 19. A PCM in the slab $0 \le x \le l$ is heated with a constant flux $q = q_L > 0$ for $0 \le t \le t_1$ at the face $x = 0$. For $t > t_1$ it is cooled with the constant flux $q = q_S < 0$. The face $x = l$ is insulated; initially the slab is solid at the melt temperature. The imposition of these fluxes creates melt and freeze fronts $X_1(t)$ and $X_2(t)$ as shown in **Figure 3.1.12.**

(a) Suppose that thermocouples are placed at the points x_0 , x_1 of the figure. Describe in qualitative terms what you would expect to see.

(b) Using the quasistationary approximation, estimate $X_1(t)$, $X_2(t)$ and the temperature histories at x_0 , x_1.

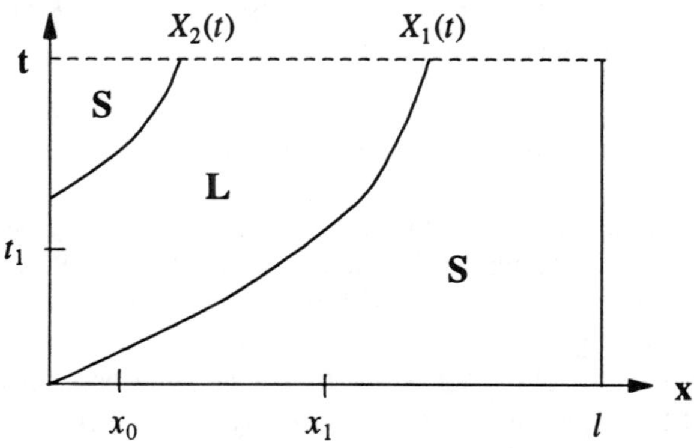

Figure 3.1.12. Melt and freeze fronts in PROBLEM 19.

3.2. QUASISTATIONARY APPROXIMATION OF AXIALLY OR RADIALLY SYMMETRIC PROCESSES

3.2.A Introduction

The quasistationary approximation may also be applied to one-phase, one-dimensional phase-change processes in cylindrical or spherical geometries. Even though such processes occur frequently and are of great practical interest, explicit solutions are not available for them, except in the few cases discussed in **§2.6**.

This makes the quasistationary solutions, to be presented here, all the more useful, but again, their utility is severly limited by the lack of any error estimates. Moreover, unlike the slab case of **§3.1**, the method does *not* yield simple representations for the temperature and interface, but rather transcendental equations with time-dependent coefficients which must be solved for the unknown interface. Nevertheless, useful information about the process can still be inferred from the approximation as we shall see.

In this section, the quasistationary approximation technique will be applied to several problems of practical interest, including inward and outward *melting* of a cylinder and inward *melting* of a sphere. The corresponding *freezing* problems will be left as exercises for the reader. One should remember that the quasistationary approximation *speeds up* the phase-change process by neglecting the sensible heat, so $R^{qs}(t)$ *overestimates* the true front $R(t)$ and therefore *underestimates* the true melt-time (i.e. $t^{qs}_{melt} < t_{melt}$).

3.2.B Outward melting of a hollow cylinder

Consider a "long" hollow cylinder, with inner radius R_{in} and outer radius R_{out}, filled with a PCM which is initially *solid* at its melt temperature T_m. Uniform heating of the inner surface will result in an *axially symmetric* melt-front $r = R(t)$ propagating *outwards* from $r = R_{in}$ and separating liquid ($R_{in} \le r < R(t)$) from solid ($R(t) < r \le R_{out}$), the latter being isothermal at T_m, see **Figure 3.2.1**. Assuming *constant* thermophysical properties, such a 1-phase Stefan Problem is described by

$$\rho c_L T_t = \frac{k}{r}(rT_r)_r, \qquad R_{in} < r < R(t), \quad t > 0 \tag{1a}$$

$$T(R(t), t) = T_m, \qquad t > 0, \tag{1b}$$

$$\rho L R'(t) = -k_L T_r(R(t)^-, t), \quad t > 0 \tag{1c}$$

$$R(0) = 0 \tag{1d}$$

together with a boundary condition at $r = R_{in}$ appropriate to the mode of heating (imposed temperature, imposed flux, or convective flux, see EXAMPLES below).

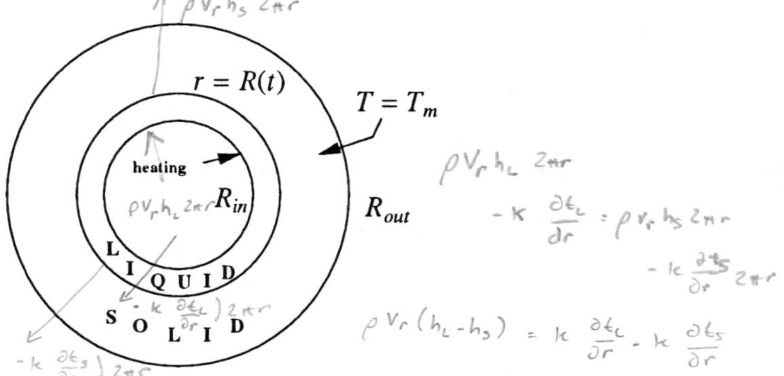

Figure 3.2.1. Outward melting of a cylinder.

The Stefan Condition, (1c), is easily derived by an energy balance (PROBLEM 24 §1.2), the total enthalpy (per unit height) here being

$$E(t) = \int_{R_{in}}^{R(t)} \{\rho c_L[T(r,t) - T_m] + \rho L\} \, 2\pi r \, dr \tag{2}$$

(referred to solid at T_m).

The quasistationary approximation replaces the heat equation (1a) by the steady-state equation $(rT_r)_r = 0$, whence

$$T^{qs}(r,t) = A \ln r + B , \tag{3}$$

with A, B independent of r, and determines $A(t)$, $B(t)$ from the remaining conditions. The calculations are routine, so we simply list the solutions corresponding to the various boundary conditions.

EXAMPLE 1. Outward melting due to imposed temperature:

$$T(R_{in}, t) = T_L(t) > T_m, \qquad t > 0.$$

Then (PROBLEM 1),

$$T^{qs}(r,t) = T_m + [T_L(t) - T_m] \frac{\ln \dfrac{r}{R^{qs}(t)}}{\ln \dfrac{R_{in}}{R^{qs}(t)}}, \qquad R_{in} \le r \le R^{qs}(t), \qquad t > 0, \tag{4a}$$

and $R^{qs}(t)$ is determined from the transcendental equation

$$2\, R^{qs}(t)^2 \ln \frac{R^{qs}(t)}{R_{in}} = R^{qs}(t)^2 - R_{in}^2 + \frac{4\, k_L}{\rho L} \int_0^t [T_L(s) - T_m] \, ds \tag{4b}$$

The melt-time of the entire cylinder is found by setting $R^{qs} = R_{out}$; in particular, for the case $T_L(t) \equiv T_L$, we find

$$t^{qs}_{melt} = \frac{\rho L}{4\, k_L (T_L - T_m)} \left[R_{in}^2 - R_{out}^2 + 2\, R_{out}^2 \ln \frac{R_{out}}{R_{in}} \right], \tag{5}$$

which *underestimates* the true t_{melt} (since $R^{qs}(t) > R(t)$).

EXAMPLE 2. Outward melting due to imposed flux:

$$-k_L T_r(R_{in}, t) = q_L(t) > 0, \qquad t > 0 \ . \tag{6}$$

In this case the solution is simpler (PROBLEM 2):

$$R^{qs}(t) = \left(R_{in}^2 + 2\frac{R_{in}}{\rho L} \int_0^t q_L(s) ds \right)^{1/2}, \qquad t > 0 \, , \tag{7a}$$

$$T^{qs}(r, t) = T_m - \frac{q_L(t)}{k_L} R_{in} \ln \frac{r}{R^{qs}(t)}, \quad R_{in} \le r \le R^{qs}(t), \ t > 0 \, , \tag{7b}$$

and when $q_L(t) \equiv q_L$, the melt-time is

$$t_{melt}^{qs} = \frac{\rho L}{2 R_{in} q_L} (R_{out}^2 - R_{in}^2) \qquad (< t_{melt}) \, . \tag{8}$$

Taking $R_{in} \to 0$ and q_L such that $2\pi R_{in} q_L \to Q$, problem (1,6) becomes that of melting due to a line source of strength Q at $r = 0$, which admits an exact solution (see **§2.6.B**). Then (7) becomes

$$R^{qs} = \sqrt{\frac{Q}{\pi \rho L} t}, \quad T^{qs} = T_m - \frac{Q}{2\pi k_L} \ln \frac{r}{R^{qs}}, \quad 0 < r < R^{qs} \, , \quad t > 0 \, , \tag{9}$$

which is precisely the approximation to the exact solution, derived in PROBLEM 4, **§2.6**, when $\frac{Q}{\rho L \alpha_L}$ is small. Note that $\frac{Q}{\rho L \alpha_L} = \frac{c_L(Q/k_L)}{L}$ may be considered as the effective Stefan Number for this melting process, with Q/k_L playing the role of a ΔT here. Thus, we see that the quasistationary solution is in fact the lowest order approximation to the exact solution for small St.

EXAMPLE 3. Outward melting due to convective flux:

$$-k_L T_r(R_{in}, t) = h[\, T_L(t) - T(R_{in}, t)\,], \qquad t > 0 \, , \tag{10}$$

with $T_L(t) > T_{in}$. We find (PROBLEM 3)

$$T^{qs}(r, t) = T_m + [T_L(t) - T_m] \frac{\ln \frac{r}{R^{qs}(t)}}{\ln \frac{R_{in}}{R^{qs}(t)} - \frac{k_L}{hR_{in}}}, \quad R_{in} \le r \le R^{qs}(t), \quad t > 0 \, , \tag{11a}$$

with $R^{qs}(t)$ determined from the transcendental equation

$$2R^{qs}(t)^2 \ln \frac{R^{qs}(t)}{R_{in}} = \left[1 - \frac{2k_L}{hR_{in}} \right] [R^{qs}(t)^2 - R_{in}^2] + \frac{4\, k_L}{\rho L} \int_0^t [T_L(s) - T_m] ds \, . \tag{11b}$$

Taking $h \to \infty$, we see that (11) reduces to (4), as it should (see **§3.1.D**).

The formulation and solution of the corresponding *outward freezing* problems is straightforward. In fact, they can be formally deduced from the above by replacing every subscript "L" by "S" and the latent heat L by $-L$, as usual (PROBLEMS 4-6).

3.2.C Inward melting of a cylinder

Now we consider axially symmetric melting when the *outer* surface of a cylinder (hollow or not) is being heated. Thus, the melt-front starts at $r = R_{out}$ and propagates *inwards*. Again we assume that the PCM is initially solid at T_m, so the mathematical problems are still as in §**3.2.B** but with R_{in} replaced by R_{out}, the melt region now being $R(t) < r < R_{out}$ at each time, and $-k_L T_r(R_{in}, t)$ replaced by $+k_L T_r(R_{out}, t)$ in (6) and (10).

The reason for the sign switch in the flux is the following: we wish to specify the flux *into* the PCM, namely $q_{in} := (-k_L \nabla T) \cdot \vec{n}_{in}$, with $\vec{n}_{in}$ the unit normal to the surface pointing *into* the PCM. At $r = R_{out}$, with the PCM occupying the region $0 \le R_{in} \le r \le R_{out}$, the inward *radial* unit normal is $n_{in} = -1$, and therefore the inward radial flux there is given by $q_{in}|_{r=R_{out}} = +k_L T_r(R_{out}, t)$ (see (15, 16) below).

It follows that the outward melting problems of §**3.2.B** will appear formally identical with the *inward* ones under the replacements

$$R_{in} \to R_{out}, \qquad q_L \to -q_L, \qquad h \to -h\,, \tag{12}$$

and the quasistationary solutions can simply be read off those of §**3.2.B.**.

EXAMPLE 4. Inward melting due to imposed temperature:

$$T(R_{out}, t) = T_L(t) > T_m, \qquad t > 0 \tag{13}$$

The only change necessary in (4) is: $R_{in} \to R_{out}$ (PROBLEM 7). To find the melt-time of a non-hollow cylinder when $T_L(t) \equiv T_L$ we interchange the subscripts "in" and "out" and then take $R_{in} \to 0$ in (5), which yields

$$t_{melt}^{qs} = \frac{\rho L R_{out}^2}{4k_L(T_L - T_m)} \qquad (< t_{melt})\ . \tag{14}$$

EXAMPLE 5. Inward melting due to imposed flux:

$$+k_L T_r(R_{out}, t) = q_L(t) > 0, \qquad t > 0\ . \tag{15}$$

The approximate solution is found from (7) by changing $R_{in} \to R_{out}$, $q_L \to -q_L$ (PROBLEM 8). For $q_L(t) \equiv q_L$, we find from (8) that a non-hollow cylinder will melt *no sooner* than $t_{melt}^{qs} = \rho L\, R_{out} / 2q_L$.

EXAMPLE 6. Inward melting due to convective flux:

$$+k_L T_r(R_{out}, t) = h[\,T_L(t) - T(R_{out}, t)\,], \qquad T_L(t) > T_m, \qquad t > 0\ . \tag{16}$$

Here the necessary changes in (11) are: $R_{in} \to R_{out}$ and $h \to -h$ (PROBLEM 9). When $T_L(t) \equiv T_L$, the melt-time of a non-hollow cylinder will be *at least*

$$t_{melt}^{qs} = \frac{\rho L}{4k_L(T_L - T_m)} \left[1 + \frac{2k_L}{hR_{out}} \right] R_{out}^2 . \tag{17}$$

The solutions to the corresponding *inward freezing* problems are formally deduced by the standard changes $L \to S$, $L \to -L$ as usual (PROBLEMS 10-12).

3.2.D Inward melting of a sphere

Consider a sphere of radius R_o, initially solid at T_m. Uniform heating of its surface will result in a spherically symmetric melt-front, $r = R(t)$, propagating *inwards* from $r = R_o$ with solid (at T_m) in $0 \le r < R(t)$ and liquid in $R(t) < r \le R_o$. Assuming *constant* properties, the heat equation in the liquid has the form

$$\rho c_L T_t = \frac{k_L}{r^2} [r^2 T_r]_r, \qquad R(t) < r < R_o, \qquad t > 0, \tag{18}$$

whose steady-state solution is simply

$$T(r, t) = \frac{A}{r} + B, \qquad A, B \text{ independent of } r . \tag{19}$$

The quasistationary approximation assumes such a temperature distribution at each time and seeks $A(t)$, $B(t)$ such that the interface and boundary conditions to be satisfied. The interface conditions here have the standard form:

$$T(R(t), t) = T_m, \qquad \rho L R'(t) = -k_L T_r(R(t)^+, t), \qquad t > 0, \tag{20}$$

as may be shown (PROBLEM 25 §**1.2**) by balancing the energy

$$E(t) = \int_{R(t)}^{R_o} \{\rho c_L[T(r, t) - T_m] + \rho L\} 4\pi r^2 \, dr . \tag{21}$$

EXAMPLE 7. Inward melting due to imposed temperature:

$$T(R_o, t) = T_L(t) > T_m, \qquad t > 0 . \tag{22}$$

From (19, 20, 22) we find (PROBLEM 13)

$$T^{qs}(r, t) = T_m + [T_L(t) - T_m] \frac{1 - \dfrac{R^{qs}(t)}{r}}{1 - \dfrac{R^{qs}(t)}{R_o}}, \qquad R^{qs}(t) \le R_o, \qquad t > 0 \tag{23a}$$

and $R^{qs}(t)$ satisfies the cubic equation

$$2\left[\frac{R^{qs}(t)}{R_0}\right]^3 - 3\left[\frac{R^{qs}(t)}{R_0}\right]^2 + 1 = \frac{6\,k_L}{\rho\,L\,R_0^2}\int_0^t [T_L(s) - T_m]ds\ . \tag{23b}$$

Hence, when $T_L(t) \equiv T_L$, the sphere of radius R_o will melt no sooner than

$$t_{melt}^{qs} = \frac{\rho\,L}{6\,k_L(T_L - T_m)}\,R_o^2.$$

EXAMPLE 8. Inward melting due to imposed flux:

$$+\,k_L T_r(R_o, t) = q_L(t) > 0, \qquad t > 0\ . \tag{24}$$

We easily obtain (PROBLEM 14)

$$T^{qs}(r, t) = T_m + \frac{q_L(t)}{k_L}\,R_o^2\left[\frac{1}{R^{qs}(t)} - \frac{1}{r}\right], \quad R^{qs}(t) \le r \le R_o, \quad t > 0\,, \tag{25a}$$

$$R^{qs}(t) = R_o\left(1 - \frac{3}{\rho\,L\,R_o}\int_0^t q_L(s)ds\right)^{1/3}, \qquad t > 0\ . \tag{25b}$$

For $q_L(t) \equiv q_L$, the melt-time estimate is $t_{melt}^{qs} = \rho\,L\,R_o\,/\,3\,q_L$.

EXAMPLE 9. Inward melting due to convective flux:

$$+\,k_L T_r(R_o, t) = h[\,T_L(t) - T(R_o, t)\,], \qquad T_L(t) > 0, \qquad t > 0\ . \tag{26}$$

In this case we have (PROBLEM 15)

$$T^{qs}(r, t) = T_m + [T_L(t) - T_m]\,\frac{1 - \dfrac{R^{qs}(t)}{r}}{1 - \left(1 - \dfrac{k_L}{h\,R_o}\right)\dfrac{R^{qs}(t)}{R_o}}, \quad R^{qs}(t) \le r \le R_o, \quad t > 0\,, \tag{27a}$$

with $R^{qs}(t)$ satisfying the cubic equation

$$2\left[1 - \frac{k_L}{h\,R_o}\right]\left[\frac{R^{qs}(t)}{R_o}\right]^3 - 3\left[\frac{R^{qs}(t)}{R_o}\right]^2 + 1 + \frac{2\,k_L}{h\,R_o} = \frac{6\,k_L}{\rho L R_o^2}\int_0^t [T_L(s) - T_m]ds\,, \tag{27b}$$

which reduces to (23b) as $h \to \infty$, as expected (see **§3.1.D**).

3.2.E Simple approximations of a heat storage process

Often the need for simple approximations to temperature, melt-depth, heat stored in a system, etc, extends to more complex systems than those examined so far. As an example of typical engineering calculations in obtaining "sizing estimates" of more complex processes, we examine a charge/discharge process coupled to convective heat transfer.

Consider a hollow cylinder along the z-axis, of inner radius R_{in} and outer radius R_{out}, filled with a PCM having melt temperature T_m. Heat is transferred to and from the PCM by pumping a heat-transfer fluid through the inner tube at desired velocity v and inlet temperature T_{inlet} at $z=0$, (**Figure 3.2.2**). If $T_{inlet} > T_m$ then the PCM is being charged, whereas if $T_{inlet} < T_m$ the PCM is being discharged. We wish to estimate the extent of melting/freezing, the temperature in the PCM and the transfer fluid, and the energy stored in the system, as functions of velocity v and inlet temperature T_{inlet}.

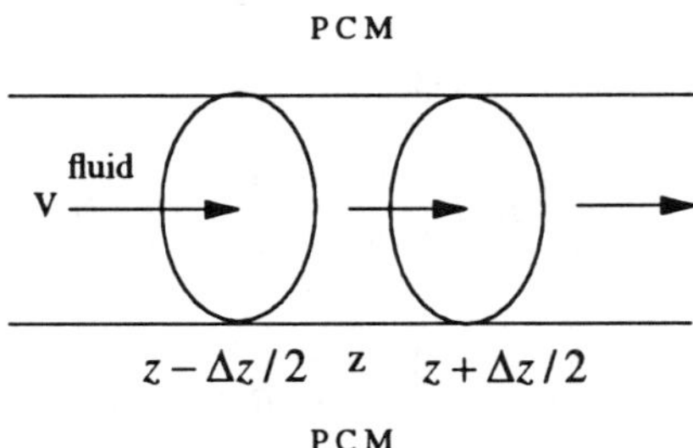

Figure 3.2.2. Heat-exchange pipe surrounded by PCM.

Suppose that initially the PCM is solid at T_m and that, starting at time $t=0$, fluid of temperature $T_{inlet} > T_m$ is pumped through the tube from $z=0$ at velocity v. As the fluid loses heat to the PCM, its temperature will decrease further up the tube and so will the extent of melting. Hence both the fluid temperature and the melt-front depend on z. The heat is exchanged across the tube wall convectively, with a heat-transfer coefficient h (see below).

We shall make the following simplifying assumptions:

(a) Heat conduction is negligible in the fluid, which is at steady-state thermally, so that its temperature depends only on z: $T_f(z)$.

(b) Radial heat transfer dominates in the PCM, so the axial one can be ignored.

(c) The latent heat and (radial) thermal conductivity of the PCM are so large that the surface temperature of the PCM at the tube wall can be taken as essentially equal to T_m during the entire melting process.

Assumption (a) is reasonable for fairly rapid flow, after an initial transient, if the tube is not too narrow and not too long. Assumption (b) is reasonable if the axial temperature drop is not too large. The third and major assumption is shakier, since it implies a *uniform* temperature of T_m throughout the melt, with all the heat going into melting; (without it however no simple approximation would be possible). Large latent heat (small Stefan Number) implies that the front will not penetrate very far into the PCM, and a large conductivity will tend to level off the temperature, so the assumption is plausible at least.

The unknowns now are the steady-state transfer-fluid temperature $T_f(z)$, and the interface $r = R(z,t)$. To find them, we set up heat balances in the fluid and in the PCM as follows.

Consider a small section of the flow channel corresponding to the interval $[z - \frac{\Delta z}{2}, z + \frac{\Delta z}{2}]$, i.e. a disk of radius R_{in} and thickness Δz, **Figure 3.2.2.** The rate of heat flowing into it is

$$Rate\ in = \pi R_{in}^2 \, v \, \rho_f \, c_f \left[T_f(z - \frac{\Delta z}{2}) - T_{ref} \right] + 2\pi R_{in} \Delta z \, h \, [T_m - T_f(z)] \, , \tag{28a}$$

where T_{ref} is an arbitrary reference temperature and ρ_f, c_f the density and specific heat of the fluid. The first term accounts for the heat convected with the flow and the second term for the heat entering the fluid through the tube walls (negative if the fluid is hotter than T_m). The rate of heat flowing out (across the face at $z + \Delta z / 2$) is

$$Rate\ out = \pi R_{in}^2 \, v \, \rho_f \, c_f \left[T_f(z + \frac{\Delta z}{2}) - T_{ref} \right] . \tag{28b}$$

Equating the two rates and letting $\Delta z \to 0$, we obtain the ordinary differential equation

$$R_{in} \, v \rho_f \, c_f \frac{\partial T_f}{\partial z}(z) = 2 h \, [T_m - T_f(z)] \, . \tag{29}$$

Using $T_f(0) = T_{inlet}$, the solution is given by

$$T_f(z) = T_m + e^{-\gamma z} [T_{inlet} - T_m], \qquad \gamma := \frac{2h}{R_{in} \, v \, \rho_f \, c_f} , \qquad z > 0 \, . \tag{30}$$

By assumption (c), all of the heat leaving the transfer-fluid will be transferred radially directly to the melt-front. Hence, the

$$Rate\ of\ heat\ gain\ at\ the\ front = 2\pi R \Delta z \frac{\partial R}{\partial t} \rho L \tag{31a}$$

must equal the

$$Heat\ flow\ rate\ into\ PCM = 2\pi R_{in} \Delta z \, h \, [T_f(z) - T_m] \, , \tag{31b}$$

whence

$$R^2 = R_{in}^2 + \frac{2 R_{in}}{\rho L} h [T_f - T_m] \, t \, , \tag{32a}$$

or, using (30),

$$R(z, t) = \left[R_{in}^2 + \frac{2 R_{in}}{\rho L} h \, e^{-\gamma z} [T_{inlet} - T_m] \, t \right]^{1/2} . \tag{32b}$$

The heat transfer coefficient, h, generally depends strongly on the fluid properties, velocity, flow conditions and geometry. There are plenty of correlations one may choose from [CHAPMAN], [ECKERT-DRAKE], [McADAMS], [PITTS-SISSOM], which essentially express h as a function of v for a given flow. A typical example of such a formula, valid for pipe flow under certain conditions, is the following [CHAPMAN p. 249]:

$$\mathbf{Nu}_D = \frac{0.0396\,(\mathbf{Re}_D)^{3/4}\ \mathbf{Pr}}{1 + 1.5\,\mathbf{Pr}^{-1/6}\ \mathbf{Re}_D^{-1/8}\,(\mathbf{Pr} - 1)} \ . \tag{33}$$

It relates three dimensionless quantities, the **Nusselt, Reynolds** and **Prandtl** numbers, defined respectively as

$$\mathbf{Nu}_D := \frac{h\,D}{k_f}, \qquad \mathbf{Re}_D := \frac{\mathrm{v}\,D}{\nu}, \qquad \mathbf{Pr} := \frac{\nu}{\alpha_f}, \tag{34}$$

where D is the pipe diameter ($= 2\,R_{in}$ here), k_f, α_f the conductivity and diffusivity of the fluid, and ν its kinematic viscosity. Hence, h can be found as a function of v.

The energy stored in the PCM (as latent heat) and in the fluid (as sensible heat) may be found by integration. Thus, for a cylinder of height l, the PCM energy will be

$$E_{PCM}(t) = \int_0^l \int_{R_{in}}^{R(z,t)} \rho\, L\, 2\pi\, r\, l\, dr\, dz \ = \ \pi R_{in}^2\, l\, \mathrm{v}\, \rho_f\, c_f\, [1 - e^{-\gamma l}][T_{inlet} - T_m]\, t\,, \tag{35a}$$

(note its independence from L), and the energy in the fluid will be (PROB. 21)

$$E_{fluid} = \int_0^l \rho_f\, c_f\, [T_f(z) - T_m]\, \pi\, R_{in}^2\, dz = \pi R_{in}^3\, \frac{\mathrm{v}}{2h}\, \rho_f^2\, c_f^2\, [1 - e^{-\gamma l}] \cdot [T_{inlet} - T_m]. \tag{35b}$$

Such estimates may be used to answer basic design questions, such as how much PCM is needed, ranges of operating parameters, duration of charge/discharge cycles, etc, see CHAPTER **5**.

PROBLEMS

PROBLEM 1. Derive the quasistationary solution (4) of EXAMPLE 1 and check (5).

PROBLEM 2. Derive (7-9) of EXAMPLE 2.

PROBLEM 3. Derive (11) and check the limit as $h \to \infty$ in EXAMPLE 3. Also derive the quasistationary solution when $h = h(t)$ instead of being constant; is the O.D.E. for $R(t)$ solvable analytically ?

PROBLEM 4. Formulate the *freezing* problem corresponding to EXAMPLE 1, derive its quasistationary solution, and check that it coincides with that (4) after the usual replacements: $L \to S$, $L \to -L$.

PROBLEM 5. Repeat the previous PROBLEM for the freezing counterpart of EXAMPLE 2.

PROBLEM 6. Repeat, for the situation in EXAMPLE 3.

PROBLEM 7. Formulate the problem for inward melting of a cylinder (EXAMPLE 4), derive its quasistationary solution, and check that it coincides with (4) under the replacement: $R_{out} \to R_{in}$. Check (13).

PROBLEM 8. Repeat for the situation in EXAMPLE 5.

PROBLEM 9. Repeat for the situation in EXAMPLE 6.

PROBLEM 10. Formulate the *freezing* counterpart of EXAMPLE 4, derive its quasistationary solution and check that it coincides with the solution of PROBLEM 7 under the standard changes $L \to S$, $L \to -L$.

PROBLEM 11. Repeat what is asked for in the previous PROBLEM for the case described in EXAMPLE 5.

PROBLEM 12. Repeat for the case of EXAMPLE 6.

PROBLEM 13. Derive (23), and check the melt-time relation of EXAMPLE 7.

PROBLEM 14. Derive (25).

PROBLEM 15. Derive (27).

PROBLEM 16. Formulate the problem of inward *freezing* of a sphere due to imposed temperature $T_S(t) < T_m$ at $r = R_{out}$. Derive its quasistationary solution and compare with (23).

PROBLEM 17. Repeat, for the freezing counterpart of EXAMPLE 8.

PROBLEM 18. Repeat, for the freezing counterpart of EXAMPLE 9 but with $h = h(t)$, and check the limit as $h \to \infty$.

PROBLEM 19. Formulate and solve (approximately) the problem of *outward melting* of a hollow sphere due to imposed temperature $T_L(t)$ at $r = R_{in}$. How can its solution be obtained directly from (23)?

PROBLEM 20. A sphere of water $(r < R_{in})$ is enclosed in a spherical shell of a PCM $(R_{in} < r < R_{out})$ which is transparent to microwave radiation. The system is placed in a microwave oven which delivers energy at a volumetric rate Q to the water. Assuming the system is initially at the PCM's melting temperature and that the resulting (outward) melting process is radially symmetric, estimate the time needed to melt the PCM completely. [see **§3.1.E** and PROBLEM 19].

PROBLEM 21. Justify the integral expressions for the energies in (35a, b) and verify their evaluation.

PROBLEM 22. A thin copper pipe of length $1\,m$ and diameter $2\,cm$ is covered with a layer of ice $2\,cm$ thick. We run water through the pipe at a constant speed of $10\,m/s$ and inlet temperature of $25°C$. Estimate how long it will take to defrost the pipe. [use $\nu = 0.01\,cm^2/s$, **Pr** ≈ 7 and the values in **§2.1.E**] [Answer: about 11 minutes for the far end].

3.3. PERTURBATION METHODS FOR ONE-PHASE PROBLEMS

3.3.A Introduction

Perturbation methods utilize series expansions in small (or large) parameters to reduce a problem to a sequence of simpler problems which, hopefully, may be explicitly solvable. Typically only the first few terms in the expansion can actually be found, due to the increasing complication of the problems. Thus, by necessity, the few terms we succeed in finding are taken as an *approximate solution.*

In some fields, such as fluid dynamics, quantum physics, etc, the perturbation approach has proved to be extremely successful and, as a result, the theory and practice of perturbation methods in those fields is very advanced.

In Stefan-type problems, there are few naturally occurring small parameters. One possibility is the **Stefan number**, $St = c\Delta T / L$ (see §**2.2**), and expansions in St have been used in a variety of problems, e.g. [DUDA-VRENTAS], [PEDROSO-DOMOTO], [JIJI-WEINBAUM], [HUANG-SHIH], [ALEXIADES, 1983]. The methodology and applications are discussed in detail in [AZIZ-NA], and a rather complete survey of small Stefan number expansions for one-phase problems in slab, cylindrical and spherical geometries may be found in [HILL]. Here we shall only present the basic methodology and a few examples to acquaint the reader with the approach. For more details we refer to [AZIZ-NA], [HILL] and references thereof.

Like all analytical methods, perturbation expansions can be applied only to relatively simple problems, serving mainly as rough guides to *qualitative* behavior rather than for precise quantitative information, unless their accuracy can be verified. Unfortunately, there are *no* precise error estimates available, and therefore validation must be performed for each problem at hand by comparison to other methods and to experiments.

Expansions in powers of St can be considered as higher order corrections to the *quasistationary approximation* (§**3.1**, §**3.2**). Indeed, small Stefan number amounts to ignorable sensible heat, and therefore the zeroth order solution should agree with the quasistationary solution.

It is more convenient and efficient to carry out the expansions in dimensionless formulations and with the interface location immobilized by means of a suitable change of variables. A general method of fixing the front location is the *Landau transformation,* discussed in §**3.3.B**. For problems with constant data, a more convenient method is to treat the interface location as an independent variable (replacing time), to be presented in §**3.3.C**.

3.3.B Landau transformation

As model problem we treat the 1-phase Stefan Problem with imposed temperature $T_L(t)$ as in **§3.1.B**. Thus, consider a slab, $0 \le x \le l$, initially solid at its melt temperature T_m, which is melted via an imposed temperature $T_L(t) > T_m$ at $x = 0$.

MATHEMATICAL PROBLEM: Find $X(t)$, $T(x,t)$ such that

$$T_t = \alpha_L T_{xx}, \qquad 0 < x < X(t), \qquad t > 0 \tag{1a}$$

$$T(X(t),t) = T_m, \qquad \rho L X'(t) = -k_L T_x(X(t),t), \qquad t > 0 \tag{1b}$$

$$X(0) = 0, \qquad T(0,t) = T_L(t) > T_m, \qquad t > 0. \tag{1c}$$

In order to fix the interface, we introduce the *Landau Transformation*, [LANDAU],

$$\xi = \frac{x}{X(t)}, \qquad t > 0, \tag{2}$$

which maps the interval $0 \le x \le X(t)$ onto the fixed interval $0 \le \xi \le 1$. To undimensionalize the problem, we employ the same variables as in **§3.1.B**. Letting

$$\varepsilon = St_L = \frac{c_L \Delta T_L}{L}, \qquad \Delta T_L = \max\{T_L(t) - T_m\}, \tag{3a}$$

we set

$$\tau = \varepsilon \frac{\alpha_L}{\hat{x}^2} t, \qquad u(\xi,\tau) = \frac{T(x,t) - T_m}{\Delta T_L}, \qquad \Sigma(\tau) = \frac{X(T)}{\hat{x}}, \tag{3b}$$

where $\hat{x}$ is an arbitrary length, and problem (1) becomes (PROBLEM 1),

$$u_{\xi\xi} = \varepsilon[\Sigma^2 u_\tau - \xi \Sigma \Sigma' u_\xi], \qquad 0 < \xi < 1, \quad \tau > 0 \tag{4a}$$

$$u(1,\tau) = 0, \qquad \Sigma\Sigma' = -u_\xi(1,\tau), \qquad \tau > 0 \tag{4b}$$

$$\Sigma(0) = 0, \qquad u(0,\tau) = U_L(\tau) := \frac{T_L(t) - T_m}{\Delta T_L}, \qquad \tau > 0. \tag{4c}$$

The usefulness of the Landau Transformation, (2), is that it replaces the *geometric* nonlinearity of problem (1) by the explicit algebraic nonlinearity in (4). Notice that the Stefan Number has been introduced into the dimensionless time τ and appears only in the partial differential equation. We proceed with the expansions.

Assuming $\varepsilon > 0$ is small, we seek $\Sigma(\tau)$ and $u(\xi,\tau)$ as series in powers of ε:

$$\Sigma(\tau) = \Sigma_0(\tau) + \varepsilon \Sigma_1(\tau) + \varepsilon^2 \Sigma_2(\tau) + \cdots \tag{5a}$$

$$u(\xi,\tau) = u_0(\xi,\tau) + \varepsilon u_1(\xi,\tau) + \varepsilon^2 u_2(\xi,\tau) + \cdots \tag{5b}$$

Substituting into (4) and matching powers of ε, we are led to the following sequence of problems for the unknown coefficients u_0, Σ_0; u_1, Σ_1; u_2, Σ_2; $\cdots$.

$$\begin{cases} u_{0\,\xi\xi} = 0\,, \quad 0<\xi<1\,, \quad \tau>0 \\ u_0(1,\tau) = 0\,, \quad u_0(0,\tau) = U_L(\tau), \quad \tau>0 \\ \Sigma_0\Sigma_0{}' = -u_{0\,\xi}(1,\tau)\,, \quad \tau>0\,, \quad \Sigma(0) = 0; \end{cases} \tag{6a}$$

$$\begin{cases} u_{1\,\xi\xi} = \Sigma_0^2\, u_{0\tau} - \xi\Sigma_0\,\Sigma_0{}'\, u_{0\,\xi}\,, \quad 0<\xi<1\,, \quad \tau>0 \\ u_1(1,\tau) = 0, \quad u_1(0,\tau) = 0\,, \quad \tau>0 \\ (\Sigma_0\Sigma_1)' = -u_{1\,\xi}(1,\tau)\,, \quad \tau>0\,, \quad \Sigma_1(0) = 0\,; \end{cases} \tag{6b}$$

$$\begin{cases} u_{2\,\xi\xi} = 2\Sigma_0\Sigma_1\, u_{0\tau} + \Sigma_0^2\, u_{1\tau} - \xi[\Sigma_0\Sigma_0{}'\, u_{1\,\xi} + (\Sigma_0\Sigma_1)' u_{0\,\xi}]\,, \quad 0<\xi<1\,, \ \tau>0 \\ u_2(1,\tau) = 0\,, \quad u_2(0,\tau) = 0\,, \quad \tau>0 \\ (\Sigma_0\Sigma_1)' + \Sigma_1\Sigma_1{}' = -u_{2\,\xi}(1,\tau)\,, \quad \tau>0\,, \quad \Sigma_2(0) = 0\,; \end{cases} \tag{6c}$$

and so on. Of course, the sub-problems become increasingly more complicated and, in practice, only the first two or three terms can be found explicitly.

Observe that the 0-th order problem, (6a), may be obtained directly from (4) by *formally* setting $\varepsilon = 0$ (called the *reduced* problem). Clearly, this is the quasistationary approximation (§**3.1.B**) but with the domain fixed to $0 < \xi < 1$. Hence, the higher order problems (6b), (6c), $\cdots$, may be viewed as higher order corrections to the quasistationary approximation (6a). The explicit solutions of (6a), (6b) are easily found (PROBLEM 2) to be, for $0<\xi<1\,,\ \tau>0$:

$$\begin{cases} u_0(\xi,\tau) = (1-\xi)\,U_L(\tau)\,, \\ \Sigma_0(\tau) = \sqrt{2W_L(\tau)}\,, \qquad W_L(\tau) = \int_0^\tau U_L(s)ds; \end{cases} \tag{7a}$$

$$\begin{cases} u_1(\xi,\tau) = [(\xi^2-\xi) - \frac{\xi^3-\xi}{3}]\,U_L{}'(\tau)\,W_L(\tau) + \frac{\xi^3-\xi}{6}\,U_L(\tau)^2\,, \\ \Sigma_1(\tau) = -\frac{1}{3\sqrt{2}}\,U_L(\tau)\sqrt{W_L(\tau)}. \end{cases} \tag{7b}$$

Continuing, one could find higher order terms which, clearly, become increasingly complicated.

As a check that we have indeed obtained a correction to the quasistationary solution we present the following

EXAMPLE (*sinusoidal imposed temperature*, c.f. EXAMPLE 4, §**3.1.B**):

For the sinusoidal temperature given in (15a) §**3.1**, we have

$$U_L(\tau) = \sin(\frac{\pi\tau}{2\tau_0})\,, \qquad W_L(\tau) = \frac{4\tau_0}{\pi}\sin^2(\frac{\pi\tau}{4\tau_0})\,, \qquad \tau_0 := \varepsilon\,\frac{\alpha_L}{\hat{x}^2}\,t_0\,, \tag{8a}$$

and from (7a), (7b) we find

$$\Sigma_0(\tau) = \sqrt{\frac{8\tau_0}{\pi}}\,\sin(\frac{\pi\tau}{4\tau_0}), \qquad \Sigma_1(\tau) = -\sqrt{\frac{2\tau_0}{9\pi}}\,\sin(\frac{\pi\tau}{2\tau_0})\,\sin(\frac{\pi\tau}{4\tau_0}). \tag{8b}$$

Hence, in physical variables the front location is

$$X(t) = \hat{x}\Sigma(\tau) = \hat{x}[\Sigma_0 + \varepsilon\Sigma_1 + \cdots] \tag{8c}$$

$$= \varepsilon^{1/2}\left(\frac{8\alpha_L t_0}{\pi}\right)^{1/2}\sin(\frac{\pi t}{4t_0}) - \varepsilon^{3/2}\left(\frac{2\alpha_L t_0}{\pi}\right)^{1/2}\sin(\frac{\pi t}{2t_0})\sin(\frac{\pi t}{4t_0}) + O(\varepsilon^{5/2}),$$

for $0 \le t \le 2t_0$. Notice that the arbitrary length scale $\hat{x}$ disappears, as it should, and the expansion of $X(t)$ turns out to be in powers of $St^{1/2}$; the first term being exactly the quasistationary front X^{qs} of (15b)§**3.1**.

The method described in this subsection is applicable to 1-phase problems in slab, cylindrical or spherical geometries with any of the standard boundary conditions. Solutions to several such problems may be found in [HILL] and in [CHO-SUNDERLAND], [DUDA-VRENTAS], [HUANG-SHIH], [SEENIRAJ-BOSE], among *many* others.

3.3.C Front location as independent variable

When the interface location is a strictly *monotone* (increasing or decreasing) function of time, then the location itself can be used to replace the time variable, thus effectively eliminating the front movement and making expansions possible.

We illustrate the method on the 1-phase Stefan Problem with constant imposed flux in slab geometry. The method, however, is applicable to the other standard geometries and boundary conditions, *as long as the boundary data are independent of time*. The problem we choose to illustrate the method is particularly interesting in the context of expansions because it contains no parameters at all, as was pointed out in §**3.1.C**; hence, a parameter must now be artificially introduced !

Consider the 1-phase problem (1) but with boundary condition

$$-k_L T_x(0,t) = q_L > 0\ . \tag{9}$$

We undimensionalize using

$$\xi = \frac{x}{\hat{x}}, \qquad \sigma = \frac{X}{\hat{x}}, \qquad Fo = \frac{\alpha_L}{\hat{x}^2}t, \qquad \bar{u} = \frac{T - T_m}{\hat{T}}, \tag{10}$$

which leads to (§**3.1.C**)

$$\bar{u}_{Fo} = \bar{u}_{\xi\xi}, \qquad \bar{u}(\sigma, Fo) = 0, \qquad \bar{u}_\xi(0, Fo) = -\frac{q_L\hat{x}}{k_L\hat{T}}, \tag{11a}$$

$$\sigma' = -\frac{c_L\hat{T}}{L}\bar{u}_\xi(\sigma, Fo), \qquad \sigma(0) = 0. \tag{11b}$$

In §3.1.C, we chose the scales $\hat{x}$, $\hat{T}$ so as to make all the coefficients equal to one. Here we choose them so as to introduce an expansion parameter ε into the problem, as follows:

$$\frac{q_L \hat{x}}{k_L \hat{T}} = 1 \quad \text{and} \quad \frac{c_L \hat{T}}{L} = \varepsilon , \tag{12a}$$

with $\varepsilon > 0$ arbitrary. Hence, the chosen scales are $\hat{x} = k_L \hat{T} / q_L$, $\hat{T} = L / c_L \varepsilon$, and $\varepsilon = c_L \hat{T} / L$ is a Stefan number again. Now set

$$\tau := \varepsilon\, Fo , \qquad u(\xi, \tau) := \tilde{u}(\xi, Fo) = \frac{T - T_m}{\hat{T}} , \tag{12b}$$

in terms of which our problem becomes

$$u_{\xi\xi} = \varepsilon\, u_\tau , \qquad 0 < \xi < \sigma(\tau) , \qquad \tau > 0 \tag{13a}$$

$$u(\sigma, \tau) = 0 , \qquad u_\xi(0, \tau) = -1 , \qquad \tau > 0 \tag{13b}$$

$$\sigma' = -u_\xi(\sigma, \tau) , \qquad \tau > 0 , \qquad \sigma(0) = 0 . \tag{13c}$$

Since there is steady heat input at $\xi = 0$, it is intuitively clear that the melt front is monotonically increasing. Hence, to each $\tau > 0$ there corresponds a unique $\sigma > 0$, the (dimensionless) front location at time τ. Now we consider σ as the independent variable, instead of τ, and let

$$U(\xi, \sigma) := u(\xi, \tau) , \tag{14}$$

in terms of which (13) becomes

$$U_{\xi\xi}(\xi, \sigma) = -\varepsilon U_\sigma(\xi, \sigma)\, U_\xi(\sigma, \sigma) \qquad 0 < \xi < \sigma , \qquad \sigma > 0 \tag{15a}$$

$$U(\sigma, \sigma) = 0 , \qquad U_\xi(0, \sigma) = -1 \qquad \sigma > 0 , \tag{15b}$$

and once $U(\xi, \sigma)$ is found, we have

$$\sigma'(\tau) = -U_\xi(\sigma, \sigma) , \qquad \sigma(0) = 0 . \tag{15c}$$

Now we seek $U(\xi, \sigma)$ in the form

$$U(\xi, \sigma) = U_0(\xi, \sigma) + \varepsilon\, U_1(\xi, \sigma) + \varepsilon^2\, U_2(\xi, \sigma) + \cdots , \tag{16}$$

and from (15) we find recursively

$$U_0(\xi, \sigma) = \sigma - \xi , \quad U_1(\xi, \sigma) = \frac{\xi^2 - \sigma^2}{2} , \quad U_2(\xi, \sigma) = \sigma^3 - \sigma\xi^2 , \quad \cdots . \tag{17}$$

Then we seek $\sigma(\tau)$ in the form

$$\sigma(\tau) = \sigma_0(\tau) + \varepsilon\sigma_1(\tau) + \varepsilon^2\sigma_2(\tau) + \cdots \tag{18}$$

and from (15c) we find

$$\sigma_0(\tau) = \tau, \qquad \sigma_1(\tau) = -\frac{1}{2}\tau^2 , \qquad \sigma_2(\tau) = \frac{\tau^3}{2} , \qquad \cdots . \tag{19}$$

Therefore, the formal expansions (16, 18) are

$$u(\xi,\tau) = U(\xi,\sigma) = (\tau-\xi) - \varepsilon\frac{\xi^2-\tau^2}{2} + \varepsilon^2\tau(\tau^2-\xi^2) + \cdots, \tag{20a}$$

$$\sigma(\tau) = \tau - \varepsilon\frac{\tau^2}{2} + \varepsilon^2\frac{\tau^3}{2} + \cdots. \tag{20b}$$

Returning to physical variables we find

$$X(t) = \hat{x}\sigma(\tau) = \hat{x}[\tau - \varepsilon\frac{\tau^2}{2} - \varepsilon^2\frac{\tau^3}{2} + \cdots]$$

$$= \frac{\varepsilon}{\hat{x}}\alpha_L t, - (\frac{\varepsilon}{\hat{x}})^3\frac{\alpha_L^2}{2}t^2 - (\frac{\varepsilon}{\hat{x}})^5\frac{\alpha_L^3}{2}t^3 + \cdots,$$

where $\frac{\varepsilon}{\hat{x}} = \frac{c_L}{L}\frac{q_L}{k_L} =: \Lambda$ is *independent* of the arbitrary scale. Therefore we have

$$X(t) = \Lambda\alpha_L t - \frac{\Lambda^3}{2}(\alpha_L t)^2 - \frac{\Lambda^5}{2}(\alpha_L t)^3 + \cdots \tag{21a}$$

and

$$T(x,t) = T_m + \hat{T}u(\xi,\tau) \tag{21b}$$

$$= T_m + \frac{q_L}{k_L}\left[\Lambda\alpha_L t - x - \frac{\Lambda}{2}x^2 + \frac{\Lambda^3}{2}(\alpha_L t)^2 + \Lambda^5(\alpha_L t)^3 - \Lambda^3(\alpha_L t)x^2 + \cdots\right].$$

We see that the arbitrary scales disappear and we are left with series expansions in powers of x and t, which provide corrections to the quasistationary solution of **§3.1.C**.

This method, of fixing the front by replacing the dimensionless time by the dimensionless front location, can be applied to any 1-phase problem (in slab, cylindrical or spherical geometry, with any of the standard boundary conditions) provided the boundary data are independent of time, see e.g. [PEDROSO-DOMOTO], [SEENIRAJ-BOSE].

In the more complicated cases of convective or radiation boundary conditions, (15c) becomes too complicated to solve explicitly so $\sigma(\tau)$ cannot be found from its expansion, (18). However, there is an obvious alternative, namely, to find τ as a function of σ by rewriting (15c) as

$$\frac{d\tau}{d\sigma} = -\frac{1}{U_\xi(\sigma,\sigma)}$$

$$= -\frac{1}{U_{0\xi}(\sigma,\sigma)}\left\{1 - \varepsilon U_{1\xi}(\sigma,\sigma) + \varepsilon^2[U_{1\xi}(\sigma,\sigma)^2 - U_{2\xi}(\sigma,\sigma)] + \cdots\right\}, \tag{22}$$

and integrating from $\tau = 0, \sigma = 0$. For example, applying this procedure to the above, we find the relationship inverse to (20b):

$$\tau(\sigma) = \sigma - \varepsilon\frac{\sigma^2}{2} - \frac{\varepsilon^2\sigma^3}{3} + \cdots, \tag{23}$$

which, in physical variables reads

$$t = \frac{1}{\Lambda\alpha_L}X - \frac{1}{2\alpha_L}X^2 - \frac{\Lambda}{3\alpha_L}X^3 + \cdots, \qquad \Lambda = \frac{c_L q_L}{L k_L}. \tag{24}$$

Thus, this approach directly produces the melt time t of location X, instead of the melt depth X at time t.

3.3.D Rapid freezing of dilute alloys

Now we apply the perturbation method to a problem modeling rapid freezing of a finite slab of dilute alloy with stirring of melt. The precise assumptions and detailed derivation of such a model were presented in **§2.5.I**, but here we shall impose a convective boundary condition, for variety.

PHYSICAL PROBLEM: Consider a finite slab, $0 \le x \le l$, of *dilute* binary alloy of concentration C_o, initially at its liquidus temperature $T_o = T_A + \frac{C_o}{\kappa_L}$ (**Figure 2.5.6**). On the face $x = 0$, a low temperature $T_\infty(t) < T_A$ is applied convectively with heat-transfer coefficient h, while the face $x = l$ is insulated (and both faces are impermeable).

As in **§2.5.I**, we assume complete mixing in the liquid ($D_L = \infty$, $\alpha_L = \infty$) and no diffusion in the solid ($D_S = 0$). Thus, the concentrations can be found (see (57) **§2.5.I**) once the freezing temperature $T_f(t)$ is found, which requires solution of the thermal problem. The latter is a 1-phase Stefan Problem but with phase-change temperature a function of interface location. For a *dilute* alloy with distribution coefficient $k = \frac{\kappa_S}{\kappa_L}$, we seek the solution of the following (c.f. (53-55, 61) **§2.5.I**, also [ALEXIADES, 1983]).

MATHEMATICAL PROBLEM: Find $T(x, t)$ and $X(t)$ such that

$$T_t = \alpha_S T_{xx}, \qquad 0 < x < X(t), \qquad t > 0 \tag{25a}$$

$$T(X(t), t) = T_f(t) = T_A + [T_o - T_A]\left[1 - \frac{X(t)}{l}\right]^{k-1}, \qquad t > 0 \tag{25b}$$

$$\rho L X'(t) = k_S T_x(X(t), t) + [l - X(t)]\rho c_S T_f'(t), \qquad t > 0 \tag{25c}$$

$$X(0) = 0, \qquad T_f(0) = T_o, \qquad T(x, 0) = T_o, \qquad 0 < x < l \tag{25d}$$

$$-k_S T_x(0, t) = h[T_\infty(t) - T(0, t)], \qquad -k_L T_x(l, t) = 0, \qquad t > 0. \tag{25e}$$

To undimensionalize the problem and fix the free boundary, we set

$$\Delta T = T_o - T_A, \qquad c = c_L / c_S, \qquad \boldsymbol{Bi} = hl / k_S, \qquad \varepsilon = c_S \Delta T / L, \tag{26a}$$

and introduce the variables (**§3.3.B**)

$$\xi = x \,/\, X(t), \qquad \tau = \varepsilon\, \alpha_S \,/\, l^2\, t\,, \qquad \Sigma(\tau) = X(t) \,/\, l\,,$$
$$u(\xi\,,\tau) = \frac{T(x\,,t) - T_A}{\Delta T}, \qquad u_f(\tau) = \frac{T_f(t) - T_A}{\Delta T}, \quad U_\infty(\tau) = \frac{T_\infty(t) - T_A}{\Delta T}\,. \tag{26b}$$

Then the expression for $T_f(t)$ becomes

$$u_f(\tau) \;=\; [1 - \Sigma(\tau)]^{k-1}, \qquad \tau > 0\,, \tag{27}$$

and the 1-phase Stefan Problem for $u(\xi\,,\tau)$, $\Sigma(\tau)$ is

$$u_{\xi\xi} = \varepsilon[\Sigma^2 u_\tau - \xi\Sigma\Sigma' u_\xi], \qquad 0 < \xi < 1, \qquad \tau > 0, \tag{28a}$$

$$u(1\,,\tau) = u_f(\tau) = (1 - \Sigma)^{k-1}, \qquad \tau > 0 \tag{28b}$$

$$u_\xi(1\,,\tau) = \Sigma\Sigma'[1 + \varepsilon\delta c(1 - \Sigma)^{k-1}], \qquad \tau > 0 \tag{28c}$$

$$\Sigma(0) = 0, \qquad u_\xi(0\,,\tau) = \boldsymbol{Bi}\,\Sigma(\tau)\,[u(0\,,\tau) - u_\infty(\tau)\,], \qquad \tau > 0\,. \tag{28d}$$

Expanding $u(\xi\,,\tau)$ and $\Sigma(\tau)$ into powers of ε, as in **§3.3.B** we find (PROBLEM 10) the 0-th order term (quasistationary solution)

$$u_0(\xi\,,\tau) = \frac{(1 - \xi)\boldsymbol{Bi}\Sigma_0\, u_\infty + (1 - \Sigma_0)^{k-1}[1 + \boldsymbol{Bi}\Sigma_0\, \xi]}{1 + \boldsymbol{Bi}\Sigma_0}, \quad 0 < \xi < 1, \quad \tau > 0\,, \tag{29}$$

with $\Sigma_0(\tau)$ the solution of

$$\Sigma_0'(\tau) = \boldsymbol{Bi}\, \frac{(1 - \Sigma_0)^{k-1} - u_\infty(\tau)}{1 + \boldsymbol{Bi}\Sigma_0}\,, \qquad \Sigma_0(0) = 0\,. \tag{30}$$

This is not explicitly solvable for $\Sigma_0(\tau)$, but in the case $k = 2$, $u_\infty \equiv$ *constant*, its inverse, $\tau(\Sigma_0)$, can be expressed as

$$\tau \;=\; -\,\Sigma_0 - [\frac{1}{\boldsymbol{Bi}} + 1 - u_\infty\,]\ln\left[1 - \frac{\Sigma_0}{1 - u_\infty}\right], \qquad 0 < \Sigma_0 < 1\,. \tag{31}$$

This constitutes the quasistationary approximation to the time at which location Σ_0 freezes. Equations for higher order terms are presented in [ALEXIADES, 1983], but are too complicated to solve explicitly. Note that when u_∞ is not a constant, even the quasistationary solution cannot be found explicitly, illustrating the limitations of these methods.

PROBLEMS

PROBLEM 1. Derive the dimensionless problem (4) from (1), using (2, 3).

PROBLEM 2. Solve (6a), (6b), (6c).

PROBLEM 3. Find at least one correction term to the quasistationary solution of
(a) EXAMPLE 2, **§3.1.B**; (b) EXAMPLE 3, **§3.1.B**;

(c) EXAMPLE 6, **§3.1.C**; (d) PROBLEM 8, **§3.1**.

PROBLEM 4. Find at least one correction term to the quasistationary solution of
(a) EXAMPLE 2, **§3.2.A**, when $q_L \equiv$ constant;
(b) EXAMPLE 4, **§3.2.B**; (c) EXAMPLE 8, **§3.2.C**.

PROBLEM 5. Derive (20) and (21).

PROBLEM 6. Derive (22-24).

PROBLEM 7. Apply the methods of **§3.3.C** to PROBLEM 3.

PROBLEM 8. Apply the method of **§3.3.C** to PROBLEM 4.

PROBLEM 9. Undimensionalize (25) in terms of (26), to verify (28).

PROBLEM 10. Show that the quasistationary solution to (28) when $k = 2$, $u_\infty(\tau) \equiv u_\infty$ is given by (29, 31).

3.4. THE MEGERLIN AND HEAT-BALANCE-INTEGRAL METHODS

3.4.A Introduction

In the preceding sections we examined two approximation approaches, the quasistationary approach, a method of low accuracy but sufficient simplicity to be useful for quick but not necessarily precise estimates, and a general perturbation expansion method, in powers of the Stefan Number, which can provide a few correction terms to the quasistationary solution.

In this section we present two additional analytical approximation methods, the Megerlin method and the Heat-Balance-Integral method of T. R. Goodman, both of which assume a parabolic profile for the temperature. They are rather ad hoc in nature, little is known about their general accuracy and reliability, and at times they may be difficult to use. Yet, they appear to provide satisfactory answers in many cases and are used widely in practice. Unfortunately, as with the previous methods, the lack of error estimates precludes any a priori assessment of their accuracy. The answers they provide should always be checked against another method or experiment.

3.4.B The Megerlin method

Consider the 1-phase Stefan Problem in a slab with imposed temperature, (1)**§3.1.B**. As we saw in **§3.1**, the quasistationary approach assumes a linear temperature profile (consistent with negligible sensible heat) and thus overestimates the front location. In general, however, we expect the temperature to look more like an error function and thus be curved. In an attempt to "improve" upon the quasistationary approximation, we may assume a *parabolic* temperature profile

$$T(x,t) \;=\; a(t) + b(t)\,x + c(t)\,x^2\,, \tag{1}$$

with $a(t)$, $b(t)$, $c(t)$ to be determined. Then we have four unknown functions (of time) to find, namely a, b, c and $X(t)$, and four conditions to satisfy, namely the heat equation, the two interface conditions and the boundary condition. Unfortunately, they lead to an *under*-determined problem (a 2nd order O.D.E. for $X(t)$ for which we lack $X'(0)$, see PROBLEM 1), and therefore the heat equation *cannot* be satisfied.

The Megerlin approach [MEGERLIN], [SOLOMON, 1978] consists of satisfying the heat equation *only* on the interface, leading to a 1st order O.D.E. for $X(t)$. We illustrate the method in the following

EXAMPLE 1. 1-phase melting of a slab with imposed temperature. Consider problem (1)§**3.1.B**, and seek the temperature in the form

$$T^{Meg}(x,t) = T_m + A(t)[x - X(t)] + B(t)[x - X(t)]^2, \quad 0 \le x \le X(t),\ t > 0, \tag{2}$$

which already satisfies $T(X(t), t) = T_m$. Then (1c, e)§**3.1.B** yield

$$A(t) = \frac{\rho L}{k_L}\,X'(t), \quad B(t) = \frac{\Delta T_L}{X(t)^2} - \frac{\rho L}{k_L}\frac{X'(t)}{X(t)}, \quad \Delta T_L(t) = T_L(t) - T_m\,; \tag{3}$$

hence only $X(t)$ remains to be found. In Megerlin's method, we try to satisfy the heat equation *only* at $x = X(t)$, which leads to the O.D.E.

$$(XX')^2 + 2\alpha_L(XX') - 2\alpha_L^2\,\frac{c_L\Delta T_L(t)}{L} = 0, \qquad X(0) = 0\;. \tag{4}$$

Keeping the positive root of this quadratic (since $XX' \ge 0$) and integrating we find

$$X^{Meg}(t) \;=\; \left\{ -\,2\alpha_L t + 2\alpha_L \int_0^t \left[1 + 2\,\frac{c_L\Delta T_L(s)}{L} \right]^{1/2} ds \right\}^{1/2} \tag{5}$$

Hence, the Megerlin solution is given by (2, 3, 5). Despite the ad hoc assumption on which the method is based, this solution in fact contains the quasistationary approximation as its lowest order term as the Stefan number $\boldsymbol{St}_L = c_L\Delta T_L / L \to 0$. Indeed, for small $\boldsymbol{St}$, expanding the square roots in (5), we find (PROBLEM 2)

$$X^{Meg}(t) \;\approx\; \left[2\,\frac{k_L}{\rho L} \int_0^t \Delta T_L(s)\,ds \right]^{1/2} - \frac{1}{2\sqrt{2}}\,\frac{c_L}{L}\,\frac{\displaystyle\int_0^t \Delta T_L(s)^2\,ds}{\left(\displaystyle\int_0^t \Delta T_L(s)\,ds\right)^{1/2}} + \cdots \tag{6}$$

for $St_L \approx 0$, the first term of which agrees precisely with (7a)§**3.1.B**. The second term is an alternative correction to that provided by the perturbation method, c.f. (7)§**3.3.B.** For $T_L(t) \equiv T_L$, ΔT_L is constant and (5) may be written as

$$X^{Meg}(t) = 2\left[\frac{\sqrt{1+2\,St_L}-1}{2}\right]^{1/2}\sqrt{\alpha_L t} =: 2\,\lambda^{Meg}\sqrt{\alpha_L t}\,, \tag{7}$$

which is an approximation to the Neumann solution $X = 2\,\lambda\sqrt{\alpha_L t}$. The values of λ, λ^{Meg} and $\lambda^{qs} = \sqrt{St_L/2}$ (quasistationary, §**3.1**) are compared in **Table 3.4.1**. As expected, λ^{Meg} provides an excellent approximation to the exact value of λ for small St_L, which progressively deteriorates as St_L increases. Note that $\lambda^{Meg} < \lambda$ for moderate St_L.

The Megerlin method may be applied to any of the standard 1-phase problems of §**3.1, 2**, but unless the boundary data are constant the resulting differential equation for $X(t)$ usually cannot be solved explicitly (PROBLEMS 3, 4). For axially symmetric problems one assumes a temperature profile quadratic in $\ln(r\,/\,R(t))$, and for spherically symmetric ones a profile quadratic in $1\,/\,r\;-\;1\,/\,R(t)$, (PROBLEMS 5, 6).

Even though the method generates good approximations in many cases, whence its rather widespread use in a variety of engineering problems (usually without the name), its validity must always be verified by independent means since

Table 3.4.1 Comparison of Quasistationary and Megerlin approximations with the Neumann solution

St_L	λ	λ^{qs}	λ^{Meg}	$\lambda - \lambda^{qs}$	$\lambda - \lambda^{Meg}$
0.000001	0.000707	0.000707	0.000707	-0.107e-06	-0.107e-06
0.00001	0.002236	0.002236	0.002236	-0.680e-07	-0.624e-07
0.0001	0.007071	0.007071	0.007071	-0.678e-07	0.109e-06
0.001	0.022357	0.022361	0.022355	-0.368e-05	0.191e-05
0.01	0.070593	0.070711	0.070535	-0.118e-03	0.576e-04
0.1	0.220016	0.223607	0.218455	-0.359e-02	0.156e-02
1.	0.620063	0.707107	0.605000	-0.870e-01	0.151e-01
2.	0.800601	1.000000	0.786151	-0.199e+00	0.144e-01
3.	0.913751	1.224745	0.907125	-0.311e+00	0.663e-02
4.	0.995727	1.414214	1.000000	-0.418e+00	-0.427e-02
5.	1.059687	1.581139	1.076249	-0.521e+00	-0.166e-01
10.	1.256972	2.236068	1.338390	-0.979e+00	-0.814e-01
20.	1.447439	3.162278	1.643643	-0.171e+01	-0.196e+00
50.	1.684005	5.000000	2.127190	-0.332e+01	-0.443e+00
100.	1.850946	7.071068	2.566851	-0.522e+01	-0.716e+00

the errors may be unpredictable (especially for large times and/or moderately large Stefan numbers). Its use is particularly risky on problems with convective boundary condition because, just like the quasistationary approximation (**§3.1.D**), it may produce qualitatively wrong solutions (PROBLEM 4).

3.4.C The Heat-Balance-Integral method

The Megerlin method commits an error by possibly violating global energy conservation, since it satisfies the heat equation *only* at the interface. The method introduced by Goodman [GOODMAN] takes its name from the fact that it seeks to satisfy a global heat balance explicitly. Like Megerlin, it assumes a parabolic profile for the temperature. We illustrate the approach in the following

EXAMPLE 2. 1-phase melting of a slab with imposed temperature. Consider again problem (1)**§3.1.B**, and seek the temperature in the form of the quadratic

$$T^{HBI}(x,t) = T_m + A(t)[x - X(t)] + B(t)[x - X(t)]^2, \quad 0 \le x \le X(t), \quad t > 0 \,, \tag{8}$$

which already satisfies $T(X(t), t) = T_m$. From $T(0, t) = T_L(t)$ we find

$$-AX + BX^2 = T_L(t) - T_m =: \Delta T_L(t) \,. \tag{9}$$

Now, instead of applying the Stefan Condition, we eliminate $X'(t)$ as follows: Taking the time-derivative of $T(X(t), t) = T_m$, we get $X' \cdot T_x(X, t) + T_t(X, t) = 0$. Assuming that the heat equation is also valid at the interface: $T_t(X, t) = \alpha_L T_{xx}(X, t)$, and using the Stefan Condition: $X' = -\dfrac{k_L}{\rho L} T_x(X, t)$, we have

$$T_x(X, t)^2 = \frac{L}{c_L} T_{xx}(X, t) \,, \tag{10}$$

which serves as replacement of the Stefan Condition. From (8, 10) we obtain $B = \dfrac{c_L}{2L} A^2$, and then (9) leads to

$$A(t) = \frac{L}{c_L X}[1 - \sqrt{1 + 2\,St_L}], \quad B(t) = \frac{L}{2c_L X^2}[1 - \sqrt{1 + 2\,St_L}]^2 \,, \tag{11}$$

where $St_L = c_L \Delta T_L(t)/L$. Having expressed everything in terms of $X(t)$, we seek to satisfy the *global heat balance*:

$$\frac{d}{dt}\int_0^{X(t)} \rho c_L [T(x,t) - T_m]\, dx = -\rho L X'(t) - k_L T_x(0, t) \,, \tag{12}$$

which easily follows by integration of the heat equation over $0 < x < X(t)$ and use of the Stefan Condition. With $T(x, t)$ as in (8), the integral in (12), which represents the sensible heat at time t, evaluates to $\rho c_L \left[\dfrac{BX^3}{3} - \dfrac{AX^2}{2}\right]$.

Also evaluating the right-hand side, (12) yields the O.D.E.

$$\frac{d}{dt}\left[\frac{BX^3}{3}-\frac{AX^2}{2}\right] = -\frac{L}{c_L}X' - \alpha_L[A-2BX], \qquad X(0)=0 \ . \tag{13}$$

Its solution (PROBLEM 7) may be written as

$$X(t) = \frac{2\sqrt{3}}{W(t)}\left[\alpha_L \int_0^t [1+2\,St_L - \sqrt{1+2\,St_L}\,]\,W(s)\,ds\right]^{1/2} , \tag{14a}$$

where

$$W(t) = 5+2\,St_L+\sqrt{1+2\,St_L}, \qquad St_L = \frac{c_L \Delta T_L(t)}{L} \ . \tag{14b}$$

Hence, the Heat-Balance-Integral solution is given by (8, 11, 14).

In particular, if $T_L(t) \equiv T_L$ then St_L and W are constants and

$$X^{HBI}(t) = 2\left(3\ \frac{1+2\,St_L-\sqrt{1+2\,St_L}}{5+2\,St+\sqrt{1+2\,St_L}}\right)^{1/2}\sqrt{\alpha_L t} \ =:\ 2\,\lambda^{HBI}\sqrt{\alpha_L t} \tag{15}$$

is an approximation to the Neumann solution $X = 2\,\lambda\sqrt{\alpha_L t}$. **Table 3.4.2** compares the values of λ^{HBI} and λ^{Meg} (from (7)) with those of the exact λ. Looking at the errors, we see that λ^{Meg} *outperforms* λ^{HBI} for all but very

Table 3.4.2 Comparison of Megerlin and Heat-Balance-Integral methods with the Neumann solution

St	λ	λ^{Meg}	λ^{HBI}	$\lambda - \lambda^{Meg}$	$\lambda - \lambda^{HBI}$
0.000001	0.000707	0.000707	0.000707	0.296E-07	0.294E-07
0.00001	0.002236	0.002236	0.002236	-0.248E-06	-0.253E-06
0.0001	0.007071	0.007071	0.007071	0.403E-07	-0.136E-06
0.001	0.022357	0.022355	0.022361	0.204E-05	-0.354E-05
0.01	0.070593	0.070535	0.070709	0.579E-04	-0.116E-03
0.1	0.220016	0.218455	0.223213	0.156E-02	-0.320E-02
1.	0.620063	0.605000	0.660014	0.151E-01	-0.400E-01
2.	0.800601	0.786151	0.859047	0.144E-01	-0.584E-01
3.	0.913751	0.907125	0.978405	0.663E-02	-0.647E-01
4.	0.995727	1.000000	1.060660	-0.427E-02	-0.649E-01
5.	1.059687	1.076249	1.121796	-0.166E-01	-0.621E-01
10.	1.256972	1.338390	1.290313	-0.814E-01	-0.333E-01
20.	1.447439	1.643643	1.420968	-0.196E+00	0.265E-01
50.	1.684005	2.127190	1.539995	-0.443E+00	0.144E+00
100.	1.850946	2.566851	1.599106	-0.716E+00	0.252E+00

small and very large St_L. We note also that $\lambda^{HBI} > \lambda$ in the range $10^{-5} \le St_L \le 10$, and $\lambda^{HBI} > \lambda^{Meg}$ for $St_L < 15.75$.

For other boundary conditions and geometries, the Heat-Balance-Integral method is complicated and cumbersome to apply (PROBLEMS 8, 9, 10). In the absence of any error estimates, the choice between this method and Megerlin's becomes a matter of personal preference.

PROBLEMS

PROBLEM 1. Substitute (2, 3) into the heat equation and show that this leads to an underdetermined problem for $X(t)$. Conclude that a parabolic temperature profile cannot possibly satisfy the phase-change problem exactly.

PROBLEM 2. Derive (6) from (5) and, for $T_L(t) \equiv T_L$, compare the first two terms with the two-term perturbation solution (7)§**3.3.B**.

PROBLEM 3. Apply Megerlin's Method to the 1-phase melting of a slab with imposed flux $q_L(t) > 0$ at $x = 0$. Compare with the quasistationary solution when the "effective Stefan number" $c_L q_L X / Lk_L$ is small.

PROBLEM 4. Apply Megerlin's method to the 1-phase melting of a slab with convective heating at $x = 0$ (assume constant h and T_L). Check the limiting cases $h \to 0$ and $h \to \infty$. For what values of Biot number does the solution become unphysical (see §**3.1.D**)?

PROBLEM 5. Apply Megerlin's method, using a temperature profile quadratic in $\ln r / R(t)$, to the following 1-phase problems (c.f. §**3.2**): (a) outward melting of a hollow cylinder due to imposed temperature at $r = R_{in}$; (b) inward freezing of a cylinder due to imposed flux at $r = R_{out}$.

PROBLEM 6. Apply Megerlin's method, using a temperature profile quadratic in $1/r - 1/R(t)$, to the following 1-phase problem (c.f. §**3.2**):
(a) outward melting of a hollow sphere due to imposed flux at $r = R_{in}$;
(b) inward freezing of a sphere due to imposed temperature at $r = R_{out}$.

PROBLEM 7. Derive the heat balance statement, (12), and using (8), show that it leads to the differential equation (13). Solve it to obtain (14).

PROBLEM 8. Apply the Heat-Balance-Integral method to the 1-phase melting of a slab with convective heating at $x = 0$ (assume constant h and T_L. For what values of Biot number does the solution become unphysical ?
(see §**3.1.D**).

PROBLEM 9. Repeat PROBLEM 5 for the Heat-Balance-Integral method.

PROBLEM 10. Repeat PROBLEM 6 for the Heat-Balance-Integral method.

3.5. SOME MELTING TIME RELATIONS

3.5.A Introduction

Combining the analytical methodology of the previous sections with numerical simulation approaches of the next chapter and curve fittings, one may derive a number of easily computable relations for various quantities of interest in melting and solidification processes. In this Section we gather together a number of these relations for the prediction of melt-times of simple bodies subject to various boundary conditions. This list is by no means exhaustive; the literature contains many such relations, whose purpose is to provide us with tools for making rapid judgements about sizings of experiments (both physical and numerical) and systems.

Our discussion in this section is divided into four subsections. We present, in order, an easily computable relation for the melt time of a simple body of phase changing material subject to a constant imposed temperature at its boundary, the melt time of a simple body with a convective boundary condition, the melt time for a rectangular body, and an approximation of use in predicting the behavior of an array of cylinders of a phase changing material. In each case we will state the relation, refer to an appropriate reference for its derivation and validation, and give an example of its use.

3.5.B Melt-time for a simple PCM body with imposed temperature

Consider a PCM body initially solid at its melt temperature. At time $t = 0$, a temperature $T_L > T_m$ is imposed at its boundary. A melting front originating at the boundary will intrude into the PCM until at some finite time t_{melt} the body is completely melted.

We assume that the melting process can be described in terms of a *single geometric variable* r; it is cylindrically symmetric for a PCM cylinder and spherically symmetric for a PCM sphere. Assume that the PCM body corresponds to the interval $0 \leq r \leq l$. If A is the surface area across which heat is transferred into the body and V is the volume of the body, let ω be a dimensionless parameter defined by

$$1 + \omega = \frac{l\,A}{V}. \tag{1}$$

Then

$$\omega = \begin{cases} 0 & \text{for a PCM } \textit{slab} \text{ insulated at one end} \\ 1 & \text{for a PCM } \textit{cylinder} \\ 2 & \text{for a PCM } \textit{sphere} \end{cases} \tag{2}$$

Thus a value $\omega = 1.5$ would correspond to a football-like shape, under the assumption that radial heat transfer is dominant, while $\omega = 0.5$ would represent heat transfer into one face of a flat cylinder with a small ratio of depth to diameter. Note that $0 \le \omega \le 2$ always.

Under these assumptions, the 1-phase melting process may be formulated, for any "shape factor" $0 \le \omega \le 2$, as follows:

$$T_t = \frac{\alpha_L}{r^\omega}\frac{\partial}{\partial r}\left(r^\omega \frac{\partial T}{\partial r}\right), \qquad R(t) \le r \le l, \qquad t > 0 \tag{3a}$$

$$T(R(t), t) = T_m, \qquad \rho L \frac{dR}{dt} = -k_L \frac{\partial T}{\partial r}(R(t)^+, t), \qquad t > 0 \tag{3b}$$

$$R(0) = l, \qquad T(l, t) = T_L, \qquad t > 0 \tag{3c}$$

Letting

$$\Delta T_L = T_L - T_m, \qquad \mathbf{St}_L = \frac{c_L \Delta T_L}{L},$$

a good approximation to the melt-time is given by [SOLOMON, 1979b] the expression

$$t_{melt} \approx \frac{l^2}{2\alpha_L(1+\omega)\mathbf{St}_L}\left\{1 + [0.25 + 0.17\omega^{0.7}]\,\mathbf{St}_L\right\}, \quad \text{valid for } 0 \le \mathbf{St}_L \le 4. \tag{4a}$$

Moreover, the *average* heat flux at $r = l$ during the duration of the melting process is

$$\bar{q} \approx \frac{-2k_L \Delta T_L}{L}\left\{1 + [0.1210 + 0.0424\,\omega]\,\mathbf{St}_L^{0.7645-0.2022\,\omega}\right\} \tag{4b}$$

and the *total* heat input, per unit surface area, during the process is

$$Q \approx \bar{q}\, t_{melt}, \tag{4c}$$

under the same restriction on $\mathbf{St}_L$ as in (4a).

Note that for $\mathbf{St}_L \approx 0$, the bracketed term in (4a) is ≈ 1 and we obtain the quasistationary value (c.f. (7a)**§3.1**, (13)**§3.2**)

$$t_{melt}^{qs} = \frac{l^2}{2(1+\omega)\alpha_L \mathbf{St}_L} = \frac{l^2}{2(1+\omega)}\frac{\rho L}{k_L \Delta T_L}, \tag{4d}$$

EXAMPLE 1. *Melting and freezing of paraffin wax:* Consider a lump of a paraffin-wax-like phase changing material with thermal properties

$$T_m = 27.67°C, \qquad L = 241.2\ kJ/kg, \qquad \rho = 814\ kg/m^3$$

$$c_L = c_S = 2.14\ kJ/kg°C, \quad k_L = 0.190\times10^{-3}, \quad k_S = 0.145\times10^{-3}\ kJ/m\,s°C.$$

Suppose the PCM is a somewhat elongated football shape with representative radius of $0.05\,m$ and a shape factor $\omega = 1.15$. We consider two situations: melting due to an imposed temperature T_L and freezing due to a T_S, in both cases starting at the melt-temperature T_m. Over a range of values for T_L and T_S, the melting and freezing times predicted by (4a) are shown in **Table 3.5.1**.

Table 3.5.1. Melting and freezing times for a paraffin-wax lump

$T_L(°C)$	$t_{melt}(hr)$	$T_S(°C)$	$t_{freeze}(hr)$
100	3.748	20	23.700
90	4.217	10	10.710
80	4.867	0	7.092
70	5.824		
60	7.373		
50	10.311		
40	18.027		
30	92.838		

EXAMPLE 2. *The performance of a PCM brick:* One of the techniques for dealing with large temperature variability from day to night in areas of the Southwest has traditionally been to enhance the thermal inertia of the structure by massive use of dense materials. This is the reasoning behind the use of adobe walls that during daylight hours can store large amounts of heat which is then available for warming the structure at night. The same philosophy has fueled a search that has gone on for many years for phase-change materials that can be incorporated into construction bricks. The hope is that the latent heat effect can serve the same role that the massiveness for sensible heat storage serves in the adobe house, namely to enhance the thermal inertia of the structure. Our next example is drawn from an attempt in this direction.

We consider a concrete block into which a phase change material has been incorporated. We assume the resulting structure to have the thermal properties of the concrete with an additional latent heat L and melt temperature T_m of phase change. We wish to compare the result of employing PCMs of various melt-temperatures (see **Table 3.5.2**), in a cubic brick of linear dimension either $0.1\,m$ or $0.025\,m$, discharging heat via direct convection into a 20 °C room with a heat transfer coefficient so high that we may assume the room temperature to be imposed on the surface of the brick.

Using (4) we may do this for the following simplified situation. Assume that initially the brick is at uniform temperature T_m in the solid phase. We take the brick to be insulated at one face and exposed to a charging temperature $T_L > T_m$ at the other; a charging time t_c is required for the melting to be completely effected. Upon being charged we assume the brick to reach a situation in which the temperature is T_m (through a small loss of sensible heat) while its phase is in the higher energy (melted) state. One of its faces is then insulated while the other is exposed to a constant (room) temperature $T_{room} = 20°C < T_m$; after a discharge time t_d the brick is assumed to have again completely changed phase. This scenario can be considered as "ideal" since actual heat transfer processes would be of longer duration.

Table 3.5.2. Charge and discharge times for the PCM brick of EXAMPLE 2

$l\,(m)$	$T_m(°C)$	$T_L(°C)$	$t_c(hrs)$	$T_S(°C)$	$t_d(hrs)$
0.025	20.62	21.41	8.	20.	10.
0.025	20.62	65.	0.15		
0.025	25.	25.79	8.	20.	1.26
0.025	25.	65.	0.17		
0.025	40.	40.79	8.	20.	0.32
0.025	40.	65.	0.26		
0.1	25.	37.83	8.	20.	20.25
0.1	25.	65.	2.68		
0.1	30.10	42.93	8.	20.	10.
0.1	30.10	65.	3.05		
0.1	40.	52.83	8.	20.	5.19
0.1	40.	65.	4.19		

As representative properties of the brick we take $\rho = 894\ kg/m^3$, $L = 174\ kJ/kg$, $c = 1.17\ kJ/kg°C$, $k = 2.16 \times 10^{-3}\ kJ/m\,s\,°C$. Then, the Stefan numbers for the charging and discharging processes are

$St_c = 0.0067\Delta T_c$, $\Delta T_c := T_L - T_m$, $St_d = 0.0067$, $\Delta T_d := T_m - T_{room}$,

and $\omega = 0$ here, since we are working in slab geometry. For St_c, $St_d \le 4$, (4a) yields the following expressions for the charge and discharge times:

$$t_c = l^2 \left\{ \frac{1674}{\Delta T_c} + 1.563 \right\}, \qquad t_d = l^2 \left\{ \frac{1674}{\Delta T_d} + 1.563 \right\}. \tag{5}$$

The resulting values for $l = 0.025$ or $0.1\,m$, and various T_m and T_L combinations, are shown in **Table 3.5.2.**

We see that the thin wall ($l = 0.025\,m$) has very short charge and discharge times; this may be desirable if we wish to simply enhance the thermal inertia of, say, floor or wall paneling. For the $0.1\,m$ brick with the thermal properties assumed here, the choice of $T_m = 30.1°C$ seems "best" among the alternatives listed, since it results in $t_c = 8\,hr$, $t_d = 10\,hr$ and $T_L = 42.93°C$, an easily attainable temperature. Let us also note that the properties of concrete vary considerably with composition, and (4) can serve as a very convenient tool for comparing various alternatives.

3.5.C Melt-time for a simple PCM body with convective boundary condition

Consider again a simple PCM body of effective length l and shape factor ω (see §**3.5.B**), but now heated or cooled convectively with a heat transfer coefficient h. To within 10% (at worst), the melt-time is given by [SOLOMON, 1980a]

$$t_{melt} \approx \frac{l^2}{2\alpha_L(1+\omega)St_L}\left\{1+\frac{2}{Bi}+[0.25+0.17\omega^{0.7}]St_L\right\} \text{ for } 0 \le St_L \le 4, \quad Bi \ge 0.1, \tag{6}$$

where $Bi = hl/k_L$ is the Biot number of the process. Note that for $St_L \approx 0$ this reduces to the quasistationary approximation value found in (35d) **§3.1.D**.

EXAMPLE 3. *Melting of Glauber's Salt*: A phase-change material appropriate for various Latent Heat Thermal Energy Storage (LHTES) applications is a mixture of Glauber's and other salts, packaged in sausage-like cylinders of length $50.8\,cm$ and diameter $5.08\,cm$, produced at the University of Delaware. The thermal properties of such a "chub", as it is called, are as follows:

$$T_m = 12.78\,°C, \quad L = 116.3\,kJ/kg, \quad \rho_L = \rho_S = 1457.6\,kg/m^3,$$
$$c_L = 3.31, \ c_S = 1.76\,kJ/kg°C, \ k_L = 0.00062, \ k_S = 0.00225\,kJ/m\,s°C. \tag{7}$$

A chub, initially solid at T_m, is placed in heated-air cross flow of ambient temperature T_L. The heat transfer coefficient is estimated to be in the range $0.0284 \le h \le 0.0568\,kJ/m^2s°C$, which is reasonable for thermal storage applications. The melt times of the chub, for each combination of values of h and T_L listed in **Table 3.5.3**, are easily found from (6) (with $\omega = 1$, $l = 0.0254\,m$). The parameter ω of (2) allows us to explore the effect of the shape on the melt time. For example, choosing $h = 0.0284$ and $T_L = 23.89$, the melt times corresponding to $\omega = 1$, 1.5 and 2 are, respectively, 3.15, 2.53 and 2.12 hours.

In convective heating or cooling, the temperature attained by the surface, across which the heat is transferred, is always of interest. For the 1-phase Stefan Problem for a slab, initially solid at T_m, melting due to convective heating at $x = 0$ with a heat transfer coefficient $h = constant$ and ambient temperature $T_L = constant$, the following relationship holds, to within a relative error of at most 10%, [SOLOMON, 1979a]:

$$t = \frac{\rho c_L k_L}{h^2}\frac{1}{2St_L}\left\{1.18\,St_L\left[\frac{T_{sfs}-T_m}{T_L-T_{sfc}}\right]^{1.83}+\left[\frac{T_L-T_m}{T_L-T_{sfc}}\right]^2-1\right\}, \tag{8}$$

Table 3.5.3. Chub melt-times (hrs) for various h and T_L

h \ T_L	18.33	23.89	40.56
0.0284	6.11	3.15	1.37
0.0568	4.24	2.21	0.99

Table 3.5.4. Surface temperature, time and melt-depth in the situation of EXAMPLE 4		
$T_{sfc}(°C)$	$t(hrs)$	$X(m)$
30	0.96	0.017
35	3.90	0.041
38	19.02	0.099
40	400.64	0.486

where $\boldsymbol{St}_L = c_L(T_L - T_m)/L$. It relates the surface temperature, $T_{sfc} := T(0, t)$, to the time.

EXAMPLE 4. Consider a slab of Glauber's salt (see (7)), heated convectively at $x = 0$ with $h = 0.0568\ kJ/m^2 s°C$ and $T_L = 40.56°C$. Then $\boldsymbol{St}_L = 0.79$ and the times at which the temperature at $x = 0$ reaches the value T_{sfc}, according to (8), are shown in **Table 3.5.4.** The third column shows the melt dept at those times, found by solving (6) for l. We see that it would take 400 *hours* for the surface temperature to rise to 40°C, by which time the melt front will penetrate only 0.5 *meters* into the slab.

3.5.D Melt-time for a rectangular body under imposed temperature

A simple, but effective, approximation (with relative error below 10%) may also be found for the melt-time of a rectangle (two-dimensional body). Assuming the rectangle, of dimensions a by b $(a < b)$, is initially solid at T_m, and that a constant temperature $T_L > T_m$ is imposed on all four sides, its melt-time is approximately given by, [SOLOMON, 1979b],

$$t_{melt} \approx \begin{cases} \dfrac{a^2[1 + 0.25\boldsymbol{St}_L]}{8\alpha_L \boldsymbol{St}_L} & \text{if } \dfrac{a}{b} < \dfrac{2[1 + 0.42\boldsymbol{St}_L]}{\pi[1 + 0.25\boldsymbol{St}_L]}, \\ \dfrac{ab[1 + 0.42\boldsymbol{St}_L]}{4\pi\alpha_L \boldsymbol{St}_L} & \text{if } \dfrac{a}{b} > \dfrac{2[1 + 0.42\boldsymbol{St}_L]}{\pi[1 + 0.25\boldsymbol{St}_L]}, \end{cases} \tag{9}$$

where $\boldsymbol{St}_L = c_L(T_L - T_m)/L$.

EXAMPLE 5. A long rectangular slab of paraffin-wax (see EXAMPLE 1 **§3.5.B** for properties), of width $10\,cm$ and height $5\,cm$, is placed in a rapid cross flow of ambient temperature $T_L = 100°C$. If the slab is initially solid at $T_m \approx 28°C$ and the heat-transfer coefficient is high, so that we may assume the T_L as being imposed on all four of the lateral faces of the slab, then the estimated melt-time of the slab, from (9a) will be $t_{melt} \approx 1.58\,hours$. For a square, 10 by $10\,cm$ slab, we find from (9b), $t_{melt} \approx 4\,hours$.

3.5.E Freezing of a PCM cylinder array

To illustrate how approximations for the behavior of a single PCM body could be linked together to deal with a complex system consisting of an array of PCM containers placed in a channel, we discuss a problem which arose in an actual heat-storage application.

An in-line array of PCM cylinders is placed in the cross flow of a transfer fluid in a duct (**Figure 3.5.1**). At time $t = 0$ the PCM is liquid at temperature T_m, and transfer fluid is pumped into the duct at temperature $T_S < T_m$. The cylinders begin to freeze and cool, while the transfer fluid gains heat as it flows through the array.

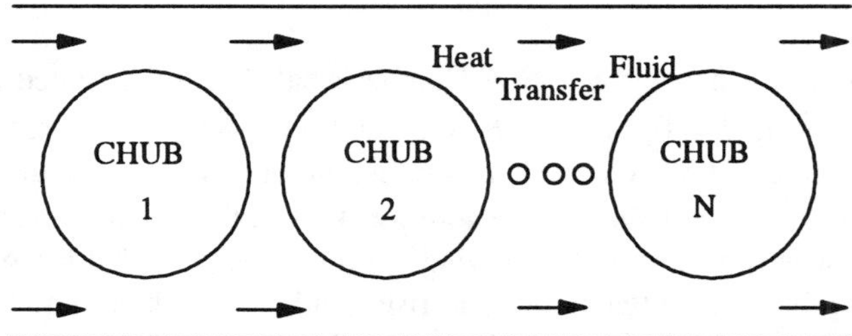

Figure 3.5.1. PCM cylinders in cross flow of a heat-transfer fluid.

This scenario is basic to the use of phase-change materials in many heat storage applications. One example is the LILCO system which was under study on Long Island, New York, in the early 1980's, and was aimed at air conditioning support [SOLOMON, 1981c]; this system was based on the use of University of Delaware "chubs" (see EXAMPLE 3, §3.5.C) and stored "cold" obtained from a low power air conditioner. Questions basic to its design were:

(a) how long does it take for each PCM cylinder in the array to freeze?

(b) how do the outlet temperature and stored energy vary in time?

The approximations that we present below permit us to address these questions without the need for detailed computation. The derivation of the approximations will be given later.

The approximations

We assume that, in each cylinder, heat transfer is by conduction and radially symmetric. We focus on a *single* row of N cylinders in a channel, and number the cylinders as #1, 2, ..., N starting at the transfer-fluid inlet (**Figure 3.5.1**).

Let t_j be the time at which cylinder $\#j$ freezes, $j = 1, 2, \ldots, N$. Reasonable approximations to these times are given by (their derivation is discussed later):

$$t_j = \tau[1 + (j-1)\gamma], \qquad j = 1, 2, \ldots, N, \tag{10}$$

where

$$\tau := \frac{R_0^2 \rho L}{2 k_S \Delta T}\left(\frac{1}{2} + \frac{k_S}{h R_0}\right), \qquad \gamma := \frac{2\pi R_0 l h}{\rho_f c_f F}, \tag{11}$$

with R_0, l and h, the radius, length, and heat-transfer-coefficient across the boundary, of each cylinder; ρ, L and k_S, the density, latent heat, and solid conductivity of the PCM; ρ_f, c_f and F the density, specific heat, and volumetric flow rate of the heat-transfer fluid; and $\Delta T = T_m - T_S$ the overall temperature drop. Then, the fluid temperature $T_j(t)$ after cylinder #j is well approximated by

$$T_j(t) = \begin{cases} T_j^*, & 0 < t < t_1 \\ T_{j-1}^*, & t_1 < t < t_2 \\ T_{j-2}^*, & t_2 < t < t_3, \quad j = 1, 2, \ldots, N, \\ \ldots & \\ \ldots & \\ T_S, & t_j < t \end{cases} \tag{12}$$

where

$$T_j^* = T_m - \Delta T(1-\gamma)^j, \quad j = 1, 2, \ldots, N. \tag{13}$$

Let "zero energy" correspond to solid PCM at the melt temperature T_m. Then the initial energy present in a cylinder is $\pi R_0^2 l \rho L$, and so, for the N cylinders the total energy initially is

$$E_0 = N \pi R_0^2 l \rho L. \tag{14}$$

The total energy E_j of the row at time t_j (of (10)) is approximately given by the expressions

$$E_1 = E_0 - 2\pi R_0 l h \Delta T \frac{\tau}{\gamma}[1-(1-\gamma)^N] \tag{15}$$

$$E_j = E_{j-1} - 2\pi R_0 l h \Delta T \tau [1-(1-\gamma)^{N-j+1}], \quad j = 2, \cdots, N. \tag{16}$$

An application

Relations (10) - (16) have been applied to the LILCO system of 1152 "chubs" in 48 channels of 24 chubs with an air cross flow. The PCM is a mix of Glauber's and other salts with $T_m \approx 12.8°C$. A chub is a sausage-like package that is $5.08 \times 10^{-2}\ m$ in diameter and $5.08 \times 10^{-1}\ m$ in length. The PCM properties are given in (7).

The chubs are arrayed in-line, with a space of $1.12 \times 10^{-2}\ m$ between them. Hence the air "channel" height above and below each chub is $5.6 \times 10^{-3}\ m$. For a pumping rate of $0.33\ m^3/s$ ($700\ ft^3/minute$) over the 48 channels, the pumping rate is $6.88 \times 10^{-3} m^3/s$ in each channel. The heat transfer coefficient is approximately [McADAMS] $h = 2.27 \times 10^{-2}\ kJ/m^2 s°C$. Moreover, $\rho_f = 1.144\ kg/m^3$ and $c_f = 1\ kJ/kg°C$.

Let us examine the predicted freezing process when air at $4.44°C$ is pumped over the chubs which are initially liquid at $12.8°C$, in which case, from (11), $\tau = 3.56$ hours, $\gamma = 0.23$, and from (14), $E_0 = 4198\ kJ$. The values of the times

t_j, energies E_j, and temperatures T_j^*, obtained from (10), (16) and (13), for $j = 1, 2, \ldots, 24$, are shown in **Table 3.5.5** under the columns labeled t_j^{apprx}, E_j^{apprx}, and T_j^*, respectively. The values shown under t_j^{comp} and E_j^{comp} were obtained from a detailed simulation code (modeling the phase-change in each chub by the enthalpy method of CHAPTER 4), which was developed for the purpose of testing the effectiveness of these approximations [GEIST et al, 1984]. The agreement is indeed very satisfactory, and using (12) we can predict hourly air temperatures anywhere down the channel. **Figure 3.5.2** shows such temperatures behind chubs #6, 12 and 24; the curves obtained from (12) and from the detailed simulation are indistinguishable at the scale of the figure. The accuracy of the approximation seen in this example is typical of that for a large number of test cases that we have examined.

Table 3.5.5 Comparison of approximations and numerical simulation for the LILCO System

j	t_j^{comp}	t_j^{apprx}	E_j^{comp}	E_j^{apprx}	T_j^*
1	3.73	3.56	3279	3336	6.36
2	4.48	4.38	3096	3140	7.84
3	5.28	5.20	2900	2944	8.97
4	6.06	6.02	2709	2747	9.85
5	6.81	6.84	2527	2551	10.47
6	7.67	7.66	2318	2357	11.04
7	8.48	8.48	2121	2162	11.44
8	9.23	9.30	1939	1968	11.75
9	10.11	10.12	1727	1774	11.98
10	10.81	10.94	1559	1530	12.17
11	11.67	11.76	1355	1390	12.31
12	12.35	12.58	1194	1200	12.42
13	13.23	13.40	990	1011	12.50
14	14.03	14.22	806	826	12.56
15	14.81	15.04	631	644	12.61
16	15.67	15.86	444	465	12.65
17	16.35	16.68	300	293	12.68
18	17.23	17.50	121	128	12.70
19	18.03	18.32	-30	-27	12.72
20	18.81	19.14	-164	-171	12.73
21	19.67	19.96	-293	-299	12.74
22	20.42	20.78	-387	-405	12.75
23	21.23	21.60	-463	-485	12.76
24	22.03	22.42	-510	-531	12.76

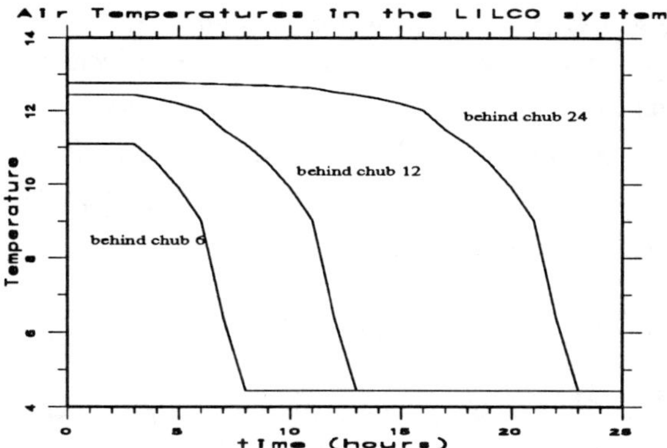

Figure 3.5.2. Air temperatures predicted by (12).

Derivation of the approximations

For small St ($St < 2$), when a phase change front is present in a PCM cylinder, the temperature is essentially constant in time, which is the basis for the use of the quasistationary approach. Thus the transfer fluid temperature is also slowly varying with time except when a cylinder completely freezes. At that time the cylinder has effectively lost its thermal inertia, its temperature will quickly fall to the ambient fluid temperature at its location, and the fluid will quickly be responding to one less cylinder.

Consider a row of N PCM cylinders after the onset of transfer-fluid flow. A phase-change front appears in each cylinder. Let T_{in}, T_{out} be the transfer fluid temperature before reaching a typical cylinder and after leaving its vicinity; let T_{sfc} be the surface temperature. Then by an elementary heat balance (PROB. 15)

$$(T_{in} - T_{out})\,\rho_f\, c_f\, F = 2\pi\, R_0\, l\, h\,(T_{in} - T_{sfc})\,. \tag{17}$$

For small Biot number, $Bi = hR_0 / k_S$, $T_{sfc} \approx T_m$, and we write

$$T_{out} = T_{in} - \frac{2\pi\, R_0\, l\, h}{\rho_f\, c_f\, F}(T_{in} - T_m) = T_{in} - \gamma\,(T_{in} - T_m). \tag{18}$$

For T_j^* the "steady state" fluid temperature after cylinder $\# j$ (using $T_0^* = T_S$), we find

$$T_{j+1}^* = \gamma\, T_m + T_j^*(1-\gamma)\,, \tag{19}$$

from which we obtain (13).

Relation (12) is derived as follows. For $0 < t < t_1$ the transfer fluid state is determined by the fact that **every** cylinder is changing phase and releasing a large and uniform amount of latent heat to the fluid. But for $t = t_1$, cylinder #1 is totally frozen, and for $t_1 < t < t_2$ it has effectively ceased heating the fluid. Thus its outlet temperature is T_S, and cylinders # 2,3,4,..., play the roles that cylinders # 1,2,3,..., did in the previous time interval, until $t = t_2$. At this time the transfer fluid is heated by cylinders # 3,4,5,...; in this way we obtain (12).

The freezing-time of chub $\#j$ is approximately given by (6) with ω=1, $\mathbf{Bi} = hR_0 / k_S$ and $\boldsymbol{St} = c_S(T_m - T^*_{j-1}) / L$; Since $\boldsymbol{St}$ is small, we drop this term from inside the brackets in (6), use (13) to replace T^*_{j-1} and thus obtain the following approximation to the freezing-time of cylinder $\#j$ exposed to temperature $T^*_{j-1} = T_m - \Delta T(1-\gamma)^{j-1}$:

$$\hat{t}_j \approx \frac{R_0^2 \rho L}{2k_S \Delta T(1-\gamma)^{j-1}} \left[\frac{1}{2} + \frac{k_S}{hR_0} \right] = \frac{\tau}{(1-\gamma)^{j-1}} . \tag{20}$$

Thus a fraction θ , $0 \le \theta \le 1$, of cylinder $\#j$ requires $\theta \hat{t}_j$ to freeze, roughly. Let us see how (10) is derived by looking at the first few cylinders.

Cylinder #1 is exposed to T^*_0 and freezing in time $t^*_1 = \hat{t}_1$, which by (20) is $t^*_1 = \tau$. During this time, a fraction $\hat{t}_1 / \hat{t}_2$ of cylinder #2 freezes, leaving unfrozen fraction $\theta = 1 - \hat{t}_1 / \hat{t}_2$. For $t > \hat{t}_1$, cylinder #2 is exposed to T^*_0, and the unfrozen portion freezes in time $\theta \hat{t}_1$; hence its total freezing time is

$$t^*_2 = \hat{t}_1 + \theta \hat{t}_1 = \tau(1+\gamma) . \tag{21}$$

Continuing this argument yields relation (10). Relations (16) are derived similarly.

Validity and extensions

On the basis of many comparisons with the results of computer simulations over a broad range of possibilities, we conclude that relations (10)-(16) are effective for $\boldsymbol{St} < 4$, $Bi < 1$ as prediction tools for the design and simulation of such systems.

If the cylinders are initially at a temperature $T_L > T_m$ we have found these relations to still be accurate if we replace the latent heat L by the "modified" latent heat

$$\bar{L} = L + \tfrac{1}{2} c_L (T_L - T_m) .$$

PROBLEMS

PROBLEM 1. For $\omega = 0, 1, 2$, derive (4d) directly from the quasistationary approximation, (**§3.1, 3.2**).

PROBLEM 2. Explore the role of the parameter ω in equation (4a). In particular, rewrite (4a) as a relation for the Fourier number and examine it as a function of ω.

PROBLEM 3. Assuming that its body has the thermophysical properties of water/ice, estimate how long it should take to defrost a frozen chicken by leaving it on your kitchen table at ambient temperature 25°C. Estimate how long it would take to refreeze it in a freezer whose temperature is -10°C.

PROBLEM 4. Using relations (4a-c) examine the validity of the assertion that states: "a material may be melted while remaining essentially at its melt temperature."

PROBLEM 5. The PCM brick of EXAMPLE 2, §3.5.B, is to be .04 m thick, and is to be exposed to 8 hours of a surface temperature of 120°C, and then is to totally give up its latent heat in 10 hours to a sink imposing a surface temperature of 30°C. Can you determine a melt temperature T_m for the PCM that will provide the desired performance (assuming all other thermophysical properties are as in EXAMPLE 2) ?

PROBLEM 6. Explain why the melt-time relation of (6) tends to that of (4a) as the Biot Number ***Bi*** tends to infinity.

PROBLEM 7. Show that (6) is the average of (4a) and the estimate of the quasi-stationary approximation.

PROBLEM 8. Compare the estimates of (4a) and (9) for the melt times of a "body" and of a rectangle. Is there an ω for which they will approximately agree?

PROBLEM 9. You wish to make a block of ice by pouring water into a thin-walled metal mold, and immersing the mold in rapidly circulating −2°C brine. Estimate the time needed to freeze a long block of ice whose cross section is a rectangle of width 0.5m and height 0.3m. Derive your estimate using (4a) and using (9).

PROBLEM 10. Given a PCM block of side-lengths a , b , c, you are asked to estimate (even very roughly!) how long it will take to melt the block via an imposed temperature on its entire surface. How would you do this?

PROBLEM 11. A dumbbell-shaped block of ice is dropped in warm water. How would you estimate how long it will take to a) divide into two distinct pieces of ice, and b) totally melt?

PROBLEM 12. Convert (8) into dimensionless form involving the Biot and Fourier Numbers.

PROBLEM 13. Using the result of PROBLEM 12 examine the question of how long you can convectively heat a body before its surface temperature reaches a given fraction θT_L of the ambient temperature of the transfer fluid.

PROBLEM 14. Verify equation (17).

PROBLEM 15. Apply the approach of §3.5.E to a system identical to that described except having rectangular containers instead of cylindrical.

CHAPTER 4

NUMERICAL METHODS– THE ENTHALPY FORMULATION

As we have repeatedly remarked, explicit and approximate solutions are obtainable only for simple problems and only in one space dimension. As most realistic phase-change processes do not neatly fall in this category, the mathematical problems modeling such processes may only be attacked numerically.

A mathematical model of a physical process may be thought of as a simulation of the process, i.e. an imitation using mathematical tools. In the same spirit as a laboratory-scale experiment of an industrial process is an imitation of the process by the means and capabilities of the laboratory, a numerical (computer) simulation is an imitation of the process by the means and capabilities of the computer.

Digital computers are capable of representing only a finite number of rational (finite decimal) numbers and therefore can only deal with discrete approximations of continuum concepts such as time and length. Moreover, memory sizes are also finite and small, thus restricting the amount of data that can be processed. Such limited capabilities of computers impose certain limitations and restrictions on the numerical simulation of a physical process. Thus, the physical region must be approximated by a small number of "control volumes," time may vary only in discrete steps, and idealized mathematical concepts, such as derivatives, integrals and limits must be re-approximated by finite-differences, sums and approximate values.

In **§4.1** we explain how such discrete approximations are set up (via finite-differences) for the simplest case of heat conduction without phase change. After a brief discussion of front-tracking methods in **§4.2**, we then quickly turn to the most general and versatile method available for the numerical simulation of phase-change processes, the so-called **enthalpy method**. Its numerical implementation is presented in **§4.3**. The mathematical ideas underlying weak formulations of PDE problems, and the mathematical formulation on which the enthalpy method is based are presented in **§4.4**. Finally, in **§4.5** we establish existence of the weak solution and convergence of the enthalpy scheme to the weak solution.

4.1. NUMERICAL HEAT TRANSFER

4.1.A Introduction

Simulation of a system means imitation of the system by a convenient replacement or "stand-in," whose performance can be studied in detail. The motivation is to use an inexpensive "stand-in" to tell us what we want to know about the original system. The simulation might consist of a field trial in place of the actual unmonitored process, a laboratory experiment in place of a field trial, or a pencil-and-paper mathematical model in place of the laboratory experiment.

A numerical, or computer, simulation is one in which the "stand-in" for the system is a computer code, whose runs simulate the system's performance. If the processes taking place are time-dependent, then the computer code must accordingly tell us what is going on with the progress of time. Such a code is often referred to as a "marching" code, with the implication being "with increasing time."

The computer simulation of a time-dependent process rests upon a **discretized** version of a mathematical model of the actual physical process. Thus, continuous quantities, such as energy and temperature, are replaced by their values at discrete points. Time itself is discretized, and the marching process takes place through discrete time steps. Just as the individual frames of a movie must be taken at close enough times, the time steps for a computer simulation must be small enough for us not to lose the impression of continuity.

The truly dramatic advances in digital computer technology achieved over the last 30 years have already elevated Numerical Simulation to the status of a third scientific method, complementing the two traditional methods of Theory and Experiment. Increasingly complicated processes may be realistically simulated numerically, often more effectively and at lower cost than actual experiments, enabling us to better predict, understand and control them. Thus, numerical simulation is fast becoming an indispensable tool in technological discovery and development and a strong driving force in the quantification and mathematization of science and technology. An excellent overview of Numerical Heat Transfer may be found in the Handbook [MINKOWYCZ et al].

There are four basic steps involved in the development of a computer simulation of a physical process:

1. **Determination of the physical problem**. Decide which physical phenomena are important enough to be taken into account, which physical variables define the system, what are the inputs (data), and what is to be found.

2. **Formulation and analysis of the mathematical model**. "Translate" the *physical* problem into a precise *mathematical* problem, identify the data and the unknowns, and convince ourselves that the resulting mathematical problem is well-posed or, at least, that it "makes sense".

3. **Discretization of the problem.** Approximate the problem by a discrete one, i.e. replace all "continuous" entities by corresponding "discrete" ones, and construct a numerical algorithm for its solution.

4. **Development and implementation of algorithms in a computer code.** Develop algorithms and code embodying them. Check them out and validate the resulting programs.

Consider, for example, heat transfer in a body occupying a region Ω in space. In Step 1, we must decide if heat is transferred by conduction, convection, or radiation; if a phase-change is involved; if temperature alone suffices to describe the thermal state; if the process is transient or steady-state; what are the initial and boundary conditions, etc. In a "real-life" situation, many of these decisions may not be as simple as they sound, and various simplifying physical assumptions may be required in order to formulate a "reasonable" problem (c.f. **§1.2**). Step 2 is achieved when we determine the equations expressing the physical laws and conditions identified in Step 1. Several examples of this process were presented in CHAPTER **2**. When the resulting mathematical problem is not amenable to analytical treatment, Step 3 becomes necessary, at which time the problem is approximated by a discrete one and algorithms are devised to compute its solution. At Step 4, we write computer programs implementing the algorithms in some convenient computer language, e.g., FORTRAN, and apply them to some simple problems with known solutions (benchmark problems), in order to check that the simulation performs as expected. Clearly, this is an inter-disciplinary endeavor, requiring knowledge from several fields: the scientific discipline pertaining to the process under study, mathematics, numerical analysis, and computer science.

In this section we are concerned with Step 3, for the case of a simple heat-transfer process. Thus, we assume that a well-posed mathematical model of heat-transfer in a region has been formulated, and we discuss its discretization and the construction of effective numerical algorithms for its solution.

Discretization begins with the subdivision of the (spatial) region into "small" subregions (**control volumes**), by an imposed spatial grid. The term "small" is relative: heat transfer in the ground around a pipe may involve a region tens of feet long; then "small" may be inches or feet. On the other hand, for heat transfer in the pipe itself, "small" may be a tenth or a hundredth of an inch. Two factors help to determine the size of the control volume. On the one hand, it should be *small* enough to capture essential variations in the computed quantities and to permit us to represent average or typical values as point values. Thus, large temperature gradients require small control volumes, and conversely, small gradients can be captured even by relatively large control volumes. On the other hand, the expense of the resulting numerical computation is the primary limiting factor in how fine a mesh one may use. If unlimited time on a Cray Supercomputer is available to run the code, then the mesh may be a hundred or a thousand times finer than if the code is to be run on a personal computer !

Control volumes are thought of as regions in which "**local equilibrium**" is achieved at a time scale considerably shorter than the computational time step;

hence, the value of a field quantity at a nodal point at the center of a control volume may be thought of as representing the **average** of the quantity over the volume.

Having chosen an "appropriate" spatial grid, we simulate heat transfer by updating the state of the discrete system through discrete time increments $\Delta t > 0$, using discrete versions of the conservation laws.

There are several actors and inter-related objectives in this play. We want the numerical scheme to be

i) **consistent,** meaning that the discrete *equations* used in the scheme tend to the correct conservation laws as the spatial and temporal grid sizes tend to zero;

ii) **convergent,** meaning that the approximations that it provides to the *solutions* of the (continuum) conservation laws, actually tend to these solutions as the spatial and temporal grid sizes tend to zero;

iii) **stable,** meaning that the computed values at each time-step are relatively insensitive to unavoidable input and roundoff errors;

iv) **effective,** meaning that the scheme achieves the above objectives with minimal computational expense, so that its use in the desired context is affordable.

Certainly, whether or not these objectives can be achieved depends on the discretization method as well as on the method used to solve the discrete equations (and even on the coding itself). The Art and Science of Numerical Computation provides us with several tools and guidelines, and the great advances in computational power and methodology during the last few years already allow us to realistically simulate fairly complicated processes. A useful principle to keep in mind is that simulation is imitation and as such it should try to follow the physical laws as closely as possible.

There are several approaches to the discretization of conservation laws: *finite differences, finite elements, collocation*, and *spectral* methods. Excellent surveys are given in [ALLEN-HERRERA-PINDER] [MINKOWYCZ et al]; see also [LAPIDUS-PINDER], [DUCHATEAU-ZACHMANN], [SEWELL]. The method that is by far the simplest, easiest to understand and implement, most amenable to direct physical interpretation, and still most widely used is that of finite-differences, especially when derived via **control-volume** discretizations. This is the method that we shall use in this book.

In order to introduce and explain the basic methodology, we begin with the simplest process of heat conduction in a finite slab. As a model problem we treat the following

PHYSICAL PROBLEM: Consider a finite slab, $0 \le x \le l$, with known initial temperature distribution, $T_{init}(x)$. Starting at time $t = 0$, the slab is heated convectively at $x = 0$ (with ambient temperature $T_\infty(t)$ and heat transfer coefficient h), while the back face $x = l$ is kept insulated. We exclude the presence of any volumetric heat sources (see **§4.1.G**). We want to predict the

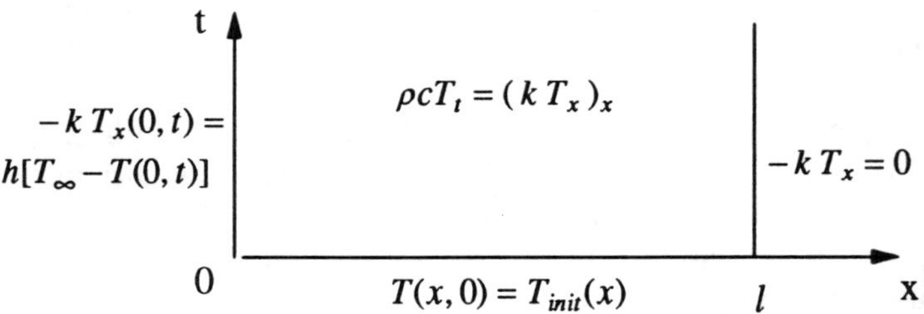

Figure 4.1.1. Model heat transfer problem.

evolution of the temperature field over time. The mathematical formulation is the following.

MATHEMATICAL PROBLEM: Find $T(x,t)$ such that (**Figure 4.1.1**)

$$\rho c T_t = (kT_x)_x, \qquad 0 < x < l, \qquad t > 0 \tag{1a}$$

$$T(x,0) = T_{init}(x), \qquad 0 \le x \le l \tag{1b}$$

$$-kT_x(0,t) = h\,[T_\infty(t) - T(0,t)], \qquad -kT_x(l,t) = 0, \qquad t > 0 \tag{1c}$$

The specific heat, c, thermal conductivity k and heat-transfer coefficient, h, may be known, temperature dependent functions.

4.1.B Control volume discretization of the conservation law

We partition the region of interest into M subregions, called **control volumes,** $V_1, V_2, \ldots, V_M$. With each subregion V_j we associate a **node** x_j, a point inside V_j. We let ΔV_j = volume of V_j, and $A_{ij} = A_{ji}$ = surface area of the face common to V_i and V_j. For the slab of length l and (constant) cross-sectional area A, we have simply

$$\Delta V_j = A \cdot \Delta x_j \quad \text{and} \quad A_{ij} \equiv A, \quad i, j = 1, \ldots, M, \tag{2a}$$

where Δx_j = length of the jth subinterval, containing node x_j. If we choose to locate nodes at the midpoints of intervals, then the endpoints of the jth subinterval are (**Figure 4.1.2**)

$$x_{j-½} = x_j - \frac{\Delta x_j}{2} \quad \text{and} \quad x_{j+½} = x_j + \frac{\Delta x_j}{2}, \quad j = 1, \ldots, M, \text{ with } x_{½} = 0, \; x_{M+½} = l. \tag{2b}$$

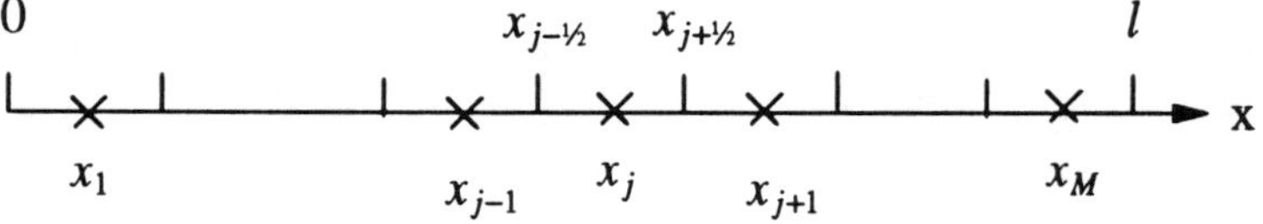

Figure 4.1.2. Nodes and faces of the spatial mesh.

In particular, if the partition is uniform, then $\Delta x_j = \Delta x = l/M$, the nodes x_j are equidistant and

$$x_{1/2} = 0, \quad x_{j-1/2} = (j-1)\Delta x, \quad j = 1, \ldots, M, \quad x_{M+1/2} = M\Delta x = l\,. \tag{2c}$$

For various other common 1- and 2-dimensional meshes see PROBLEMS 3-6.

Let $\Delta t_n > 0$ be time increments and define the discrete time-steps

$$t_0 = 0\ , \quad t_1 = \Delta t_0\ , \cdots , \quad t_{n+1} = t_n + \Delta t_n\ , \cdots , \quad n = 0, 1, 2, \ldots . \tag{2d}$$

If $\Delta t_n = \Delta t$ for all n, then: $t_n = n\Delta t,\ n = 0, 1, 2, \ldots$.

With $T(x, t)$ denoting the exact solution of (1), $T(x_j, t_n)$ represents its value at node x_j at time t_n, and its numerical approximation will be denoted by

$$T_j^n \approx T(x_j, t_n), \quad j = 1, \ldots, M, \quad n = 0, 1, \ldots\ . \tag{3a}$$

We regard T_j^n as also an *approximation* to the *mean temperature of* V_j *at time* t_n, see PROBLEM 8. In addition, we introduce approximations to the *boundary* temperatures

$$T_0^n \approx T(0, t_n) \quad \text{and} \quad T_{M+1}^n \approx T(l, t_n), \quad n = 0, 1, \ldots\ . \tag{3b}$$

From the initial condition (1b),

$$T_j^0 := T_{init}(x_j), \quad j = 1, \ldots, M, \tag{4}$$

is known; for $n = 0, 1, \cdots$, we want to define an algorithm for determining the values T_j^{n+1} at the next time-step, when we know the values T_j^n at the current time-step.

Discrete heat balance

Finite-difference discretizations of the heat equation (1) may be derived in various ways (see [LAPIDUS-PINDER], [PATANKAR], [MINKOWYCZ et al]). We prefer the one that has direct physical meaning, the *discrete heat balance*, that originally formed the basis for the conservation law (1) itself (**§1.2**). Thus, we think of (1) in its primitive form :

$$E_t + q_x = 0\ , \tag{5}$$

with

$$E = \text{thermal energy density per unit volume} = \int_{T_{ref}}^{T} \rho c(\bar{T})d\bar{T} \approx \rho c[T - T_{ref}], \tag{6}$$

T_{ref} being some convenient reference temperature, and

$$q = \text{heat flux} = -kT_x \qquad (\text{Fourier's law}). \tag{7}$$

Note that we may use either the volumetric enthalpy E (per unit volume), or the specific enthalpy e (per unit mass), $E = \rho e$. Integrating (5) over the control volume V_j (**Figure 4.1.3**), and over the time interval $[t_n, t_n + \Delta t_n]$, we find

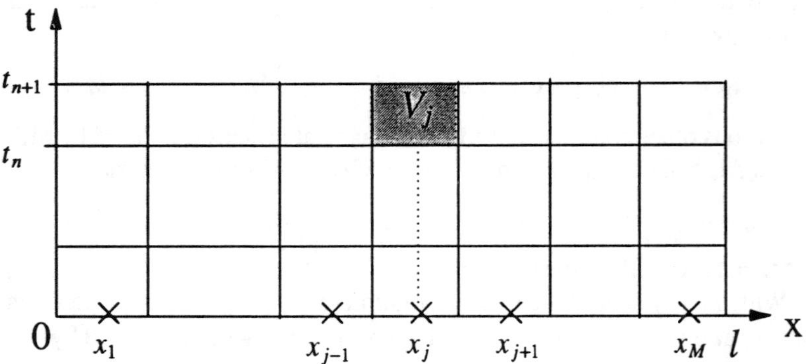

Figure 4.1.3. Space - time grid.

$$\int_{t_n}^{t_{n+1}} \frac{\partial}{\partial t}\left(A \int_{x_{j-1/2}}^{x_{j+1/2}} E(x\,,t)\,dx \right) dt \;=\; -\int_{t_n}^{t_{n+1}} A \int_{x_{j-1/2}}^{x_{j+1/2}} q_x(x\,,t)\,dx\,dt. \tag{8a}$$

Dividing out the constant cross-sectional area A and integrating the derivatives yields

$$\int_{x_{j-1/2}}^{x_{j+1/2}} E(x\,,t)\,dx \,\Big|_{t=t_n}^{t=t_{n+1}} \;=\; \int_{t_n}^{t_{n+1}} [\,q(x_{j-1/2}\,,t) - q(x_{j+1/2}\,,t)\,]\,dt\,. \tag{8b}$$

Assuming V_j is small enough for $E(x_j, t)$ to be approximately the mean energy (density) inside V_j, i.e. assuming E is approximately uniform in V_j we have

$$\int_{x_{j-1/2}}^{x_{j+1/2}} E(x\,,t)\,dx \;\approx\; E(x_j\,,t)\,\Delta x_j\,,$$

and (8) becomes

$$[\,E(x_j\,,t_{n+1}) - E(x_j\,,t_n)\,]\;\;\Delta x_j \;=\; \int_{t_n}^{t_{n+1}} [\,q(x_{j-1/2}\,,t) - q(x_{j+1/2}\,,t)\,]\,dt\,. \tag{9}$$

This simply expresses the heat balance in V_j during (t_n , t_{n+1}), namely, the gain of heat during this time is equal to the amount of heat entering the volume (from the left), minus the heat leaving it (at the right, per unit cross-sectional area).

Next, we assume that the time-increment Δt_n may be so brief that during the time (t_n , t_{n+1}) the fluxes are approximately constant and arbitrarily close to their values at any intermediate time in this interval. Let

$$t_{n+\theta} \;:=\; t_n + \theta\Delta t_n \;=\; (1-\theta)t_n + \theta t_{n+1}\,, \tag{10}$$

be some intermediate time with $0 \le \theta \le 1$. The usual choices are $\theta = 0$, ½ or 1, and these will be discussed later. We can then approximate (9) by

$$[E(x_j, t_{n+1}) - E(x_j, t_n)]\,\Delta x_j = \Delta t_n\,[q(x_{j-1/2}, t_{n+\theta}) - q(x_{j+1/2}, t_{n+\theta})], \quad j = 1, \ldots, M, \tag{11}$$

which constitutes a *complete discretization of the conservation law* (5). To obtain a numerical scheme, we introduce the discrete approximations

$$E_j^n \approx E(x_j, t_n), \qquad q_{j\pm 1/2}^{n+\theta} \approx q(x_{j\pm 1/2},\ t_n + \theta\Delta t_n), \qquad 0 \le \theta \le 1,$$

and write (11) as

$$E_j^{n+1} - E_j^n = \frac{\Delta t_n}{\Delta x_j}\,[q_{j-1/2}^{n+\theta} - q_{j+1/2}^{n+\theta}], \qquad j = 1, \ldots, M, \qquad n = 0, 1, \ldots \tag{12}$$

This is the discretization of the energy conservation law that will be extended to phase change processes as the "enthalpy method" in **§4.3**. The thermal state of the control volume V_j at the time t_n is completely determined by the enthalpy E_j^n. Relation (12) provides us with the means for updating that thermal state to the next discrete time t_{n+1}.

Let us now discuss the choice of the parameter θ. For $\theta = 0$ the fluxes are evaluated at the old time, t_n, and (12) constitutes an **explicit** determination of the enthalpy approximation E^{n+1} of E at the advanced time step t_{n+1} in terms of the state of the material at t_n :

explicit scheme:
$$E_j^{n+1} = E_j^n + \frac{\Delta t_n}{\Delta x_j}\,[q_{j-1/2}^n - q_{j+1/2}^n], \quad j = 1, \ldots, M, \quad n = 0, 1, \ldots. \tag{13}$$

For $\theta = 1$ we have the

fully implicit scheme:
$$E_j^{n+1} = E_j^n + \frac{\Delta t_n}{\Delta x_j}\,[q_{j-1/2}^{n+1} - q_{j+1/2}^{n+1}], \quad j = 1, \ldots, M, \quad n = 0, 1, \ldots. \tag{14}$$

For $0 < \theta < 1$, the scheme is also implicit, using intermediate values of the flux which we define as

$$q^{n+\theta} := (1 - \theta)\,q^n + \theta\,q^{n+1}.$$

The most common choice is $\theta = 1/2$, in which case the resulting numerical method is known as the **Crank – Nicolson scheme**, about which more will be said later.

Discrete fluxes

These schemes require approximations of the fluxes across the faces located at $x_{j-1/2}$ and $x_{j+1/2}$. Let us consider *interior* control-volume faces first; the boundary cases will be discussed in **§4.1.C**. The conductive flux is given by

$$q = -kT_x \approx -k\,\frac{\Delta T}{\Delta x}. \tag{15}$$

Since the temperature is represented discretely by nodal values T_j, we may use first-order finite differences to approximate q discretely. Thus,

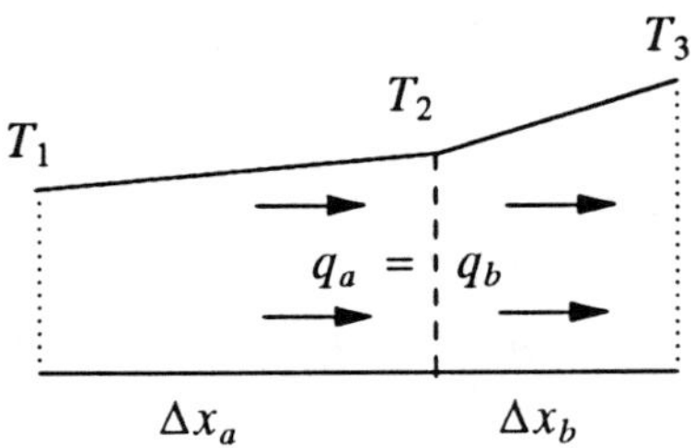

Figure 4.1.4. Steady-state profile.

$$q_{j-1/2} = -k_{j-1/2}\frac{T_j - T_{j-1}}{x_j - x_{j-1}}, \qquad j = 2, \ldots, M, \tag{16}$$

is the amount of heat flowing from V_{j-1} into V_j across a unit cross-sectional area per unit time. But what does $k_{j-1/2}$ represent? Generally, the conductivity is *not* constant but a function of *location* (when V_{j-1}, V_j consist of different materials, e.g. a wall and a phase change material or liquid and solid phases of the same material), and of *temperature*, $k = k(x, T)$. So, in general, the flux must represent heat flow through media of different conductivities, k_{j-1} for V_{j-1} and k_j for V_j, and we need to assign, in a consistent manner, an **effective conductivity** $k_{j-1/2}$. A reasonable definition of effective conductivity for a *layered* structure is obtained as follows.

Consider steady-state heat conduction ($T_{xx} = 0$) through two adjacent layers of thicknesses $\Delta x_a, \Delta x_b$ and conductivities k_a, k_b. Then the temperature profiles are straight lines (**Figure 4.1.4**) and at the common wall the flux from the left must equal the flux from the right:

$$-k_a\frac{T_2 - T_1}{\Delta x_a} = q = -k_b\frac{T_3 - T_2}{\Delta x_b}.$$

Solving the first equality for $T_2 - T_1$, the second for $T_3 - T_2$ and adding we obtain

$$T_3 - T_1 = -q\left(\frac{\Delta x_a}{k_a} + \frac{\Delta x_b}{k_b}\right).$$

Hence, the flux across the common wall is

$$q = -\frac{T_3 - T_1}{\dfrac{\Delta x_a}{k_a} + \dfrac{\Delta x_b}{k_b}}.$$

We refer to the ratio of length to conductivity as the **thermal resistance**. Hence, the relationship between flux q and resistance R is $q = -\dfrac{\Delta T}{R}$ where the temperature drop ΔT is often referred to as the **thermal driving force.**

It should be noted that the common definition of thermal resistance is

$$\frac{\textit{length of resistance path}}{(\textit{crosssectional area})\,(\textit{conductivity})},$$

making the formula

$$\text{heat flow rate} \;=\; qA \;=\; -k\,\frac{\Delta T}{\Delta x}\,A \;=\; -\frac{\Delta T}{\Delta x\,/\,Ak} \;=\; -\frac{\Delta T}{R}$$

correct. However, when the cross sectional area A is constant and $\Delta V = A\Delta x$, the A divides out in the discretization of the conservation law:

$$\Delta E \;=\; \frac{\Delta t}{\Delta V}\,[\![qA]\!]_+^- \;=\; \frac{\Delta t}{A\Delta x}\,[\![q]\!]_+^- A \;=\; \frac{\Delta t}{\Delta x}\,[\![q]\!]_+^- ,$$

so we only need an expression for the flux and not the flow rate. Hence, for 1-dimensional Cartesian geometry, it is more convenient to take as resistance the quantity $\dfrac{\Delta x}{k}$ instead of the standard $\dfrac{\Delta x}{Ak}$ (See also **§4.1.F**).

From the above analysis we see that the effective overall resistance of the composite layer is $R = R_a + R_b$. Hence it is **not** the conductivities that add up but the resistances of the two layers, in this *serial* arrangement. We conclude that the total flux through a composite layer equals the overall temperature drop divided by the sum of the resistances of the layers.

With this in mind, we set (**Figure 4.1.5**)

$$R_{j-\frac{1}{2}} \;=\; \frac{\frac{1}{2}\Delta x_{j-1}}{k_{j-1}} + \frac{\frac{1}{2}\Delta x_j}{k_j} \;=\; \text{resistance of the path } [x_{j-1}, x_j] \tag{17}$$

and express the interior fluxes, (16), as

$$q_{j-\frac{1}{2}} \;=\; -\frac{T_j - T_{j-1}}{R_{j-\frac{1}{2}}}\,, \qquad j = 2, \ldots, M \;. \tag{18}$$

In particular, if $\Delta x_{j\pm1} = \Delta x_j \;=\; \Delta x$ and $k_{j\pm1} = k_j = k$ then their common resistance is

$$R \;=\; \frac{\Delta x}{2}\left(\frac{1}{k} + \frac{1}{k}\right) \;=\; \frac{\Delta x}{k}\,,$$

as expected.

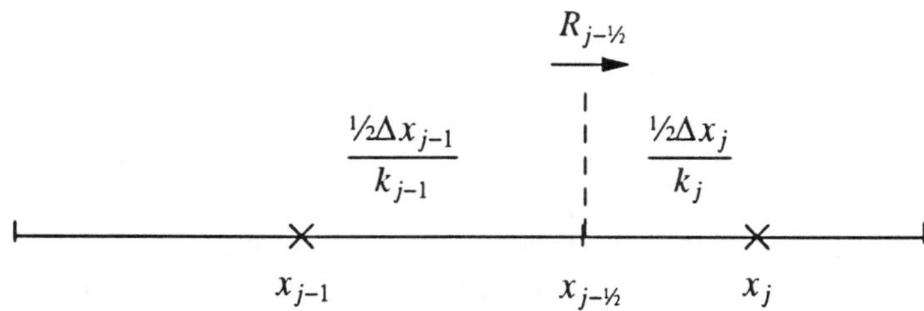

Figure 4.1.5. Resistances of adjacent control volumes.

Discrete heat equation

Substituting the flux expression (18) into the discrete heat balance (12) we obtain

$$E_j^{n+1} = E_j^n + \frac{\Delta t_n}{\Delta x_j} \left[\frac{T_{j+1}^{n+\theta} - T_j^{n+\theta}}{R_{j+½}} - \frac{T_j^{n+\theta} - T_{j-1}^{n+\theta}}{R_{j-½}} \right]$$

$$= E_j^n + \frac{\Delta t_n}{\Delta x_j} \left[\frac{1}{R_{j-½}} T_{j-1}^{n+\theta} - \left(\frac{1}{R_{j-½}} + \frac{1}{R_{j+½}}\right) T_j^{n+\theta} + \frac{1}{R_{j+½}} T_{j+1}^{n+\theta} \right] . \quad (19)$$

In particular, if $\Delta x_j = \Delta x$, and $k_j = k$, then $R_{j\pm½} = \frac{\Delta x}{k}$ and (19) becomes

$$E_j^{n+1} = E_j^n + \frac{k\Delta t_n}{\Delta x^2} [T_{j-1}^{n+\theta} - 2T_j^{n+\theta} + T_{j+1}^{n+\theta}] ; \quad (20)$$

the bracketed expression is, of course, the standard centered finite-difference discretization of T_{xx}.

For *plain heat conduction*, the energy is simply the sensible heat, (6), which, when the specific heat is independent of temperature, becomes

$$E_j^n \approx E(x_j, t_n) = \rho c_j [T_j^n - T_{ref}] .$$

This can be used to eliminate E_j^n, and thus (12) takes the form

$$T_j^{n+1} = T_j^n + \frac{\Delta t_n}{\rho c_j \Delta x_j} [q_{j-½}^{n+\theta} - q_{j+½}^{n+\theta}] , \qquad j = 1, \ldots, M, \qquad n = 0, 1, 2, \ldots \quad (21)$$

with $0 \leq \theta \leq 1$ to be chosen. This is often a convenient discretization, the fluxes being given by (18) for interior faces, and as described in **§4.1.C** for the boundary faces. Alternatively, the fluxes may be eliminated completely, using (18), to obtain

$$T_j^{n+1} = T_j^n + \frac{\Delta t_n}{\rho c_j \Delta x_j} \left[\frac{1}{R_{j-½}} T_{j-1}^{n+\theta} - \left(\frac{1}{R_{j-½}} + \frac{1}{R_{j+½}}\right) T_j^{n+\theta} + \frac{1}{R_{j+½}} T_{j+1}^{n+\theta} \right] , \quad j = 1, \ldots, M; \quad n = 0, 1, 2, \ldots . \quad (22)$$

which is a complete discretization of the heat conduction equation (1) in terms of temperatures only, with $T_0^{n+\theta}$ and $T_{M+1}^{n+\theta}$ determined by the boundary conditions.

In particular, for a uniform grid, $\Delta x_j = \Delta x$, uniform time steps, $\Delta t_n = \Delta t$ and constant thermophysical properties ($\alpha = k/\rho c$) we have

$$T_j^{n+1} = T_j^n + \frac{\alpha \Delta t}{\Delta x^2} [T_{j-1}^{n+\theta} - 2T_j^{n+\theta} + T_{j+1}^{n+\theta}] , \quad j = 1, \ldots, M; \quad n = 0, 1, 2, \ldots . \quad (23)$$

For $\theta = 0$, this is the usual explicit discretization of the heat equation, $T_t = \alpha T_{xx}$, obtained by forward Euler discretization of T_t and centered differencing of T_{xx}. For any $0 < \theta \leq 1$, the discretization is implicit. Their pros and cons are discussed in **§4.1.E** and **§4.1.F**.

4.1.C Discretization of boundary conditions

For equations (12) (or (19) or (22)) to constitute a closed system allowing the state of the system to be advanced from t_n to t_{n+1}, values are needed for the boundary fluxes $q_{½}$ and $q_{M+½}$, representing the fluxes through the walls $x = 0$ and $x = l$ (or, values for T_0 and T_{M+1}). We discuss the treatment of boundary conditions at $x = 0$, the treatment at $x = l$ being completely analogous. The concept of thermal resistance makes the treatment of boundary conditions simple.

Case I. Imposed temperature: $T(0, t) = T_0(t)$

Here the wall temperature is specified, so the value of T_0^n is known at each time t_n,

$$T_0^n = T_0(t_n), \qquad n = 0, 1, 2, \ldots \tag{24}$$

Then from (18), the boundary flux is

$$q_{½}^n = -\frac{T_1^n - T_0^n}{R_{½}}, \quad \text{with} \quad R_{½} = \frac{½\Delta x_1}{k_1}. \tag{25}$$

Case II. Imposed Flux: $-kT_x(0, t) = q_0(t)$

Here the boundary flux is specified, so

$$q_{½}^n = q_0(t_n), \qquad n = 0, 1, 2, \ldots. \tag{26}$$

Then, the surface temperature T_0^n is obtained from $-\dfrac{T_1^n - T_0^n}{R_{½}} = q_0(t_n)$, whence

$$T_0^n = T_1^n + R_{½} q_0(t_n), \qquad \text{with} \quad R_{½} = \frac{½\Delta x_1}{k_1}. \tag{27}$$

Case III. Convective Flux: $-kT_x(0, t) = h[T_\infty(t) - T(0, t)]$

Setting $T_\infty^n := T_\infty(t_n)$, and employing the standard discretization

$$q_{½}^n = -\frac{T_1^n - T_0^n}{R_{½}}, \qquad R_{½} = \frac{½\Delta x_1}{k_1}, \tag{28}$$

of the conductive flux $-kT_x(0, t)$, we see that the boundary condition requires $-\dfrac{T_1^n - T_0^n}{R_{½}} = h[T_\infty^n - T_0^n]$, from which T_0^n is expressed as a weighted average of T_∞^n and T_1^n :

$$T_0^n = \frac{T_1^n + hR_{½}T_\infty^n}{1 + hR_{½}}. \tag{29}$$

Substituting this value of T_0^n into (28), we find

$$q^n_{1/2} = -\frac{T^n_1 - T^n_\infty}{1/h + R_{1/2}}. \tag{30}$$

Comparison with (28) reveals that the ambient temperature, T^n_∞, can play the role of the face temperature T^n_0 provided the conductive resistance, $R_{1/2}$, is replaced by the total effective resistance $\frac{1}{h} + R_{1/2}$, (the sum of the convective and conductive resistances).

As usual, the *imposed* temperature case, (25), corresponds to $h \to \infty$ in (29), (30).

4.1.D The discrete problem

Having derived discretizations of both the partial differential equation and the boundary conditions, we can now present algorithms for finding the unknown nodal temperatures $T^{n+1}_1, T^{n+1}_2, \cdots, T^{n+1}_M$, at time t_{n+1}, from their values at the old time t_n for our model heat conduction problem (1). Indeed, combining (4), (30), (26) but applied to $x = l$, and (21), (18), (17), the updating equations for the T^{n+1}_j's are as follows:

initial values:
$$T^0_j = T_{init}(x_j), \quad j = 1, \ldots, M, \tag{31a}$$

boundary condition at $x = 0$:
$$q^{n+\theta}_{1/2} = -\frac{T^{n+\theta}_1 - T^{n+\theta}_\infty}{\frac{1}{h} + R_{1/2}}, \qquad R_{1/2} = \frac{\frac{1}{2}\Delta x}{k_1}, \tag{31b}$$

boundary condition at $x = l$:
$$q^{n+\theta}_{M+1/2} = 0, \tag{31c}$$

interior values:
$$T^{n+1}_j = T^n_j + \frac{\Delta t_n}{\rho c_j \Delta x_j}\left[q^{n+\theta}_{j-1/2} - q^{n+\theta}_{j+1/2} \right], \quad j = 1, \ldots, M, \tag{31d}$$

where

$$q^{n+\theta}_{j-1/2} = -\frac{T^{n+\theta}_j - T^{n+\theta}_{j-1}}{R_{j-1/2}} \quad \text{with } R_{j-1/2} = \frac{\frac{1}{2}\Delta x_{j-1}}{k_{j-1}} + \frac{\frac{1}{2}\Delta x_j}{k_j}, \quad j = 2, \ldots, M, \tag{31e}$$

and

$$T^{n+\theta} = (1-\theta)T^n + \theta T^{n+1}, \qquad 0 \le \theta \le 1. \tag{31f}$$

The solvability of this system for the choices $\theta = 0$, $0 < \theta \le 1$, is discussed in the following subsections.

Note that neither the spatial steps Δx_j nor the time steps Δt_n need be uniform. A finer mesh may be needed near boundaries, or wherever steep gradients are expected, to resolve rapid variations, etc. However, unless there is specific reason to use non-uniform spatial grids, uniform ones are preferred because they are simpler and they yield better accuracy. Moreover, if the heat transfer coefficient is not constant but a function of t, $T_\infty(t)$ and $T(0,t)$ as in the radiation boundary condition (§1.2), then in (31b) h is actually

$$h^{n+\theta} = h(t_{n+\theta}\,,\, T_\infty(t_{n+\theta})\,,\, T_0^{n+\theta})\,, \tag{32a}$$

where, by (29),

$$T_0^{n+\theta} = \frac{T_1^{n+\theta} + \dfrac{h^{n+\theta}\Delta x_1}{2k_1} T_\infty^{n+\theta}}{1 + \dfrac{h^{n+\theta}\Delta x_1}{2k_1}}\,, \tag{32b}$$

making the system highly nonlinear if $\theta > 0$. Similarly, if the specific heat and/or conductivity are functions of location and temperature, then c_j, and k_j actually change with time because of the temperature change, and must be evaluated at $t = t_{n+\theta}$. The resulting nonlinear system must be solved by some iterative method (see §4.1.F).

Discretization replaces a Partial Differential Equation, $\mathbf{PDE}[u] = 0$, by a Finite-Difference Equation, $\mathbf{FDE}[U_j^n] = 0$. The amount by which the exact solution u of the **PDE** fails to satisfy the **FDE** is called the

local truncation error: $\qquad \mathbf{te}_j^n := \mathbf{FDE}[u(x_j, t_m)]$.

Since $\mathbf{PDE}[u(x_j, t_n)] = 0$, the truncation error may be viewed as the difference between **FDE** and **PDE** applied to $u(x_j, t_n)$. The discretization is **consistent** if $\mathbf{te}_j^n \to 0$ as $\Delta x, \Delta t \to 0$, which signifies that the **FDE** is indeed an approximation to the given **PDE** (instead of to some other PDE); see PROBLEM 14. On the other hand, the distance between the continuous and discrete solutions is measured by the

local discretization error: $\qquad \mathbf{de}_j^n := U_j^n - u(x_j, t_n)$.

The method is **convergent** if $\mathbf{de}_j^n \to 0$ as $\Delta x, \Delta t \to 0$, which signifies that the discrete solution does indeed approximate the exact solution, see PROBLEM 15. Note that U_j^n denotes the *exact* solution of the **FDE**. The actual *computed* solution $\vec{U}_j^n$, however, may be contaminated by **roundoff errors** $\mathbf{re}_j^n = \vec{U}_j^n - U_j^n$. These may be introduced at any point in the computation (because of unavoidable rounding of data or computed values). Such errors then propagate to subsequent time-steps and to neighboring points. Even though a single rounding error is typically negligibly small, the concern is that it may grow so fast as it propagates that substantial accuracy in the computed solution is lost (see §4.1.E and PROBLEM 16). The best we can hope for is that the numerical scheme does not amplify errors so that they grow faster than the exact solution of the **FDE**. In particular, if the exact solution does not grow, then errors should not be amplified. In this case the numerical method is called **stable**. Clearly, the actual (local) error in the numerical solution is the sum $\mathbf{de}_j^n + \mathbf{re}_j^n = \vec{U}_j^n - u(x_j, t_n)$. In a convergent method, we can reduce $\mathbf{de}_j^n$ by taking smaller $\Delta x, \Delta t$ but then $\mathbf{re}_j^n$ increases (see PROBLEM 21); hence, in practice, there is always an error in the computed results.

For any $0 \le \theta \le 1$, (31) is a consistent scheme with $\mathbf{te}_j^n = O(\Delta t + \Delta x^2)$, see PROBLEM 14. Stability is discussed in §**4.1.E, F** and convergence follows from the ([ISAACSON-KELLER], [LAPIDUS-PINDER])

Lax Equivalence Theorem: A *consistent* finite-difference method for a *well-posed* (linear) problem is *convergent* if and only if it is *stable*.

Also see PROBLEM 15.

4.1.E Explicit time updating

Choosing $\theta = 0$ in (31), the fluxes are evaluated at the old time t_n and therefore they are completely known. This amounts to assuming that the values of the fluxes do **not** change appreciably during the time interval $[t_n, t_{n+1}]$, so that the process at time t_{n+1} is still driven by the fluxes at time t_n. The time discretization is then the standard forward Euler discretization, and the new values, T_j^{n+1}, are obtained directly, simply by evaluating the right-hand sides.

Written in terms of temperatures only, the explicit scheme consists of (PROBLEM 13)

$$T_j^0 = T_{init}(x_j), \qquad j = 1, \ldots, M, \tag{33a}$$

$$T_0^n = \frac{T_1^n + h\, R_{½} T_\infty^n}{1 + h\, R_{½}}, \qquad \text{where} \qquad R_{½} = \frac{\Delta x_1}{2k_1}, \tag{33b}$$

$$T_{M+1}^n = T_M^n - 0 \cdot R_{M+½}, \qquad \text{where} \qquad R_{M+½} = \frac{\Delta x_M}{2k_M} \tag{33c}$$

(since $q_{M+½} = 0$ in our example problem), and

$$T_j^{n+1} = T_j^n + \frac{\Delta t_n}{\rho c_j \Delta x_j}\left[\frac{1}{R_{j-½}} T_{j-1}^n - \left(\frac{1}{R_{j-½}} + \frac{1}{R_{j+½}}\right) T_j^n + \frac{1}{R_{j+½}} T_{j+1}^n\right], \quad j = 1, \ldots, M \tag{33d}$$

with

$$R_{j+½} = \frac{\Delta x_j}{2k_j} + \frac{\Delta x_{j+1}}{2k_{j+1}}, \qquad j = 1, 2, 3, \ldots, M-1, \text{ and}$$

$$R_{j-½} = \frac{\Delta x_{j-1}}{2k_{j-1}} + \frac{\Delta x_j}{2k_j}, \qquad j = 2, 3, \ldots, M. \tag{33e}$$

The **local truncation error** is of order Δt in time and Δx^2 in space ([SMITH], [LAPIDUS-PINDER], [SEWELL], PROBLEM 14). Clearly, if the thermal conductivity and specific heat are constants, $k_j \equiv k$, $c_j \equiv c$, and the mesh is uniform, $\Delta x_j \equiv \Delta x$, then $R_{j+½} = R_{j-½} \equiv \Delta x / k$, so setting

$$\mu = \frac{\alpha \Delta t}{\Delta x^2}, \qquad \alpha = \frac{k}{\rho c}, \tag{34}$$

we see that (33d) simplifies to (c.f. (23))

$$T_j^{n+1} = T_j^n + \mu [T_{j-1}^n - 2T_j^n + T_{j+1}^n], \quad j = 2, \ldots, M, \tag{35}$$

for the internal nodes ($j = M$ is also included here since $T_{M+1}^n = T_M^n$ by (33c)). Boundary nodes are discussed later.

The extreme simplicity and convenience of the explicit scheme however is partially offset by the necessity of restricting the time step size to ensure the *numerical stability* of the scheme. This is easiest to explain in the simplest case of (35), which may be re-written as

$$T_j^{n+1} = (1 - 2\mu) T_j^n + \mu (T_{j-1}^n + T_{j+1}^n), \quad j = 2, \ldots, M. \tag{36}$$

The condition for stability is that $1 - 2\mu \geq 0$, known as the

$$\textbf{Courant – Friedrichs – Lewy } (CFL) \textbf{ Condition:} \quad \Delta t \leq \frac{1}{2} \frac{\Delta x^2}{\alpha}, \tag{37}$$

after its discoverers [COURANT-FRIEDRICHS-LEWY]. It guarantees that the exact solution of the numerical scheme will obey a *Maximum Principle*, as does the solution of the Heat Equation itself, namely that

$$\min \{T_{j-1}^n, T_j^n, T_{j+1}^n\} \leq T_j^{n+1} \leq \max \{T_{j-1}^n, T_j^n, T_{j+1}^n\}.$$

Indeed, if condition (37) is violated, then (36) can produce physically unrealistic values, for example, a negative T_j^{n+1} from positive $T_{j-1}^n, T_j^n, T_{j+1}^n$. The numerical consequence is that errors would grow exponentially with n. Indeed, if errors are introduced at any time, from whatever source (say, roundoff), then (36) will compute contaminated values, $\bar{T}_j^n = T_j^n + e_j^n$, instead of the desired values T_j^n in later steps; since both the T_j^n's and the $\bar{T}_j^n$'s satisfy (36), the errors e_j^n also do:

$$e_j^{n+1} = (1 - 2\mu) e_j^n + \mu [e_{j-1}^n + e_{j+1}^n], \quad j = 2, \ldots, M; \tag{38}$$

as a simple illustration, assuming $e_j^0 = (-1)^j e$, we find

$$e_j^1 = (1 - 4\mu) e_j^0, \quad e_j^2 = (1 - 4\mu)^2 e_j^0, \ldots, \quad e_j^n = (1 - 4\mu)^n e_j^0,$$

whence the error amplification factor is $1 - 4\mu$ at each step; it follows that, unless $|1 - 4\mu| \leq 1$, i.e. $0 \leq \mu \leq \frac{1}{2}$, the errors will be amplified exponentially fast and will destroy the computation after a few steps ! On the other hand, the requirement that Δt be greater than the relaxation time $\tau = \Delta x^2 / \pi^2 \alpha$ (PROBLEM 11) provides a lower bound $\mu > 1/\pi^2$.

More generally, in the **von Neumann stability analysis** approach, errors are represented by Fourier expansions and amplification factors of typical Fourier terms are determined, which leads to (37) as the condition for no-growth in the propagated errors, see [ALLEN-HERRERA-PINDER, p. 86, p. 206], [LAPIDUS-PINDER, p. 170]. Another approach to numerical stability is the "matrix method" [LAPIDUS-PINDER, p. 179].

A simple and effective way to guarantee stability is the "*positive-coefficient rule*": when T_j^{n+1} is written as a linear combination of its neighbors $T_{j-1}^n, T_j^n, T_{j+1}^n$ (see (33d)), the coefficients must all be positive; this has been shown to be sufficient for stability by [FORSYTHE-WASOW], (see [PATANKAR] for a discussion). Thus, in the more general case of (33d), *stability at internal nodes* is guaranteed by

$$\Delta t_n \leq \min_{2 \leq j \leq M-1} \frac{\rho c_j \Delta x_j}{\dfrac{1}{R_{j-1/2}} + \dfrac{1}{R_{j+1/2}}} \quad \text{or simply} \quad \Delta t_n \leq \frac{1}{2} \frac{\Delta x_{min}^2}{\alpha_{max}}, \tag{39}$$

where $\Delta x_{min} = \min \Delta x_j$ and $\alpha_{max} = \max \alpha_j$.

We now consider boundary nodes for each type of boundary conditions. It is easy to see (PROBLEM 17) that if $T_0^n = T_0(t_n)$ is imposed at $x = 0$, then the coefficient of T_1^n in (33d) for $j = 1$ is $1 - \dfrac{\Delta t_n}{\rho c_1 \Delta x_1}(\dfrac{1}{R_{1/2}} + \dfrac{1}{R_{1+1/2}})$, whose positivity is guaranteed by choosing

$$\Delta t_n \leq \frac{1}{3} \frac{\Delta x^2}{\alpha}. \tag{40a}$$

Note that this restricts the time-step even more than (39). On the contrary, if the flux $q_{1/2}^n = q_0(t_n)$ is prescribed at $x = 0$, then the coefficient of T_1^n is $1 - \dfrac{\Delta t_n}{\rho c_1 \Delta x_1 R_{1+1/2}}$, whence it suffices to choose

$$\Delta t_n \leq \frac{\Delta x^2}{\alpha} \tag{40b}$$

which will automatically hold under (39). Finally, the convective boundary condition case leads to the restriction

$$\Delta t_n \leq \frac{1 + h\dfrac{\Delta x}{2k}}{1 + 3h\dfrac{\Delta x}{2k}} \cdot \frac{\Delta x^2}{\alpha}, \tag{40c}$$

for which (40a) is sufficient.

We see that it suffices to restrict the time-step according to (PROBLEM 18)

$$\Delta t_n < \frac{1}{3} \frac{\Delta x_{min}^2}{\alpha_{max}} \quad \text{for } \textit{imposed temperature} \text{ or } \textit{convective boundary conditions}, \tag{41}$$

or

$$\Delta t_n < \frac{1}{2} \frac{\Delta x_{min}^2}{\alpha_{max}} \quad \text{for } \textit{imposed flux boundary conditions}. \tag{42}$$

We have replaced $\leq$ with $<$ as a precaution, against roundoff.

Such restrictions on the time-step size may be rather severe, making computations with an explicit scheme expensive. When the material properties vary with temperature, Δt_n needs to be re-adjusted (re-computed) before a new time-step is taken. In such a case it is good programming practice *not* to let Δt become smaller than a pre-set minimum; for if it does, no practical time-advancing will be observed and computation will be wasted; instead, halt the computation and carefully examine what has caused the time step size to become so small.

Note also that (41) may be significantly more restrictive than (42) which is all that is required at internal nodes. Fortunately, there is a simple way of avoiding (41) entirely: for the boundary node(s) only, use the fully implicit discretization. For example, if $T_0^n = T_0(t_n)$ is imposed at $x = 0$, (also, see PROBLEM 20), then the implicit discretization at the boundary node ($j = 1, \theta = 1$ in (31d), assuming uniform Δx and constant k for simplicity) is

$$(1 + 3\mu)T_1^{n+1} = T_1^n + \mu[2T_0^{n+1} + T_2^{n+1}] ; \tag{43}$$

with Δt_n as in (42), we update the internal nodes $T_2^{n+1}, T_3^{n+1}, \ldots,$ explicitly, $T_0^{n+1} = T_0(t_{n+1})$ is given, and therefore we can find T_1^{n+1} from (43).

4.1.F Implicit time updating

Choosing the parameter θ to be greater than zero in (31) results in a system of simultaneous equations for the unknowns $T_1^{n+1}, T_2^{n+1}, \cdots, T_M^{n+1}$, which must be solved, usually by some iterative method. The advantage of implicit schemes over explicit ones is their possible unconditional stability dependent on the choice of θ. The price to be paid is having to solve a system of equations, instead of just evaluations; we shall discuss some commonly used methods below.

The common choices for the value of θ are ½ and 1. Choosing $\theta = 1$, the fluxes are computed at the latest time, t_{n+1}, and the scheme is referred to as **fully implicit**. It results from the backward Euler time discretization and its local error is again of order Δt in time and Δx^2 in space. In this regard, the **Crank-Nicolson** scheme, resulting from taking $\theta = ½$ in (31) is preferable. The fluxes, $q^{n+½}$ at the mid-point of the time-interval $[t_n, t_{n+1}]$ are taken as the averages of the values at t_n and t_{n+1}. This amounts to employing a centered-difference formula for the time derivative, resulting in a local error of order Δt^2 in time and Δx^2 in space.

Written entirely in terms of temperatures, the implicit scheme for any $0 < \theta \le 1$ takes the form

$$-\frac{\theta \Delta t_n}{\rho c_j \Delta x_j}\frac{T_{j-1}^{n+1}}{R_{j-½}} + \left[1 + \frac{\theta \Delta t_n}{\rho c_j \Delta x_j}\left(\frac{1}{R_{j-½}} + \frac{1}{R_{j+½}}\right)\right]T_j^{n+1} - \frac{\theta \Delta t_n}{\rho c_j \Delta x_j}\frac{T_{j+1}^{n+1}}{R_{j+½}} \tag{44a}$$

$$= \frac{(1-\theta)\Delta t_n}{\rho c_j \Delta x_j}\frac{T_{j-1}^{n}}{R_{j-½}} + \left[1 - \frac{(1-\theta)\Delta t_n}{\rho c_j \Delta x_j}\left(\frac{1}{R_{j-½}} + \frac{1}{R_{j+½}}\right)\right]T_j^{n} + \frac{(1-\theta)\Delta t_n}{\rho c_j \Delta x_j}\frac{T_{j+1}^{n}}{R_{j+½}},$$

$$j = 1, \ldots, M,$$

while the boundary conditions (31b, c) contribute the equations

$$T_0^{n+1} = \frac{T_1^{n+1} + hR_{½}T_\infty^{n+1}}{1 + hR_{½}}, \qquad T_{M+1}^{n+1} = T_M^{n+1} - 0 \cdot R_{M+½} . \tag{44b}$$

These constitute a linear system of $M+2$ equations for the $M+2$ unknowns $T_0^{n+1}, T_1^{n+1}, \ldots, T_M^{n+1}, T_{M+1}^{n+1}$.

Let us examine this system in the simplest case of uniform $\Delta t_n = \Delta t$, $\Delta x_j = \Delta x$, and constant $c_j = c$ and $k_j = k$. Then, setting

$$\mu = \frac{\Delta t}{\rho c \Delta x} \frac{1}{R_{j\pm½}} = \frac{k\Delta t}{\rho c \Delta x^2} = \frac{\alpha\, \Delta t}{\Delta x^2}, \tag{45}$$

the system consists of the $M+2$ equations

$$(1 + \frac{h\Delta x}{2k})T_0^{n+1} - T_1^{n+1} = \frac{h\Delta x}{2k} T_\infty^{n+1}, \tag{46a}$$

$$-\theta\mu T_{j-1}^{n+1} + (1+2\theta\mu)T_j^{n+1} - \theta\mu T_{j+1}^{n+1} = \tag{46b}$$
$$(1-\theta)\mu T_{j-1}^n + [1 - 2(1-\theta)\mu]T_j^n + (1-\theta)\mu T_{j+1}^n$$

$$-T_M^{n+1} + T_{M+1}^{n+1} = 0 \cdot \frac{\Delta x}{2k} \tag{46c}$$

The last equation simply says $T_{M+1}^{n+1} = T_M^{n+1}$, so we omit it, and write the remaining $M+1$ equations for the unknowns $T_0^{n+1}, T_1^{n+1}, \cdots, T_M^{n+1}$ in matrix form:

$$\begin{bmatrix} (1+\frac{h\Delta x}{2k}) & -1 & 0 & \cdots & \cdots & 0 \\ -\theta\mu & (1+2\theta\mu) & -\theta\mu & \cdots & \cdots & 0 \\ 0 & -\theta\mu & (1+2\theta\mu) & \cdots & \cdots & 0 \\ \cdots & 0 & \cdots & \cdots & \cdots & 0 \\ 0 & \cdots & \cdots & -\theta\mu & (1+2\theta\mu) & -\theta\mu \\ 0 & \cdots & \cdots & 0 & -\theta\mu & (1+2\theta\mu) \end{bmatrix} \begin{bmatrix} T_0^{n+1} \\ T_1^{n+1} \\ \cdots \\ \cdots \\ T_{M-1}^{n+1} \\ T_M^{n+1} \end{bmatrix}$$

$$= \begin{bmatrix} \frac{h\Delta x}{2k} T_\infty^{n+1} \\ (1-\theta)\mu T_0^n + [1-2(1-\theta)\mu]T_1^n + (1-\theta)\mu T_2^n \\ \cdots \\ \cdots \\ (1-\theta)\mu T_{M-2}^n + [1-2(1-\theta)\mu]T_{M-1}^n + (1-\theta)\mu T_M^n \\ (1-\theta)\mu T_{M-1}^n + [1-(1-\theta)\mu]T_M^n \end{bmatrix} \tag{47}$$

The coefficient matrix has several important properties. It is **tridiagonal** and **strictly diagonally dominant**, meaning that the magnitude of each diagonal entry is greater than the sum of the absolute values of the off-diagonal entries, $|1+2\theta\mu| > |-\theta\mu| + |-\theta\mu|$. Moreover, multiplying the first equation by $\theta\mu$ makes the coefficient matrix symmetric. It is known, (see, for example, [SEWELL]) that such a matrix is positive-definite and the elements of its inverse are all positive. It follows that the linear system always has a unique solution, which may be obtained by the very efficient **tridiagonal algorithm** (a variant of Gaussian elimination, see [PRESS et al], [MINKOWYCZ et al], [SEWELL]).

In more general cases, the tridiagonal system (47) may be solved by the Gauss-Seidel iterative method or, more efficiently, by the SOR iterative method ([YOUNG-GREGORY], [LAPIDUS-PINDER], [SEWELL]). We shall discuss these later for phase change problems (see **§4.3**)

The advantage of implicit schemes lies in their improved stability properties. Indeed, the scheme (46) with $\frac{1}{2} \le \theta \le 1$ is known to be *unconditionally stable* [ISAACSON-KELLER], thus imposing no restriction on the time-step. Yet in practice, the Crank-Nicolson scheme ($\theta = \frac{1}{2}$) may exhibit oscillations for large time-steps (see [PATANKAR] for a discussion). In fact, the "positive coefficient rule" mentioned earlier, when applied to (46), requires $1 - 2(1-\theta)\mu \ge 0$, i.e.,

$$\mu \le \frac{1}{2(1-\theta)}\,, \tag{48}$$

which imposes a restriction on the time step for any $0 \le \theta < 1$. Only the fully implicit scheme ($\theta = 1$) is truly unconditionally stable in this stronger sense!

4.1.G Heat conduction in 2 or 3 dimensions

All of the previous developments generalize naturally to 2 or 3 space dimensions. We shall outline the treatment for a 3-dimensional analogue of the model heat conduction problem (1).

For simplicity, we consider a box, $\Omega : 0 \le x \le l_1\,,\ 0 \le y \le l_2\,,\ 0 \le z \le l_3$, initially at temperature $T_{init}(\vec{x})$. The face $x = 0$ is heated convectively, from a source at ambient temperature $T_\infty(t)$, the face $y = 0$ is kept at a fixed temperature T_{fixed} (for variety!) and the other faces are insulated (**Figure 4.1.6**). The parameters ρ , c , k will be assumed to be constant.

MATHEMATICAL PROBLEM : Find $T(\vec{x}, t) = T(x, y, z, t)$ such that

$$\rho c T_t = \nabla \cdot (k \nabla T) \qquad \text{in } \Omega\,, \quad t > 0\,, \tag{49a}$$

$$T(\vec{x}, 0) = T_{init}(\vec{x}), \qquad \vec{x} \in \Omega\,, \tag{49b}$$

$$-kT_x \big|_{x=0} = h\,[\,T_\infty(t) - T(0,y,z,t)]\,, \qquad T(x,0,z,t) = T_{fixed}\,, \tag{49c}$$

$$-kT_x \Big|_{x=l_1} = -kT_y \Big|_{y=l_2} = -kT_z \Big|_{z=0} = -kT_z \Big|_{z=l_3} = 0\,. \tag{49d}$$

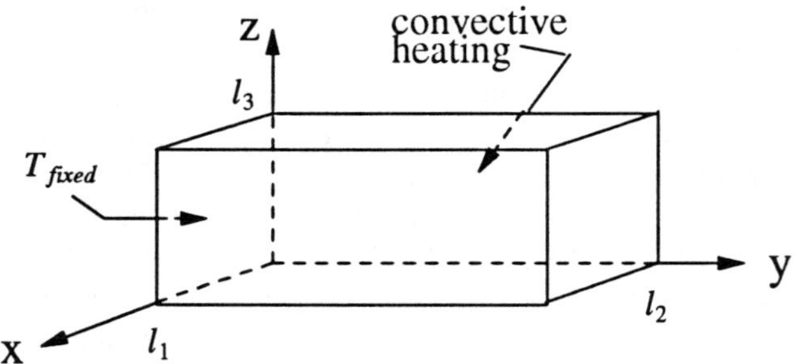

Figure 4.1.6. Melting of a box.

We subdivide $[0, l_1]$ into M_1 subintervals, $[0, l_2]$ into M_2 subintervals and $[0, l_3]$ into M_3 subintervals. For simplicity we take uniform grids in each direction, so that

$$\Delta x = \frac{l_1}{M_1}, \qquad \Delta y = \frac{l_2}{M_2}, \qquad \Delta z = \frac{l_3}{M_3},$$

and $\Delta V = \Delta x \Delta y \Delta z$. Thus, the box Ω is subdivided into $M_1 M_2 M_3$ boxes V_{ijk} of uniform volume ΔV with centers (x_i, y_j, z_k) and bounding surface ∂V_{ijk}. Approximations to the temperature $T(x_i, y_j, z_k, t_n)$ will be denoted by T^n_{ijk}, and to the energy density $E(x_i, y_j, z_k, t_n)$ by E^n_{ijk}, considered as mean values over V_{ijk}. Integrating the conservation law

$$E_t + \nabla \cdot \vec{q} = 0 \tag{50}$$

over the control volume V_{ijk} and $[t_n, t_{n+1}]$, we obtain similarly to (6)-(12), the discrete conservation law

$$\begin{aligned} E^{n+1}_{ijk} - E^n_{ijk} &= -\frac{\Delta t}{\Delta V_{ijk}} \int_{\partial V_{ijk}} \vec{q}^{\,n+\theta} \cdot \vec{N}\, dS \\ &= \frac{\Delta t}{\Delta V_{ijk}} [q^{n+\theta}_{i-½\, j\, k} \cdot A_{i-½\, j\, k} - q^{n+\theta}_{i+½\, j\, k} \cdot A_{i+½\, j\, k} + q^{n+\theta}_{i\, j-½\, k} \cdot A_{i\, j-½\, k} \\ &\quad - q^{n+\theta}_{i\, j+½\, k} \cdot A_{i\, j+½\, k} + q^{n+\theta}_{i\, j\, k-½} \cdot A_{i\, j\, k-½} - q^{n+\theta}_{i\, j\, k+½} \cdot A_{i\, j\, k+½}] \end{aligned} \tag{51}$$

for $i = 1, \ldots, M_1$, $j = 1, \ldots, M_2$, $k = 1, \ldots, M_3$ and $0 \le \theta \le 1$,

the A's denoting the areas of the corresponding faces.

In the rectangular geometry chosen, the areas of pairs of opposite faces are the same and $\Delta V = \Delta x \Delta y \Delta z$, so (51) simplifies to

$$\begin{aligned} E^{n+\theta}_{ijk} = E^n_{ijk} &+ \frac{\Delta t}{\Delta x} [q^{n+\theta}_{i-½\, j\, k} - q^{n+\theta}_{i+½\, j\, k}] + \frac{\Delta t}{\Delta y} [q^{n+\theta}_{i\, j-½\, k} - q^{n+\theta}_{i\, j+½\, k}] \\ &+ \frac{\Delta t}{\Delta z} [q^{n+\theta}_{i\, j\, k-½} - q^{n+\theta}_{i\, j\, k+½}] . \end{aligned} \tag{52}$$

In general, however, some pairs of opposite faces may have different areas (e.g. in the radial direction for the case of cylindrical geometry) and the above simplification will be unfeasible. In such a case, we need to express the heat flow rate, $q \cdot A$ and not just the flux (**§4.1.B**) in terms of temperature gradients, and use of the standard resistance becomes more convenient. So, in general, we define

$$\tilde{R}_{i-½\,j\,k} = \frac{1}{A_{i-½\,j\,k}} \left(\frac{½\Delta x_{i-1}}{k_{i-1\,j\,k}} + \frac{½\Delta x_i}{k_{i\,j\,k}} \right) \equiv \frac{1}{A_{i-½\,j\,k}} \cdot R_{i-½\,j\,k} , \tag{53}$$

and similarly for $\tilde{R}_{i+½\,j\,k}$, $\tilde{R}_{i\,j-½\,k}$, etc, so that the heat flow rate may be expressed as

$$(qA)^{n+\theta}_{i-½\,j\,k} = - \frac{T^{n+\theta}_{i\,j\,k} - T^{n+\theta}_{i-1\,j\,k}}{\tilde{R}_{i-½\,j\,k}} , \quad \text{etc} . \tag{54}$$

In the simpler case of (52), we have

$$q^{n+\theta}_{i-½\,j\,k} = \frac{(qA)^{n+\theta}_{i-½\,j\,k}}{A_{i-½\,j\,k}} = - \frac{T^{n+\theta}_{ijk} - T^{n+\theta}_{i-1\,j\,k}}{\Delta y \Delta z \tilde{R}_{i-½\,j\,k}} = - \frac{T^{n+\theta}_{ijk} - T^{n+\theta}_{i-1\,j\,k}}{\frac{\Delta x}{2} \left(\frac{1}{k_{i-1\,j\,k}} + \frac{1}{k_{ijk}} \right)} , \quad \text{etc} . \tag{55}$$

The boundary conditions are discretized as in the 1-dimensional case (**§4.1.C**). For example, the convective flux boundary condition on the face $x = 0$ becomes

$$(qA)^{n+\theta}_{1-½\,j\,k} = - \frac{T^{n+\theta}_{1jk} - T^{n+\theta}_{\infty}}{\frac{1}{h\,A_{½\,j\,k}} + \tilde{R}_{½\,j\,k}} \equiv - \frac{T^{n+\theta}_{1jk} - T^{n+\theta}_{\infty}}{\left(\frac{1}{h} + R_{½\,j\,k} \right) \frac{1}{A_{½\,j\,k}}} \tag{56}$$

with $R_{½\,j\,k} = \frac{½\Delta x}{k_{1jk}}$; the imposed temperature on the face $y = 0$ becomes

$$(qA)^{n+\theta}_{i\,1-½\,k} = - \frac{T^{n+\theta}_{i1k} - T_{fixed}}{\tilde{R}_{i\,½\,k}} \equiv - \frac{T^{n+\theta}_{i1k} - T_{fixed}}{\frac{1}{A_{i\,½\,k}} R_{i\,½\,k}} \quad \text{with } R_{i\,½\,k} = \frac{½\Delta y}{k_{i1k}} ; \tag{57}$$

and the zero flux on the face $x = l_1$ becomes

$$(qA)^{n+\theta}_{M_1+½\,j\,k} = 0. \tag{58}$$

For the heat conduction process we are examining here, the energy is simply the sensible heat measured relative to some convenient T_{ref} :

$$E_{ijk} = \rho c_{ijk} [T_{ijk} - T_{ref}]. \tag{59}$$

When everything is expressed in terms of the temperatures, the equation for T^{n+1}_{ijk} involves the temperatures of the 6 adjacent nodes. Choosing $\theta = 0$ (explicit scheme), these neighboring temperatures will be at time t_n and thus known. The stability condition becomes

$$\Delta t_n < \frac{\min(\Delta x_i^2, \Delta y_j^2, \Delta z_k^2)}{6 \cdot \max \alpha_{ijk}}, \tag{60}$$

often making computations lenghty and prohibitively expensive.

In the implicit case ($\theta = ½$ or 1), the resulting linear system will be hepta-diagonal and diagonally- dominant, so again it may be solved efficiently, especially if the ADI method is used to reduce the system to three tridiagonal ones, see [ALLEN-HERRERA-PINDER], [LAPIDUS-PINDER].

Cylindrical and spherical geometries are examined in PROBLEMS 5, 6, 25, 26, 28, 29.

4.1.H Internal heat source

The presence of an internal (volumetric) heat source adds a term in the energy conservation law (**§1.2**) , which, instead of (50) will read

$$E_t + \nabla \cdot \vec{q} = f \ . \tag{61}$$

The source term $f(\vec{x}, t)$ is the power density, representing the amount of energy delivered at location $\vec{x}$ at time t per unit volume per unit time (so it may be in units of $J/s\,cm^3 = Watts/cm^3$).

Its integration over V_{ijk} and $[t_n, t_{n+1}]$ contributes the additional term

$$\frac{1}{\Delta V_{ijk}} \int_{t_n}^{t_{n+1}} \int_{V_{ijk}} f(\vec{x}, t)\, dVdt \tag{62}$$

in (51). It should be noted that this integral should *not* be discretized by the low order approximations used for the derivative terms because large errors may ensue. The integration in (62) should be performed analytically whenever possible, or high order numerical integration methods should be employed. The result may be represented as $S(\Delta V_{ijk}, \Delta t_n)$, and its discrete approximation as S_{ijk}^n. With this term added, the numerical scheme remains the same in all other aspects. In particular, the stability condition, (60), for the explicit scheme is not altered.

In some processes the power density, f, may also depend on temperature, $f(\vec{x}, t, T(\vec{x}, t))$, making its treatment difficult. Direct integration of (62) is now impossible and the low order discretization of T is imposed on this term as well. Its mean value approximation,

$$\frac{\Delta t_n}{\Delta V_{ijk}} \Delta V_{ijk} f(x_{ijk}, t_{n+\theta}, T_{ijk}^{n+\theta})$$

may introduce large errors, unless the time step Δt_n is taken to be small. Some expedient ways for handling nonlinear source terms are suggested by [PATANKAR].

4.1.I Some programming suggestions

In implementing the schemes described above there are some steps that can be taken to enhance the utility, efficiency and maintainability of the code. Let us describe some points related to the construction, output, debugging and validation of a code, for the benefit of inexperienced programmers.

For clarity, readability, and adaptability, it is a good idea to place the control logic of the algorithm into the MAIN PROGRAM and code the various tasks as subroutines, e.g. INPUT, MESH, START, FLUX, PDE, OUTPUT, which will be called by MAIN. An example of such a structure is shown in **Table 4.1.1** below, as it would pertain to the explicit scheme applied to a slab with imposed temperatures at its ends. Comments and explanations of what is done are very helpful. In coding the algorithm, care should be taken to avoid unnecessary or inefficient computation. For example, expressions should be arranged so as to minimize loss of significant digits; polynomials should be evaluated in nested form; wherever an expression is used several times, evaluate it once and then use its value; "if" statements, and especially subroutine calls are relatively expensive, so their use should be minimized; frequent output slows down execution, so unnecessary output should be avoided. The longer the runs one plans to make, the more attention should be paid to such simple programming issues.

The computation begins by calling **Subroutine INPUT**, which reads in the data file, for example:

read	*tmax*, *maxsteps*, *dtout*		
read	l, M	!	l = slab length, M = number of nodes
read	ρ, c, k	!	material properties
read	T_{init}, T_0, T_l	!	initial and boundary temperatures

where	*tmax*	=	desired duration of the simulation,
	maxsteps	=	maximum number of time-steps to be allowed for the entire computation,
	dtout	=	desired time-interval for output.

The time-stepping will be monitored both by the actual *time* and by the number of time-steps taken , *nsteps* (see **Table 4.1.1**), and will end when *time* > *tmax* **or** *nsteps* > *maxsteps*, the latter as a precaution just in case the time-step *dt* gets to be too small (or even negative!) for any reason.

After the data have been read in, **Subroutine MESH** sets up the mesh structure (defines locations of nodes $x(i)$, control-volume faces, areas, etc.), and determines the appropriate time-step Δt. At each time step, time will be advanced by Δt.

In order to obtain output at precisely the desired time intervals, we introduce the variable *tout* = output time, which will be advanced only after each output step; before each *tout* is reached we temporarily reduce Δt, if necessary, so that *time* + Δt equals *tout*; then Δt is restored to its permanent value. (see **Table 4.1.1**).

Table 4.1.1 Example of a driver code

```
c        HEAT.f : Heat conduction in a slab with imposed temperatures
c 7-17-91 : entered and debugged basic code
c------- Notation ----------
         (explanation of symbols used and their meaning)
c**************************
         Program HEAT
         (common blocks)
c---------- Initialize ----------
         call INPUT
         call MESH
           dtperm = dt
           tout = max(dtout, dt)
         time = 0.
         call START
c---------- Begin time-stepping --------
100      continue
         time = time + dt
         nsteps = nsteps + 1
         if( time .gt. tmax .OR. nsteps .gt. maxsteps) go to 1000
         call FLUX
         call PDE
         if (time .eq. tout) then
           call OUTPUT
           dt = dtperm
           tout = tout + max(tout, dt)
         else if (time + dt .gt. tout) then
           dt = tout - time
         end if
         go to 100
c---------- end of time-stepping ----------
1000     continue
         (write any exiting information, such as time)
         stop
         end
```

Subroutine START initializes variables to their values at time = 0. Then each time-step consists of calling **FLUX**, which computes resistances and heat-flow-rates, and **PDE**, which solves the PDE.

Subroutine OUTPUT writes out the current values of the quantities of interest, such as temperature. It is usually desirable to have output at certain specified locations (where thermocouples may be located, for example), but it is generally overly complex to attempt to arrange the mesh in such a way that all these output locations coincide with computational nodes; instead, one may extract values at the desired locations via interpolation of the nodal values.

To debug the code, one usually starts with short-time runs on a very coarse mesh, say with $M = 10$ nodes, and observes the behavior of the solution as various parameters are varied, watching out for any non-physical behavior, for example violation of the Maximum Principle. A crucial and necessary check is provided by an energy-balance check from time-step to time-step: the total energy of the system at time t_n is $E^n_{total} = \sum_{i=1}^{M} E^n_i \Delta V_i$, so the energy gain during $[t_n\ ,\ t_n + \Delta t_n]$, is $E^{n+1}_{total} - E^n_{total}$; this must equal the energy input from the boundaries, $\int_{t_n}^{t_n+\Delta t_n} \int_{\partial\Omega} \vec{q} \cdot \vec{N} dS$, i.e. the sum of the boundary flow-rates times Δt_n, (PROBLEM 12).

The final step in preparing a code is to validate it by running one or more *benchmark problems* on it with known solutions, and comparing the computed and exact solutions. One may start with a coarse mesh, and successively double the number of nodes (halving the mesh width Δx) to verify that various measures of the error (e.g. $\max_{1 \le j \le m} |T^n_j - T(x_j, t_n)|$ at a fixed t_n, or $\max_{0 \le t_n \le t_{max}} (\max_{1 \le j \le M} |T^n_j - T(x_j, t_n)|)$) decrease as M increases. For the algorithms described earlier, one expects to see errors of order $O(\Delta x^2)$, which is the order of the discretization error, at least for M's up to about 100 (in single precision); for larger M however, roundoff error takes over and the accuracy actually deteriorates, see PROBLEM 21.

Two- and three-dimensional simulations can easily tax the capabilities of even "large" mainframe computers, so vector or/and parallel "super-computing" becomes necessary. Then one must use various programming "tricks" to take advantage of the special features of such machines. For example, to aid vectorization one may unfold 2- or 3-dimensional arrays into 1-dimensional long vectors using "red-black ordering", and replace "if" statements with logical (boolean) equivalents inside DO loops, [WILLIAMS-WILSON], [ORTEGA-VOIGT]. Parallelism for transient problems may be achieved by "domain-decomposition" methods at a basic level [DRAKE-NARANG]. These new computing technologies, which are currently under intense development, have brought about a re-examination of the various serial algorithms to see which methods are best suited to the various machine architectures and classes of problems.

PROBLEMS

PROBLEM 1. Write a brief essay on the simulation of a thermal process, addressing the possible reasons for preparing it, the roles that it is to serve, the kinds of information available to it, the type of output it is to provide, the importance or lack of importance of computational speed and the accuracy that it is to have. Specific points that you might address are simulations in various contexts, including laboratory scale studies, industrial size studies, and real time control.

PROBLEM 2. For the model heat conduction problem of **§4.1.A**, describe *qualitatively* how you expect the temperature to evolve in time. In particular, what is the expected appearance of temperature-time curves at preassigned thermocouple locations at the faces of the slab and at, say, two interior points? If a *fluxmeter* were attached at each of the faces, what flux-time curves would you expect to see?

PROBLEM 3. Set up a 1-dimensional *radial* mesh for *axially symmetric* heat transfer in a hollow *cylinder* $R_{in} \le r \le R_{out}$ of *unit height.* That is, subdivide the interval $[R_{in}, R_{out}]$ into M subintervals and determine the nodes r_i, faces $r_{i-1/2}$, radial "areas" $A_{i-1/2} = 2\pi r_{i-1/2}$, and control "volumes" $\Delta V_i = \pi r_{i+1/2}^2 - \pi r_{i-1/2}^2 = 2\pi r_i \Delta r_i$. For further developments see PROB. 9, 22.

PROBLEM 4. Set up a 1-dimensional *radial* mesh for *spherically symmetric* heat transfer in a solid *sphere* $0 \le r \le R_{out}$, by subdividing $[0, R_{out}]$ into M subintervals, [Here $A_{i-1/2} = 4\pi r_{i-1/2}^2$, $\Delta V_i = (4/3)\pi(r_{i+1/2}^3 - r_{i-1/2}^3)$]. See PROBLEMS 10, 23.

PROBLEM 5. Set up a 2-dimensional (r, z) mesh for *axially symmetric* heat transfer in a hollow *cylinder* $R_{in} \le r \le R_{out}$, $0 \le z \le Z$, by subdividing $[R_{in}, R_{out}]$ into M_r subintervals and $[0, Z]$ into M_Z subintervals. Determine the nodes (r_i, z_j), faces $r_{i-1/2}$, $z_{j-1/2}$, areas of radial faces $A_{i-1/2\, j}$, of axial faces $A_{i\, j-1/2}$, and control volumes ΔV_{ij}. See PROBLEMS 25, 28.

PROBLEM 6. Set up a 2-dimensional (r, θ) mesh for *axially symmetric* heat transfer in a *sphere* $0 \le r \le R_{out}$, by subdividing $[0, R_{out}]$ into M_r subintervals and $[0, \pi]$ into M_θ sectors (the right-half sphere suffices, so let $x = r \sin\theta$, $z = r\cos\theta$, with θ the azimouthal angle measured off the positive z-axis). Determine the nodes (r_i, θ_j), faces $r_{i-1/2}$, $\theta_{j-1/2}$, areas of radial faces

$A_{i-1/2\, j} = 2\pi \int_{\theta_{j-1/2}}^{\theta_{j+1/2}} \int_{r_{i-1/2}}^{r_{i+1/2}} x\, r\, dr\, d\theta$, of angular faces $A_{i\, j-1/2} = 2\pi \int_{r_{i-1/2}}^{r_{i+1/2}} x\, dr$, and

control volumes $\Delta V_{ij} = 2\pi \int_{\theta_{j-1/2}}^{\theta_{j+1/2}} \int_{r_{i-1/2}}^{r_{j+1/2}} x\, r\, dr\, d\theta$.

[Check: $A_{i-1/2 j} = 2\pi r_{i-1/2}^2(\cos\theta_{j-1/2} - \cos\theta_{j+1/2})$, $A_{ij-1/2} = \pi(r_{i+1/2}^2 - r_{i-1/2}^2)\sin\theta_{j-1/2}$, $\Delta V_{ij} = (2\pi/3)\,(r_{i+1/2}^3 - r_{i-1/2}^3)\,(\cos\theta_{j-1/2} - \cos\theta_{j+1/2})$]. See PROBLEMS 26, 29.

PROBLEM 7. Discuss the factors that would lead you to use non-uniform spatial and time subdivisions. In particular, what would you do if results are desired at definite times (e.g. in accordance with the readings of some recording device), and under conditions where high temperature gradients are present in certain locations. Under what conditions would the latter actually occur?

PROBLEM 8. Using the Taylor expansion, show that for the centered-nodes mesh (2b) the error in the approximation of the mean value by the nodal value is $O(\Delta x_j^3)$.

PROBLEM 9. Derive the discrete heat balance, analogous to (11), or (12), for axially symmetric heat conduction in a cylinder of unit height, using the mesh constructed in PROBLEM 3.

PROBLEM 10. Derive the discrete heat balance for spherically symmetric heat conduction in a sphere, using the mesh of PROBLEM 4.

PROBLEM 11. The time increment Δt used in a time-stepping scheme should be so large that local equilibrium obtains in a control volume during this time, i.e. Δt should be larger than the **relaxation time** τ of the heat conduction process. This is the time required for the temperature to relax to its equilibrium (steady-state) value T_∞, relative to its initial distance from the steady state; the convenient and commonly used definition of the relaxation time τ is (e.g. see [PINSKY]): $\left| T(x, \tau) - T_\infty / T(x, 0) - T_\infty \right| = 1 / e$ or, equivalently,

$$1 / \tau := \lim_{t \to \infty} (1 / t) \ln | T(x, t) - T_\infty |.$$

Consider a control volume $0 \le x \le \Delta x$ of width Δx. Using the fact that the solution of the heat equation with vanishing boundary values (hence $T_\infty = 0$ here) is given by $T(x, t) = \exp(-\pi^2 \alpha\, t / \Delta x^2) \sin(\pi x / \Delta x)$, show that the relaxation time is $\tau = \Delta x^2 / \pi^2 \alpha$. Then, the requirement $\Delta t > \tau$ combined with the CFL condition restrict the ratio $\mu = \alpha \Delta t / \Delta x^2$ to be $1 / \pi^2 < \mu < 1 / 2$.

PROBLEM 12. Prove that with all choices of θ the numerical scheme (12) obeys a global heat balance identically.

PROBLEM 13. Choose $\theta = 0$ in (31) to derive the explicit scheme (33).

PROBLEM 14. For the simplest explicit scheme (35), we have **PDE**$[T] \equiv$

$T_t - \alpha T_{xx}$ and **FDE**$[T_j^n] \equiv \dfrac{T_j^{n+1} - T_j^n}{\Delta t} - \alpha \dfrac{T_{j-1}^n - 2T_j^n + T_{j+1}^n}{\Delta x^2}$, see **§4.1.D.**

Using Taylor expansions show that the *local truncation error* is given by

$$\mathbf{te}_j^n = \frac{\Delta t}{2} T_{tt}(x_j, t_n) - \alpha \frac{\Delta x^2}{12} T_{xxxx}(x_j, t_n) + O(\Delta t^2 + \Delta x^4) = O(\Delta t + \Delta x^2)$$

provided T_{tt} and T_{xxxx} are bounded. Hence the scheme is *consistent*. Next, using $T_{tt} = \alpha (T_{xx})_t = \alpha (T_t)_{xx} = \alpha^2 T_{xxxx}$, show that the choice $\mu = \alpha \Delta t / \Delta x^2 = 1 / 6$ reduces this error to $O(\Delta x^4)$.

PROBLEM 15. Show that if $\mu \le ½$ in (35) then the *local discretization error* satisfies $\|\mathbf{de}^{n+1}\| \le \|\mathbf{de}^n\| + \Delta t \cdot (A\Delta t + B\Delta x^2)$, where $\|\mathbf{de}^n\| = \max_{1 \le j \le M} |\mathbf{de}_j^n|$,

$A = \max |T_{tt}/2|$, $B = \max |\alpha T_{xxxx}/12|$. Deduce that $\|\mathbf{de}^n\| \le n\Delta t (A\Delta t + B\Delta x^2)$, and since $n\Delta t \le t_{max}$ conclude that $\|\mathbf{de}^n\| = O(\Delta t + \Delta x^2)$, thus establishing convergence of the scheme directly.

PROBLEM 16. (a) Show that if the CFL condition holds for (36) , then the error at any $n > 0$ due to initial roundoff error ε_j^0 is bounded by that initial error.
(b) Roundoff error may be introduced at every point that a computation is performed. Thus even if the CFL condition is met for (36) error is introduced not only at the initial step $n = 0$ but at every time step. What is its cumulative effect? Can it grow exponentially?

PROBLEM 17. (a) Derive the stability condition (40a) for imposed temperature at $x = 0$. (b) Derive the stability condition (40b) for imposed flux at $x = 0$. (c) Derive the stability condition (40c) for the convective boundary condition at $x = 0$, and show that (40a) is sufficient for it.

PROBLEM 18. Combine (39) and (40) to establish (41-42).

PROBLEM 19. Analyze carefully the effect of using the discretization (43) for the first interior node. In particular, what happens if errors originate both at the initial line and at the boundary $j = 1$? What if no error originates at the boundary line?

PROBLEM 20. Find the counterpart to the implicit equation (43) for convective and flux boundary conditions.

PROBLEM 21. (a) Implement the explicit scheme (33) in a computer code (see **§4.1.I** for helpful suggestions) for the simple case (35) of uniform mesh and constant properties. To debug and validate your code, take $h = 0$ (whence the boundary conditions are $T_x(0,t) = T_x(l,t) = 0)$ and $T_{init}(x) = 100\cos(\pi x/l)$, in which case the exact solution is $T(x,t) = \exp(-\pi^2 \alpha t / l^2 \cdot 100\cos(\pi x / x)$, $0 \le x \le l$, $t \ge 0$. For simplicity, choose $l = 1$, $\alpha = 0.1$, $M = 10$ and compare the numerical and exact solutions up to time $t_{max} = 1$.
(b) Examine convergence by making runs with $M = 10, 20, 40, 80, 160$ nodes (remember to adjust Δt so that $\mu = ½)$ and looking at the maximum error $\max\limits_{0 \le t_n \le t_{max}} (\max\limits_{1 \le j \le M} |T_j^n - T(x_j, t_n)|)$. Does it behave like $O(\Delta x^2)$? For which M do you get the least error? For that M, make a run in double precision. Does the error reduce further?
(c) Examine the effects of instability by fixing $M = 20$ and making runs with $\mu = \alpha \Delta t / \Delta x^2 = 0.4, 0.5, 0.501, 0.6, 1.0$. Discuss what you observe.
(d) According to PROBLEM 14, the choice $\mu = 1/6$ improves the error to $O(\Delta x^4)$. Test this by repeating (6) but with $\mu = 1/6$ now. Compare with the results from (6).

PROBLEM 22. For axially symmetric heat conduction in a hollow cylinder of unit height (PROBLEMS 3 and 9) with convective boundary condition at $r = R_{in}$ and imposed temperature at $r = R_{out}$: (a) set up the general $(0 \le \theta \le 1)$ algorithm (analogous to (31)); (b) find the stability conditions (analogous to (39, 40, 48)); (c) in the implicit case $(0 < \theta \le 1)$, write down the tridiagonal system (analogous to (47)).

PROBLEM 23. Do the same for spherically symmetric heat conduction in a sphere (PROBLEMS 4 and 10) with imposed temperature at $r = R_{out}$. In particular, examine carefully the discretization and stability restriction at the most internal node $[0, r_{1+½}]$. Note that the natural boundary condition at $r = 0$ is $q_{½} = 0$.

PROBLEM 24. Repeat PROBLEM 23 for the cases of imposed flux and of a convective boundary condition at $r = R_{out}$.

PROBLEM 25. Derive the discrete heat balance (see **§4.1.G**), for 2-dimensional (r, z), axially symmetric heat conduction in a cylinder, using the mesh of PROBLEM 5.

PROBLEM 26. Derive the discrete heat balance for 2-dimensional (r, θ), axially symmetric heat conduction in a sphere using the mesh of PROBLEM 6.

PROBLEM 27. Set up a 3-dimensional (r, θ, z) mesh for heat conduction in a hollow cylinder $R_{in} \le r \le R_{out}$, $0 \le \theta < 2\pi$, $0 \le z \le Z$, with $M_r \times M_\theta \times M_z$ nodes, and derive the discrete heat balance.

PROBLEM 28. For the process of PROBLEM 25, with convective boundary condition at $r = R_{in}$, imposed flux at $r = R_{out}$, and insulated axial faces $(z = 0, z = Z)$: (a) set up the computational algorithm for $0 \le \theta \le 1$; (b) find the stability conditions at internal and boundary nodes.

PROBLEM 29. Repeat, for the process of PROBLEM 26.

PROBLEM 30. Consider the initial-boundary value problem for the heat equation, $T_t = T_{xx}$, $a < x < b$, $t > 0$, with $T(x, 0) = C\cos(\gamma_0 x)$, $a < x < b$, and $T(a, t) = A\sin(\alpha_0 t)$, $T(b, t) = B\sin(\beta_0 t)$, where γ_0, α_0, β_0 are real numbers. Discuss how you would decide upon the size of the spatial and temporal mesh sizes to be used in calculating the solution to this problem.

PROBLEM 31. What would be your method for simulating heat transfer in a material whose thermal diffusivity varies by an order of magnitude or more over the range of temperatures encountered?

PROBLEM 32. (a) Discretize the heat equation $T_t = \alpha T_{xx}$ using the *centered-difference* $(T_j^{n+1} - T_j^{n-1}) / 2\Delta t$ for T_t and the standard centered-difference $(T_{j-1}^n - 2T_j^n + T_{j+1}^n) / \Delta x^2$ for T_{xx}. Show that this scheme is **unstable** for any $\mu > 0$!!! (b) In the previous scheme replace the central-term $-2T_j^n$ by $-(T_j^{n-1} + T_j^{n+1})$ to obtain the *Dufort-Frankel method* ([LAPIDUS-PINDER], [DUCHATEAU-ZACHMANN]). Show that if the ratio $\mu_1 := \Delta t / \Delta x$ is held fixed as $\Delta x, \Delta t \to 0$ then the method is consistent with the *hyperbolic* PDE $T_t + \alpha\mu_1 T_{tt} = \alpha T_{xx}$ and *not* with the heat equation.

4.2. BRIEF OVERVIEW OF NUMERICAL METHODS FOR PHASE CHANGE PROBLEMS

Consider the one-dimensional heat conduction problem ((1)-(4) of **§4.1**), but now regarding the material as a **phase change** material with a melt temperature T_m, and take

$$T_{init}(x) < T_m \quad \text{and} \quad T_\infty(t) > T_m .$$

Then melting of the material will occur, commencing at $x = 0$ at some time $t_{init} > 0$, and we want to find t_{init}, $T(x,t)$ and $X(t)$ such that

$$\rho c_S T_t = (k_S T_x)_x \quad \text{for } 0 < x < l,\ 0 < t < t_{init} \quad \text{and for } X(t) < x < l,\ t > t_{init},$$

$$\rho c_L T_t = (k_L T_x)_x \quad \text{for} \quad 0 < x < X(t) \quad , t > t_{init},$$

$$-k_S T_x(0,t) = h[T_\infty(t) - T(0,t)] \quad \text{for} \quad t < t_{init},$$

$$-k_L T_x(0,t) = h[T_\infty(t) - T(0,t)] \quad \text{for} \quad t > t_{init},$$

$$T_x(l,t) = 0 \quad \text{for} \quad t > 0,$$

$$T(x,0) = T_{init}(x), \quad X(t_{init}) = 0,$$

$$T(X(t),t) = T_m, \quad t > t_{init},$$

$$\rho L X'(t) = -k_L T_x(X(t)^-,t) + k_S T_x(X(t)^+,t), \quad t > t_{init}.$$

Clearly we have to separate the problem into two distinct problems: a pure heat conduction problem, until the face $x = 0$ reaches the melt temperature at time $t = t_{init}$, and a two-phase Stefan problem after time t_{init}. The numerical solution to the first problem was presented in **§4.1**, but that of the Stefan problem is much more difficult, due to the underlying geometric nonlinearity of the problem: the regions in which the two heat conduction equations are to hold change in time, and we have to compute the location of the interface $x = X(t)$ concurrently. Several approaches have been devised with this aim, collectively referred to as **front tracking schemes**, because they attempt to explicitly track the interface using the Stefan condition.

One approach is to fix the spatial step, Δx, but allow the time step, Δt_n to "float" in such a way that the front always passes through a node (x_j, t_n). An example of this approach is the method of [DOUGLAS-GALLIE].

Another approach is to fix the time step and allow the spatial step to "float," in fact, use two distinct and time-varying space steps for the two phases. The isotherm migration method of J. Crank is of this type, [CRANK, 1981, 1984].

Yet another approach is based on the Landau transformation, which we have used in the perturbation method (see §3.3). By a change of variables (**§3.3.B**), the regions representing the phases become fixed, the underlying geometric nonlinearity showing up algebraically now in the transformed equations. Then one solves the resulting system of nonlinear equations by some numerical method.

All such approaches work well, more or less, for simple Stefan problems (that would arise e.g., in laboratory settings) in which we know what to expect, namely a single sharp front separating the two phases. It is not difficult to realize however, that such problems are not the rule in practice, particularly when time dependence of heat input/output or thermal cycling occur. For example, in Latent Heat Thermal Energy Storage, one must deal with cases of extreme thermal cycling, multiple fronts, disappearing phases and non-predictable behavior (**§1.3**, **§5.3**). Internal heating is another source of difficulties, first documented by [ATTHEY], in which extended mushy zones may appear instead of sharp fronts. Constitutional supercooling of binary alloys results in similar effects which may not be ignored (see [ALEXIADES-WILSON-SOLOMON, 1985]). Simultaneous mass transfer by diffusion and/or convection complicate the phase change process to the point that we cannot guess a priori the qualitative picture in enough detail to even be able to *formulate* the problem in the classical fashion of a Stefan type problem with sharp front, etc. If so many complications can arise in 1-dimensional processes, what about 2- and 3-dimensional processes? Despite the difficulties, successful methods for 2-D hydrodynamic instability problems have been developed by Glimm and coworkers, [GLIMM]. Surveys of front tracking methods appear in [MEYER, 1978], [CRANK, 1981, 1984], [ALBRECHT-COLLATZ-HOFFMANN], etc.

Such reasons make front-tracking schemes unviable as general simulation tools for modeling realistic phase-change processees. The only viable general approach is the so-called **enthalpy method**, precisely because it bypasses the explicit tracking of the interface. In this approach the jump condition (Stefan condition) is *not forced* on the solution, but it is obeyed automatically by it as a "natural boundary condition" (in the sense of the Calculus of Variations). Its theoretical basis is a formulation of the Stefan problem different than the classical one, the so-called **weak** or **enthalpy formulation**, described in **§4.4**. It is similar to the weak formulations commonly used in gas dynamics for shocks (see [HYMAN] for a brief overview). Another fixed-domain method (as opposed to front-tracking), based on "variational inequalities" [DUVAUT], [ODEN-KIKUCHI] reformulation of the Stefan problem and finite elements, lacks the direct physical interpretation of the enthalpy method and has not lived up to its initial promise for Stefan-type problems.

It can be safely concluded today that the enthalpy method, to which we turn in the next section, discretized by (integrated) finite differences, is the most versatile, convenient, adaptable, and easily programmable numerical method available for phase change problems in 1, 2 or 3 space dimensions.

We hasten to add however that it does *not* solve *all* the problems. Excluded are problems which we do not know how to formulate weakly due to their special interface conditions. Such is the case with supercooling problems, where the instability of the interface must be studied. A very successful computational approach for such problems is another fixed-domain type formulation, the so-called "phase-field" approach, under intense development lately, [CAGINALP, 1989, 1991], [KOBAYASHI].

4.3. THE ENTHALPY METHOD IN ONE SPACE DIMENSION

4.3.A Introduction

The, so called, enthalpy or weak solution approach is based on the fact that the energy conservation law, expressed in terms of energy (enthalpy) and temperature, together with the equation of state contain all the physical information needed to determine the evolution of the phases. It turns out that, for the purpose of obtaining numerical schemes, the most appropriate and convenient way to state energy conservation is the primitive integral heat balance over arbitrary volumes and time-intervals, from which all other formulations can be obtained, namely

$$\int_t^{t+\Delta t} \frac{\partial}{\partial t}\left(\int_V E\,dV \right) dt = \int_t^{t+\Delta t} \int_{\partial V} -\vec{q}\cdot\vec{n}\,dS\,dt \tag{1}$$

where $E = \rho e$ is the energy density (per unit volume), and $-\vec{q}\cdot\vec{n}$ is the heat flux *into* the volume V across its boundary ∂V, $\vec{n}$ being the outgoing unit normal to ∂V.

The distinct advantage of this primitive form is that it is valid irrespectively of phase, and even if E and $\vec{q}$ experience jumps, so it is actually more general than the localized differential form

$$E_t + div\vec{q} = 0 . \tag{2}$$

The two forms are equivalent for smooth E, $\vec{q}$, thanks to the Divergence Theorem (**§1.2**). In the presence of a phase-change, the partial differential equation (2) can only be interpreted in the classical pointwise sense inside each phase separately, and then conservation across the interface must be imposed explicitly as an additional interface (Stefan) condition, making front-tracking necessary. Alternatively, the PDE (2) may be interpreted in a generalized (weak) sense globally, as described in detail in **§4.4**. It turns out that the numerical solutions obtained via the enthalpy method approximate this weak solution, as we shall show in **§4.5**.

In this section, we describe the enthalpy method for Stefan problems in one space dimension, and its numerical implementation via time-explicit or time-implicit schemes.

4.3.B The enthalpy method

The idea of the enthalpy approach is very simple, direct, and physical. We partition the volume occupied by the phase-change material into a finite number of control volumes V_j and apply energy conservation, (1), to each control volume to obtain a discrete heat balance. Note that this is the same discrete heat balance as for plain heat conduction (**§4.1.B**), and we use it to update the enthalpy, E_j, of each control volume. From the equation of state we know that $E_j \le 0 ==> V_j$ is solid, $E_j \ge \rho L ==> V_j$ is liquid, and $0 < E_j < \rho L ==> V_j$ is partially liquid and partially solid, so we call it "mushy". A mushy cell contains an interface and the fraction of the cell occupied by liquid is naturally given by the value of the

liquid fraction : $$\lambda_j = \frac{E_j}{\rho L}.$$

Note that in this scheme the phases are determined by the enthalpy alone, with no mentioning of interface location(s). It is a "volume-tracking" scheme, as opposed to "front-tracking". Since the front location may be recovered *a posteriori* from the values of the enthalpy, it may be characterized as a **front-capturing scheme**, similar in spirit to "shock-capturing" schemes of gas dynamics (see [HYMAN] for an overview of various types of schemes).

Let us see how the method works in detail, by considering the heat conduction problem of **§4.1**, except now we assume that our slab $0 \le x \le l$ is occupied by a material that changes phase at a melt temperature T_m. We assume that initially the material is solid with

$$T(x,0) = T_{init}(x) \le T_m, \quad 0 \le x \le l, \tag{3}$$

the face $x = 0$ is heated convectively by $T_\infty(t) \ge T_m$:

$$q(0,t) = h\,[\,T_\infty(t) - T(0,t)\,], \quad t > 0, \tag{4}$$

and the face $x = l$ is insulated :

$$q(l,t) = 0, \quad t > 0. \tag{5}$$

The energy conservation law in its integrated form (1) applied to the present one-dimensional control volumes $V_j = [\,x_{j-½}\,,\,x_{j+½}\,] \times A$ (see **§4.1.B**) becomes

$$\int_{t_n}^{t_{n+1}} \frac{\partial}{\partial t}\left(A \int_{x_{j-½}}^{x_{j+½}} E(x,t)dx \right) dt = -\int_{t_n}^{t_{n+1}} A \int_{x_{j-½}}^{x_{j+½}} q_x(x,t)\,dx\,dt. \tag{6}$$

We seek numerical approximations to the temperature, energy and flux obeying (3)-(6) with $q = -k\,T_x$ of course.

This problem is formally identical to the heat conduction problem (1b,c),(8a) of **§4.1**. The two differ in that now the enthalpy E is the sum of sensible and latent heat in the liquid, so that, instead of (6) of **§4.1.B**, we have

$$E(x,t) = \begin{cases} \displaystyle\int_{T_m}^{T(x,t)} \rho c_S(\tau)d\tau\,, & T(x,t) < T_m \quad \text{(solid)} \\ \displaystyle\int_{T_m}^{T(x,t)} \rho c_L(\tau)d\tau + \rho L\,, & T(x,t) > T_m \quad \text{(liquid)} \end{cases} \tag{7}$$

The phases are described by

$$E(x,t) \leq 0 \quad => \quad \text{solid at } (x,t) \tag{8a}$$

$$0 < E(x,t) < \rho L \quad => \quad \text{interface at } (x,t) \tag{8b}$$

$$E(x,t) \geq \rho L \quad => \quad \text{liquid at } (x,t)\,. \tag{8c}$$

Thus, only the relation between E and T is now different than in **§4.1**, while the discretization of (6) is still (12) of **§4.1.B**. The flexibility and generality of this approach will be further illustrated in **§4.3.H** where even a wall layer will be incorporated into the global scheme by adjusting the energy.

Consider the case in which

$$c_S\,,\quad c_L = \text{constants}\,. \tag{9}$$

Then (7) becomes

$$E = \begin{cases} \rho c_S[T - T_m]\,, & T < T_m \\ \rho c_L[T - T_m] + \rho L\,, & T > T_m \end{cases} \tag{10}$$

or, solving for T,

$$T = \begin{cases} T_m + \dfrac{E}{\rho c_S}\,, & E \leq 0 & (\text{solid}) \\ T_m\,, & 0 < E < \rho L & (\text{interface}) \\ T_m + \dfrac{E - \rho L}{\rho c_L}\,, & E \geq \rho L & (\text{liquid}) \end{cases} \tag{11}$$

Proceeding with the discretization of fluxes and boundary conditions as in **§4.1**, we arrive at the following discrete problem.

initial values:

$$T_j^0 = T_{init}(x_j), \quad j = 1, \ldots, M \tag{12a}$$

boundary condition at $x = 0$:

$$q_{½}^{n+\theta} = -\frac{T_1^{n+\theta} - T_\infty^{n+\theta}}{\frac{1}{h} + R_{½}}\,, \quad R_{½} = \frac{½\Delta x}{k_1} \tag{12b}$$

boundary condition at $x = l$:

$$q_{M+½}^{n+\theta} = 0\,, \tag{12c}$$

interior values:

$$E_j^{n+1} = E_j^n + \frac{\Delta t_n}{\Delta x_j}\left[q_{j-½}^{n+\theta} - q_{j+½}^{n+\theta} \right], \quad j = 1, \ldots, M \tag{12d}$$

where

$$q_{j-½}^{n+\theta} = -\frac{T_j^{n+\theta} - T_{j-1}^{n+\theta}}{R_{j-½}} \quad \text{with} \quad R_{j-½} = \frac{½\Delta x_{j-1}}{k_{j-1}} + \frac{½\Delta x_j}{k_j}\,, \quad j = 2, \ldots, M\,, \tag{12e}$$

and

$$T_j^n = \begin{cases} T_m + \dfrac{E_j^n}{\rho c_S}\,, & E_j^n \leq 0 & (\text{solid}) \\ T_m\,, & 0 < E_j^n < \rho L & (\text{interface}) \\ T_m + \dfrac{E_j^n - \rho L}{\rho c_L}\,, & E_j^n \geq \rho L & (\text{liquid}) \end{cases} \tag{12f}$$

The updating algorithm from any time t_n to the next $t_n + \Delta t_n$ proceeds as follows: Knowing the enthalpy, temperature and phase (see below) of each control volume, we compute the resistances and fluxes, which are then used to update the enthalpies, which in turn yield new temperatures and phase states.

The most convenient phase-indicator is the **liquid fraction** of a control volume V_j, defined as

$$\lambda_j^n = \begin{cases} 0, & \text{if} \quad E_j^n \le 0 \quad (\text{solid}) \\ \dfrac{E_j^n}{\rho L}, & \text{if} \quad 0 < E_j^n < \rho L \quad (\text{mushy}). \\ 1, & \text{if} \quad \rho L \le E_j^n \quad (\text{liquid}) \end{cases} \tag{13}$$

If $0 < \lambda_j^n < 1$ the control volume is said to be **mushy** with liquid volume $\lambda_j^n \Delta x_j$ and solid volume $(1 - \lambda_j^n)\Delta x_j$ (per unit cross sectional area).

The definitions of resistances and fluxes between control volumes are identical to their definitions in **§4.1**, with the resistance at $x_{j-1/2}$ expressed as

$$R_{j-1/2} = \frac{\Delta x_{j-1}}{2k_{j-1}} + \frac{\Delta x_j}{2k_j}.$$

The effective conductivity k_j of a mushy control volume depends on the structure of the phase-change front, and it is not always clear how to choose it, especially in 2 or 3 dimensional situations. Some alternative choices are:

Sharp front(s): A control volume containing a sharp front consists of layers of solid and liquid in a "serial" arrangement, for which the effective resistivity is the sum of the resistivities of the layers. With the layer thicknesses determined from the solid and liquid fractions, we have

$$\frac{1}{k_j^n} = \frac{\lambda_j^n}{k_L(T_m)} + \frac{1 - \lambda_j^n}{k_S(T_m)}, \qquad j = 1, 2, \ldots, M. \tag{14a}$$

Columnar front: A front consisting of columns of solid and liquid constitutes a "parallel" arrangement, so the effective conductivity is the sum of the conductivities of the phases :

$$k_j^n = \lambda_j^n k_L(T_m) + (1 - \lambda_j^n) k_S(T_m). \tag{14b}$$

Amorphous mixture of solid and liquid: The inter-phase region may be a random mixture of solid and liquid. In this case one may use the following formula which interpolates the previous two cases [CHEMICAL ENGINEERING GUIDE, p.242]:

$$k_j^n = k_S(T_m) \frac{1 + \lambda^{2/3}(\kappa - 1)}{1 + (\lambda^{2/3} - \lambda)(\kappa - 1)}, \quad \lambda = \lambda_j^n, \quad \kappa = \frac{k_L(T_m)}{k_S(T_m)}. \tag{14c}$$

In most situations we do not know which of the above cases is relevant. In 2 or 3 space dimensions even a sharp front will generally not be moving in the direction of one of the axes, so the choice is not clear. A simple expedient is to take the average of the solid and liquid conductivities,

$$k_j^n = \tfrac{1}{2}(k_S + k_L) \tag{14d}$$

However, the best alternative, when applicable, is to employ the "Kirchoff transformation" (see (7)§**4.4.D**), to replace the temperature T by the "Kirchoff temperature" u. This can be used when the conductivity is a function of temperature only. In particular, for constant k_S, k_L the "Kirchoff temperature" is

$$u = \begin{cases} k_S\,[\,T - T_m\,] & \text{if} \quad T < T_m \\ 0 & \text{if} \quad T = T_m\,. \\ k_L\,[\,T - T_m\,] & \text{if} \quad T > T_m \end{cases} \tag{15}$$

Then $q \equiv -k\,T_x = -u_x$, so the discrete flux is simply $q_{j-1/2} = (\,u_{j-1} - u_j\,)/\Delta x$. To compare this with (12e), resubstitute u in terms of T: $u_j = k_j\,[\,T_j - T_m\,]$, and write it as

$$q_{j-1/2} = \frac{T_{j-1} - T_m}{\hat{R}_{j-1}} + \frac{T_m - T_j}{\hat{R}_j}\,, \qquad \hat{R}_j := \Delta x\,/\,k_j\,. \tag{16}$$

Thus the flux neatly splits to a sum of two terms, one for each of the two adjacent nodes. Note that if one of the nodes, say node j, is mushy then $T_j = T_m$ and the node is not contributing to conduction. We see that in this prescription each node has its own resistivity $\hat{R}_j = \Delta x\,/\,k_j$, which may be found conveniently from

$$\hat{R}_j = \Delta x\,\{\,\lambda_j\,/\,k_L + (1 - \lambda_j)\,/\,k_S\,\}\,, \tag{17}$$

the value for mushy nodes being irrelevant since they are not contributing to the flux. No averaging of values is used, so this prescription results in the highest effective conductivity. It is the best choice for the enthalpy scheme since it is consistent with the mushy nodes being treated as isothermal.

Note that such issues arise only when $k_L(T_m)$ and $k_S(T_m)$ are substantially different, in which case the sensitivity of the solution to the choice of effective conductivity should be examined by comparing the results from the various choices. (14a) yields the lowest effective conductivity and (16)-(17) yields the highest, so these two pretty much bracket the system behavior.

We emphasize again that the interface location is not involved in the computation at all, this being an essential advantage of the enthalpy method. If the problem being modeled admits a sharp interface, then the enthalpy scheme ought to produce a *single mushy node* at each time step. If at time t_n the mushy node is the m-th node, then a good approximation to the interface location $X(t_n)$ is given by

$$X^n := x_{m-1/2} + \lambda_m^n \Delta x_m\,. \tag{18}$$

REMARK 1. *Missing the phase-change*

The latent heat effect is only felt in mushy nodes, so each control volume should pass through the mushy state before changing phase. The algorithm described here is "robust" in this respect, so skipping this transition indicates a bug in the code, or too large a time-step. Some other implementations may not be so robust and one must be careful not to miss the transition. Especially prone are algorithms that track the temperature and account for the latent heat via a source term.

REMARK 2. *Temperature dependent heat capacities*

If $c_S = c_S(T)$, $c_L = c_L(T)$, then the equation of state is the nonlinear relation (7), and finding T from E is not quite as simple as in the constant c_S, c_L case. If $E \leq 0$ we need to find T from the equation $E = \int_{T_m}^{T} \rho c_S(\tau)d\tau$, for which a Newton-Raphson method may be employed. Alternatively, we may rewrite the equation as $\frac{dE}{dT} = \rho c_S(T)$, or $\frac{dT}{dE} = \frac{1}{\rho c_S(T)}$, which is an ODE with "initial condition" $T = T_m$ for $E = 0$. Similarly, if $E \geq \rho L$, then we may solve the equation $E = \int_{T_m}^{T} \rho c_L(\tau)d\tau + \rho L$ via a Newton-Raphson method, or solve the ODE $\frac{dT}{dE} = \frac{1}{\rho c_L(T)}$ with "initial condition" $T = T_m$ for $E = \rho L$. Any ODE solver can be used for this purpose, e.g. forward Euler, backward Euler, Runge-Kutta, etc.

Actually, the temperature dependence of heat capacities is commonly expressed in the form

$$c_i(T) \;=\; A_i + B_i\,T + \frac{C_i}{T^2}\,, \qquad i = S, L\,, \tag{19}$$

with T in degrees Kelvin and A_i, B_i, C_i given constants. Then the integrals expressing the sensible heat can be computed analytically, and the resulting algebraic equations may be solved very effectively via a Newton-Raphson method. Note that this needs to be done for each node at each time step, adding considerably to the expense of the computation. A reasonable starting value is the temperature at the previous time step, T_j^n.

4.3.C A time-explicit scheme

Choosing $\theta = 0$ in (12), the fluxes are evaluated at the old time t_n and we assume that up to time t_{n+1} the process is driven by these fluxes. The explicit scheme proceeds as follows.

Initially, the phase and temperature of each control volume are known with

$$T_j^0 \;=\; T_{init}(x_j)\,, \qquad j = 1, 2, \ldots, M\,, \tag{20}$$

which in turn determine the enthalpies E_j^0, $j = 1, 2, \ldots, M$, via (10). Assume that we have found enthalpies, temperatures and phase-states (λ_j) through the n-th time step. From (13) we find the liquid fractions, λ_j^n, hence the phase of each node and from (12f) the (mean) temperatures. If only one node is mushy the interface location at time t_n is given by (18).

Now we compute the conductivities from (14), the resistances and fluxes from (12b,c,e), with $\theta=0$, and then E_j^{n+1} is found from (12d), $j=1,2,\ldots,M$. Note that for the boundary control volumes ($j=2, M-1$) we may use the analog to the implicit relation (43) of §**4.1**, in order to guarantee the Maximum Principle and still have the CFL condition guarantee no growth of errors. Thus, the updating of enthalpies to time t_{n+1} is complete.

The stability condition is the same as in the pure conduction case, namely

$$\Delta t_n \leq \frac{1}{2} \frac{(\min \Delta x)^2}{(\max \alpha^n)}, \tag{21}$$

where $\min \Delta x = \min\limits_{j=1,2,\ldots,M} \Delta x_j$ and

$$\max \alpha^n = \max \left\{ \frac{k_L(T_j^n)}{\rho c_j}, \frac{k_S(T_j^n)}{\rho c_j}, \quad j=1,2,\ldots,M \right\}.$$

It is good practice to take a number slightly smaller than ½ in order to avoid stability problems arising from roundoff. The value of $\max \alpha^n$ can be computed at each time step t_n and then we can use $\Delta t_n = \dfrac{1}{2} \dfrac{(\min \Delta x)^2}{\max \alpha^n}$ as the next time step. As this quantity may become impractically small, it is good programming practice to halt the computation if Δt_n becomes smaller than a prescribed minimum Δt, and carefully examine what caused it to become so small.

The great advantage of the time-explicit scheme lies in its simplicity and the ease with which it can be programmed. In situations where the time-step must be small for physical reasons (to capture rapidly moving fronts or resolve rapid changes in data, for example), the stability requirement may not impose undue restrictions, and the explicit scheme may turn out to be as efficient as implicit schemes. The extreme case of laser annealing with picosecond or nanosecond pulses is such a situation [ALEXIADES et al, 1985a].

4.3.D Performance of the explicit scheme on a one-phase problem

To see how the scheme of §**4.3.C** performs, we test it on the simplest problem with known exact solution, namely the one-phase Stefan Problem with constant imposed temperature at $x = 0$. The Neumann (similarity) solution in dimensionless variables appears in (14)-(16) of §**2.1**.

To retain direct physical meaning, we implement the enthalpy scheme on the original formulation (1)-(4) of §**2.1**, which can be made identical to the dimensionless formulation ((19)-(23), §**2.1**) by choosing

$$T_m = 0\,. \qquad \rho = c_L = k_L = 1\,, \qquad T_L = 1\,, \qquad L = \frac{1}{St}\,. \tag{22}$$

We simulate melting in the slab $0 \le x \le 1$ for two extreme values of the Stefan number: $St = .1$ and $St = 5$; the corresponding transcendental roots are found to be $\lambda = .22$ and $\lambda = 1.06$.

To exhibit the convergence of the algorithm, we discretize the slab $0 \le x \le 1$ with $M = 10$, 20 and 40 uniform subintervals. Since $\alpha = k_L/\rho c_L = 1$ (from (22)), the corresponding time steps will be (see (21)) $\Delta t \le \frac{1}{2}(\frac{1}{10})^2, \frac{1}{2}(\frac{1}{20})^2$ and $\frac{1}{2}(\frac{1}{40})^2$, showing the severe limitation on the time step imposed by the stability criterion.

Table 4.3.1 shows temperatures at several locations x, at time $t = 3$ for the problem with $St = 0.1$ and at time $t = .12$ when $St = 5$. At these times the melt fronts have not reached $x = .8$ yet, so there has been no backface influence. The second column shows the exact (Neumann) temperatures found as in **§2.1**. The numerically computed temperatures with $M = 10, 20$ and 40 nodes are listed in the other columns (linearly interpolated from nodal values for $M = 20$ and $M = 40$). Observe the progressive convergence to the exact solution as M increases. It is also interesting to compare the code execution run times for the three mesh sizes: on an IBM-PC/XT, they were 12.5, 41.5 and 260 seconds for $St = 0.1$ and 7, 9 and 24 seconds for $St = 5$, respectively for $M = 10$, 20 and 40 illustrating the dramatic slowdown the finer mesh causes.

In **Table 4.3.2** we compare the exact and numerical interface locations at a few sample times from the same runs as above.

In **Figures 4.3.1** and **4.3.4**, the computed temperature history (melting curve) at a fixed location is compared with the exact (Neumann) solution when $St = 0.1$ and $St = 5$ respectively. The "staircase" shape is characteristic of enthalpy methods and it is much more pronounced for $M = 10$ nodes (**Fig. 4.3.1(a), 4.3.4(a)**) than for

Table 4.3.1: Exact and Computed Temperature Profiles

For **St** $= 0.1$ at time $t = 3.0$					For **St** $= 5$ at time $t = .12$				
x	T_{exact}	$M = 10$	$M = 20$	$M = 40$	x	T_{exact}	$M = 10$	$M = 20$	$M = 40$
0	1.0000	1.0000	1.0000	1.0000	0	1.000	1.000	1.000	1.000
.1	.8667	.8667	.8677	.8672	.1	.8132	.8144	.8135	.8133
.2	.7336	.7333	.7359	.7346	.2	.6341	.6360	.6346	.6342
.3	.6010	.6000	.6051	.6026	.3	.4692	.4711	.4698	.4693
.4	.4691	.4666	.4755	.4712	.4	.3236	.3254	.3243	.3238
.5	.3380	.3333	.3473	.3405	.5	.2003	.2035	.2015	.2004
.6	.2080	.2000	.2203	.2105	.6	.1001	.1076	.1017	.1002
.7	.0793	.0667	.0942	.0809	.7	.0220	.0331	.0193	.0241
.8	0.0	0.0	0.0	0.0	.8	0.0	0.0	0.0	0.0
.9	0.0	0.0	0.0	0.0	.9	0.0	0.0	0.0	0.0
1.0	0.0	0.0	0.0	0.0	1.0	0.0	0.0	0.0	0.0

Table 4.3.2: Exact and Computed Interface Locations

For **St** = 0.1					For **St** = 5				
time	X_{exact}	$M = 10$	$M = 20$	$M = 40$	*time*	X_{exact}	$M = 10$	$M = 20$	$M = 40$
0	0.	0.	0.	0.	0	0.	0.	0.	0.
.5	.3111	.3089	.3101	.3108	.02	.2998	.2995	.3008	.3004
.0	.4400	.4375	.4401	.4399	.04	.4240	.4237	.4243	.4251
1.5	.5389	.5369	.5390	.5388	.06	.5193	.5227	.5197	.5206
2.0	.6223	.6207	.6218	.6225	.08	.5996	.6082	.6023	.6006
					.10	.6704	.6958	.6724	.6722

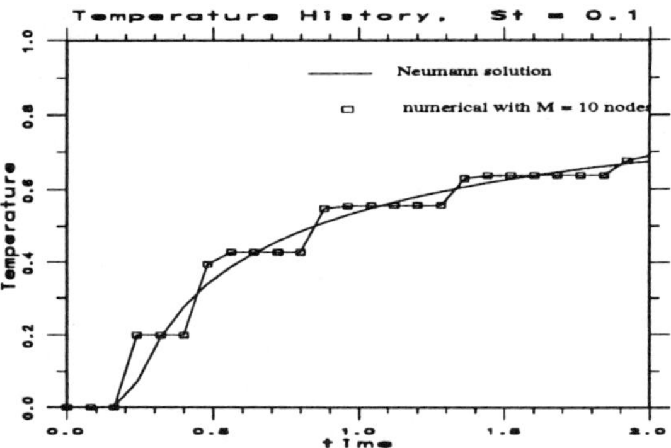

Figure 4.3.1(a). Temperature history at $x = .3$, with $M = 10$ nodes, compared with the exact solution, $St = 0.1$.

$M = 40$ nodes (**Fig. 4.3.1**(b), **4.3.4**(b)). This is due to the fact that while the interface lies anywhere inside a particular mesh interval, the temperature of that interval is held at T_m, so the temperature in the rest of the slab relaxes to a steady state corresponding to a fixed isotherm through that node. When the interface moves to the next mesh interval, the temperature adjusts rapidly and then relaxes to a new steady state. It follows that the duration of each "step" is strictly a function of the time the interface remains in each mesh interval, and therefore, the finer the mesh the shorter the "steps." Indeed, using $M = 80$ nodes the computed and exact solutions would be indistinguishable graphically. **Figures 4.3.2** and **4.3.5** show that the interface location computed with only $M = 10$ and $M = 20$ nodes, for $St = 0.1$ and $St = 5$ respectively, agrees well with the exact interface. Finally, temperature profiles with $M = 10$ and $M = 40$ nodes are shown in **Figures 4.3.3**(a),(b) (at time $t = 2.$, $St = 0.1$) and **Figures 4.3.6**(a),(b) (at time $t = .1$, $St = 5$).

These figures bare out the fact that while numerical methods for phase-change problems can easily capture interface locations and even temperature profiles, the errors show up vividly in temperature history plots.

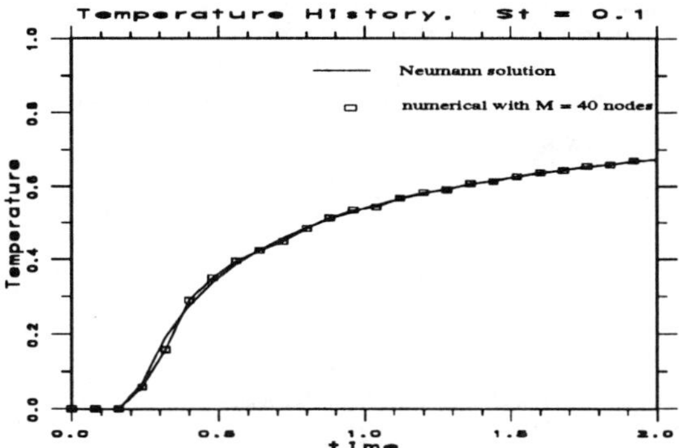

Figure 4.3.1(b). Temperature history at $x = .3$, with $M = 40$ nodes compared with the exact solution, $St = 0.1$.

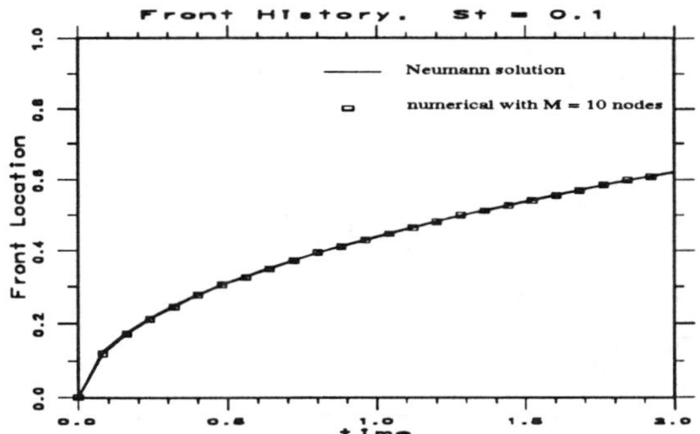

Figure 4.3.2. Melt front location with $M = 10$ nodes, compared with the exact solution, $St = 0.1$.

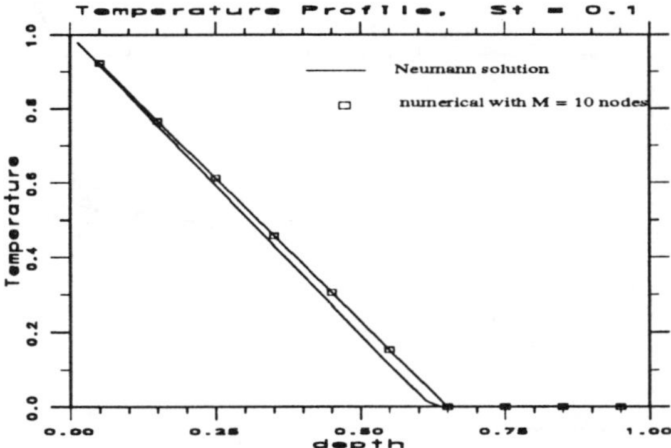

Figure 4.3.3(a). Temperature profile at time $t = 2.$, with $M = 10$ nodes, compared with the exact solution, $St = 0.1$.

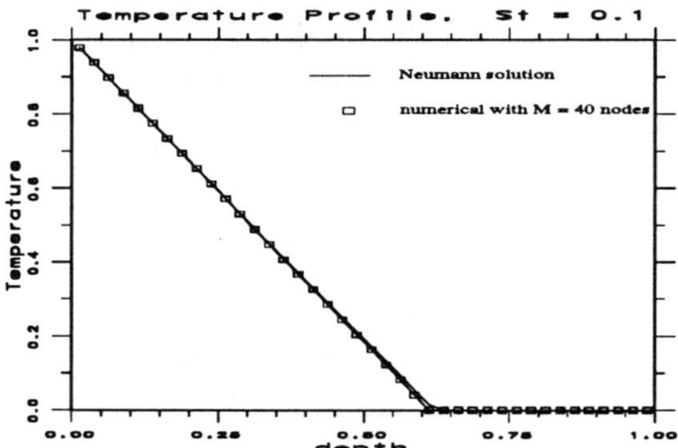

Figure 4.3.3(b). Temperature profile at time $t = 2.$, with $M = 40$ nodes, compared with the exact solution, $St = 0.1$.

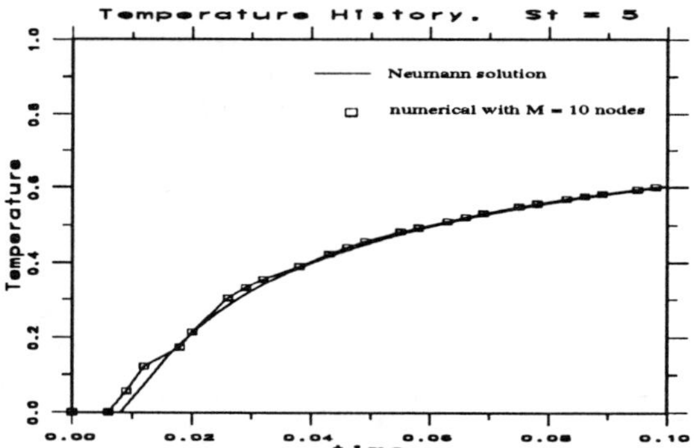

Figure 4.3.4(a). Temperature history at $x = .3$, with $M = 10$ nodes, compared with the exact solution, $St = 5$.

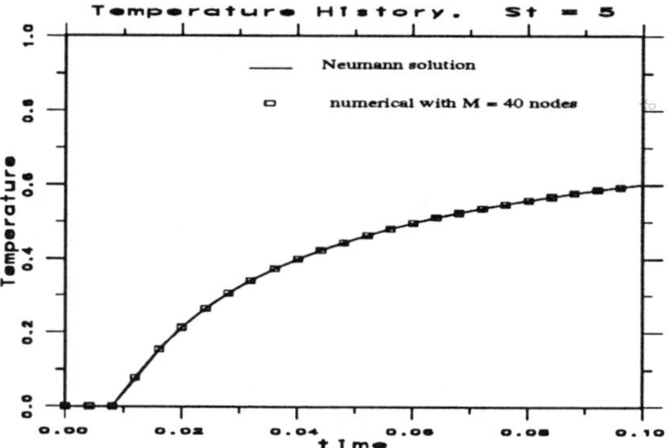

Figure 4.3.4(b). Temperature history at $x = .3$, with $M = 40$ nodes, compared with the exact solution, $St = 5$.

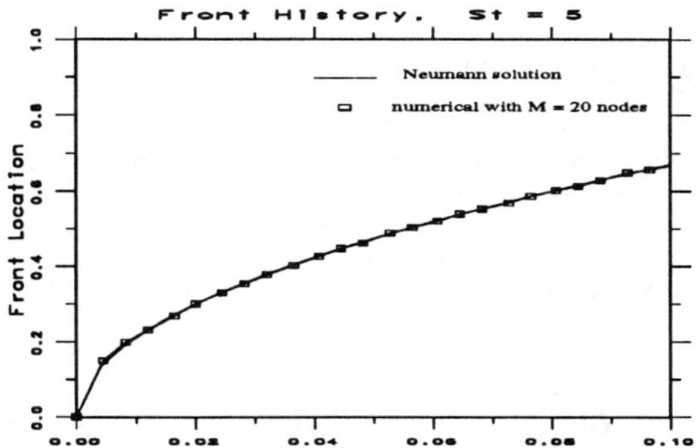

Figure 4.3.5. Melt front location with $M = 20$ nodes, compared with the exact solution, $St = 5$.

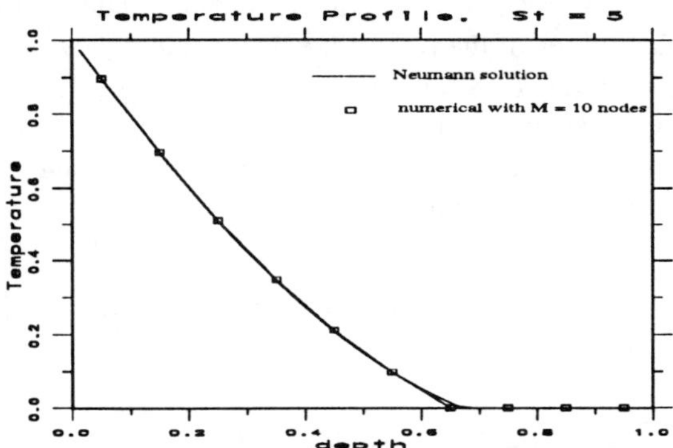

Figure 4.3.6(a). Temperature profile at time $t = .1$, with $M = 10$ nodes, compared with the exact solution, $St = 5$.

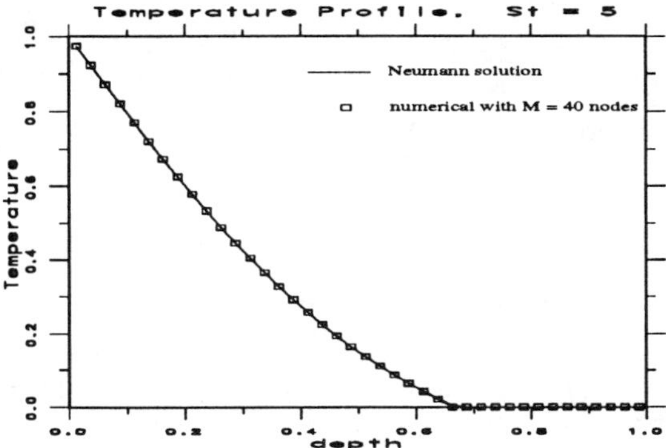

Figure 4.3.6(b). Temperature profile at time $t = .1$, with $M = 40$ nodes, compared with the exact solution, $St = 5$.

4.3.E Implicit schemes

Choosing $0 < \theta \leq 1$ in (12) with $q^{n+\theta} = \theta q^{n+1} + (1-\theta)q^n$ results in a system of equations for the temperatures $T_1^{n+1}, \cdots, T_M^{n+1}$. It is convenient to normalize the temperature scale relative to the melt temperature by introducing the new unknown

$$u = T - T_m , \tag{23}$$

and writing the fluxes $q_{j\pm½}^{n+\theta}$ in terms of the $u_j = T_j - T_m$, $j = 1,2,\ldots,M$, using (12e), as

$$q_{j-½}^{n+\theta} = \theta \frac{u_{j-1}^{n+1} - u_j^{n-1}}{R_{j-½}^{n+1}} + (1-\theta)\frac{u_{j-1}^n - u_j^n}{R_{j-½}^n} , \qquad j = 2,3,\ldots,M . \tag{24}$$

For handling unequal conductivities, a simpler and more efficient implementation results from replacing T by the "Kirchoff temperature", (15), in the conservation law and the equation of state. This is equivalent to the choice (16)-(17), and faster convergence is observed in the iterative scheme, making it more efficient. Whenever the conductivities of the phases are unequal and they are functions of T only and not of location, this transformation should be used. Here we continue the presentation of the general scheme in terms of the original physical variables, delegating the easily derivable simpler scheme to the problems (PROBLEM 9).

Substituting (24) into (12d) we obtain

$$E_j^{n+1} - \frac{\theta\,\Delta t_n}{\Delta x_j}\left[\frac{1}{R_{j+½}^{n+1}}u_{j+1}^{n+1} - (\frac{1}{R_{j+½}^{n+1}} + \frac{1}{R_{j-½}^{n+1}})u_j^{n+1} + \frac{1}{R_{j-½}^{n+1}}u_{j-1}^{n+1}\right]$$

$$= E_j^n + \frac{(1-\theta)\,\Delta t_n}{\Delta x_j}\left[\frac{1}{R_{j+½}^n}u_{j+1}^n - (\frac{1}{R_{j+½}^n} + \frac{1}{R_{j-½}^n})u_j^n + \frac{1}{R_{j-½}^n}u_{j-1}^n\right], \tag{25}$$

for $j = 1,2,\ldots,M$. This form of the equation is valid even for the boundary nodes $j = 1$ and $j = M$ if we take (for convective flux at $x = 0$)

$$u_0 = T_\infty - T_m , \qquad R_{½} = \frac{1}{h} + \frac{\Delta x_1}{2k_1} , \tag{26a}$$

and (for insulated boundary at $x = l$)

$$u_{M+1} = u_M , \qquad R_{M+½} = \frac{\Delta x_M}{2k_M} . \tag{26b}$$

The remaining resistances have the standard form (12e),

$$R_{j-½} = \frac{\Delta x_{j-1}}{2k_{j-1}} + \frac{\Delta x_j}{2k_j} , (j = 2,\ldots,M), \quad R_{j+½} = \frac{\Delta x_j}{2k_j} + \frac{\Delta x_{j+1}}{2k_{j+1}} , (j = 1,\ldots,M-1). \tag{26c}$$

The energy E and new temperature u are related by (c.f. (10))

$$E = \begin{cases} \rho c_S u \,, & u < 0 \\ \rho c_L u + \rho L \,, & u > 0 \end{cases} \tag{27}$$

or (c.f. (11))

$$u = \begin{cases} E \,/\, \rho c_S \,, & E \leq 0 & \text{(solid)} \\ 0 \,, & 0 < E < \rho L & \text{(interface)} \\ E - \rho L \,/\, \rho c_L \,, & \rho L \leq E & \text{(liquid).} \end{cases} \tag{28}$$

The right-hand side of (25) is a known quantity (everything is evaluated at the old time t_n), which we shall denote by b_j^n, namely we set

$$b_j^n = E_j^n + (1-\theta)\frac{\Delta t_n}{\Delta x_j}\left[\frac{1}{R_{j+½}^n}u_{j+1}^n - (\frac{1}{R_{j+½}^n}+\frac{1}{R_{j-½}^n})u_j^n + \frac{1}{R_{j-½}^n}u_{j-1}^n\right], \tag{29}$$

for $j = 1, 2, \ldots, M$. Then, the system for the unknowns $u_1^{n+1}, \cdots, u_M^{n+1}$ takes the form

$$E_j^{n+1} - \theta\frac{\Delta t_n}{\Delta x_j}\left[\frac{1}{R_{j+½}^{n+1}}u_{j+1}^{n+1} - (\frac{1}{R_{j+½}^{n+1}}+\frac{1}{R_{j-½}^{n+1}})u_j^{n+1} + \frac{1}{R_{j-½}^{n+1}}u_{j-1}^{n+1}\right] = b_j^n \,, \tag{30}$$

for $j = 1, 2, \ldots, M$ together with (26)-(27). This system *cannot* be solved directly (by Gauss elimination), because, due to (27) it is actually a *nonlinear* system, and may only be solved by an iteration method.

A very convenient iterative method is the **Gauss-Seidel** iteration, in which the j-th equation is solved for the j-th unknown using the *latest* values available for all other variables [YOUNG-GREGORY], [GOLUB-VanLOAN], [PRESS et al]. Let us describe the method for a standard linear system $A\vec{u} = \vec{b}$, with $A = [a_{ij}]$ an $M \times M$ matrix, $\vec{u}$, $\vec{b}$ $M \times 1$ vectors. Using a superscript (p) to denote the p-th iteration, $p = 0, 1, 2, \ldots$, the iterations begin with an initial "guess," $\vec{u}^{(0)}$, and then, for $p = 0, 1, 2, \ldots$, one solves the j-th equation for u_j, $j = 1, 2, \ldots, M$:

$$u_j^{p+1} = \frac{1}{a_{jj}}[\, b_j - a_{j1}u_1^{(p+1)} - \cdots - a_{j\,j-1}u_{j-1}^{(p+1)} - a_{j\,j+1}u_{j+1}^{(p)} - \cdots - a_{jM}u_M^{(p)}\,], \tag{31}$$

until some convergence criterion is met (typically $\max_{1\leq j\leq M} |u_j^{(p+1)} - u_j^{(p)}| <$ TOLERANCE). Then one should also check the *residual* $A\vec{u} - b$ to make sure it is acceptably small.

For our system, (30), this amounts to solving the equation

$$E_j^{p+1} + \frac{\theta\Delta t_n}{\Delta x_j}\left[\frac{1}{R_{j+½}^{(p)}} + \frac{1}{R_{j-½}^{(p)}}\right]u_j^{(p+1)} = b_j^n + \frac{\theta\Delta t_n}{\Delta x_j}\left[\frac{1}{R_{j+½}^{(p)}}u_{j+1}^{(p)} + \frac{1}{R_{j-½}^{(p)}}u_{j-1}^{(p+1)}\right] \tag{32}$$

for $u_j^{(p+1)}$, $j = 1, 2, \ldots, M$ for each $p = 0, 1, 2, \ldots$ until convergence. Note that the right-hand side of (32) and the coefficient of $u_j^{(p+1)}$ contain only known terms (b_j^n from the old time-step, (29), $u_{j-1}^{(p+1)}$ because it was just found by solving the $(j-1)$-st equation, and $u_{j+1}^{(p)}$, $R_{j\pm½}^{(p)}$ come from the previous iteration). Remember, however, that the energy, $E_j^{(p+1)}$, itself depends on the current unknown $u_j^{(p+1)}$ via (27), exhibiting the nonlinearity of system (32). In fact, E_j as a function of u_j is **multivalued** when $u_j = 0$ (meaning $T_j = T_m$), as discussed in **§4.4**, since it may be any number between 0 and ρL (we have interpreted $E/\rho L$ as the liquid fraction in **§4.3.B**). To bring out the essence of equation (32), let us give names to the known terms, namely, set

$$C_j^{(p)} = \theta \frac{\Delta t_n}{\Delta x_j} \left(\frac{1}{R_{j+½}^p} + \frac{1}{R_{j-½}^p}\right), \quad j = 1, 2, \ldots, M, \tag{33}$$

$$z_j^{(p)} = b_j^n + \theta \frac{\Delta t_n}{\Delta x_j} \left[\frac{1}{R_{j-½}^p} u_{j-1}^{(p+1)} + \frac{1}{R_{j+½}^p} u_{j+1}^{(p)} \right], \quad j = 1, \ldots, M. \tag{34}$$

Then (32) takes the form

$$E_j^{(p+1)} + C_j^{(p)} u_j^{(p+1)} = z_j^{(p)}, \quad j = 1, 2, \ldots, M, \tag{35}$$

for $p = 0, 1, 2, \ldots$, with $C_j^{(p)}$ and $z_j^{(p)}$ known quantities, and $E_j^{(p+1)}$, $u_j^{(p+1)}$ related by (28), from which we want to find the phase and temperature. Let us examine the three possibilities of node V_j being solid ($E_j \le 0$), mushy ($0 < E_j < \rho L$) or liquid ($E_j \ge \rho L$).

If the node is solid, then $u_j^{(p+1)} \le 0$, $E_j^{(p+1)} = \rho c_S u_j^{(p+1)} \le 0$ and, from (35),

$$z_j^{(p)} = \rho c_S u_j^{(p+1)} + C_j^{(p)} u_j^{(p+1)} \le 0, \tag{36}$$

whence

$$u_j^{(p+1)} = \frac{z_j^{(p)}}{\rho c_S + C_j^{(p)}}. \tag{37}$$

If the node is mushy, then $u_j^{(p+1)} = 0$, $0 < E_j^{(p+1)} < \rho L$ and, from (35),

$$0 < z_j^{(p)} = E_j^{(p+1)} < \rho L. \tag{38}$$

If the node is liquid, then $u_j^{(p+1)} \ge 0$, $E_j^{(p+1)} = \rho c_L u_j^{(p+1)} + \rho L$, and, from (35),

$$z_j^{(p)} = \rho c_L u_j^{(p+1)} + \rho L + C_j^{(p)} u_j^{(p+1)} \ge \rho L, \tag{39}$$

whence

$$u_j^{(p+1)} = \frac{z_j^{(p)} - \rho L}{\rho c_L + C_j^{(p)}}. \tag{40}$$

We observe that the (known) quantity $z_j^{(p)}$ does in fact determine the phase of the node V_j by its size according to (36), (38), (39). Hence the solution of (35) is

$$u_j^{(p+1)} = \begin{cases} \dfrac{z_j^{(p)}}{\rho c_S + C_j^{(p)}}, & \text{if } z_j^{(p)} \leq 0, \\ 0, & \text{if } 0 < z_j^{(p)} < \rho L, \\ \dfrac{z_j^{(p)} - \rho L}{\rho c_L + C_j^{(p)}}, & \text{if } \rho L \leq z_j^{(p)}, \end{cases} \tag{41}$$

and the Gauss-Seidel iteration proceeds as follows: for each $p = 0, 1, 2, \ldots$, knowing $u_j^{(p)}$, $j = 1, 2, \ldots, M$, we compute $z_j^{(p)}$ and $C_j^{(p)}$ from (34), (33), obtain $u_j^{(p+1)}$ from (41), $j = 1, 2, \ldots, M$, and repeat for the next p, until

$$\max_{1 \leq j \leq M} |u_j^{(p+1)} - u_j^{(p)}| \leq TOLERANCE. \tag{42}$$

As initial iterates we take the values from the previous time-step

$$u_j^{(0)} = u_j^n = T_j^n - T_m, \qquad j = 1, 2, \ldots, M, \tag{43}$$

and the boundary data $u_0^{n+1} = T_\infty^{n+1} - T_m$ remain fixed during the iteration.

Upon convergence, i.e. when (42) is satisfied, we have obtained the new temperatures, $u_j^{n+1} = u_j^{(p+1)}$, $j = 1, 2, \ldots, M$, (for the latest p), and then from (35) we also obtain the new energies

$$E_j^{n+1} = z_j - C_j u_j^{n+1}, \qquad j = 1, 2, \ldots, M, \tag{44}$$

z_j, C_j denoting the latest values of these quantities (upon meeting the convergence criterion).

The SOR algorithm (successive over-relaxation) is a standard method for accelerating the convergence of the Gauss-Seidel iteration by linear extrapolation, [YOUNG-GREGORY], [LAPIDUS-PINDER], [GOLUB-VanLOAN]). Elliott and Ockendon have suggested a partial SOR method for phase-change problems ([ELLIOTT], [ELLIOTT-OCKENDON]) which uses relaxation only when there is **no** phase change. The Elliott-Ockendon SOR algorithm is as follows:

Let ω be a relaxation parameter, $0 < \omega < 2$, to be chosen later. Knowing u_j^n, E_j^n, $(j = 1, 2, \ldots, M)$ at time t_n, set $u_j^{(0)} = u_j^n$ $(j = 1, \ldots, M)$ and for $p = 0, 1, 2, \ldots$ do the following:

1. Compute $z_j^{(p)}$, $C_j^{(p)}$ from (34), (33).
2. Find the Gauss-Seidel iterate $\bar{u}_j$ from (41).
3. Set $$\hat{u}_j = u_j^{(p)} + \omega [\bar{u}_j - u_j^{(p)}]. \tag{45}$$

4. Set

$$u_j^{(p+1)} = \begin{cases} \hat{u}_j & \text{if } \hat{u}_j \cdot u_j^{(p)} > 0, \\ \tilde{u}_j & \text{if } \tilde{u}_j \cdot u_j^{(p)} \leq 0, \end{cases} \tag{46}$$

i.e. over-relax only at the nodes that have **not** just changed phase.

5. Test for convergence: if (42) is satisfied then set

$$u_j^{n+1} = u_j^{(p+1)}, \qquad \text{hence } T_j^{n+1} = T_m + u_j^{n+1}, \tag{47}$$

$$E_j^{n+1} = z_j^{(p)} - C_j^{(p)} \cdot u_j^{n+1}, \tag{48}$$

otherwise go to step 1.

It has been shown, [ELLIOTT], that this SOR scheme is convergent for any $0 < \omega < 2$, to the **weak** solution of the Stefan problem. Of course, the convergence is accelerated only if $1 < \omega < 2$, $\omega = 1$ being exactly the Gauss-Seidel iteration.

The relaxation parameter, ω, is chosen according to linear SOR theory for the discrete heat equation. Replacing E_j^{n+1} by $\rho c T_j^{n+1}$ in (25) and assuming constant properties and uniform Δx, Δt, the resulting linear tri-diagonal system (**§4.1.E**) (for the internal nodes) may be written in the form

$$[D - (L + U)]\vec{u}^{n+1} = \vec{b}, \tag{49}$$

where $D = (1 + 2\mu)I$, $\mu = \theta \dfrac{\alpha \Delta t}{\Delta x^2}$, I the $M \times M$ identity matrix, and

$$L + U = \begin{bmatrix} 0 & \mu \cdots 0 & 0 \\ \mu & 0 \cdots\cdots\cdots \\ 0 & \mu \cdots\cdots\cdots \\ \cdots\cdots\cdots\cdots & \mu \\ 0 \; 0 \; 0 & \mu \; 0 \end{bmatrix}. \tag{50}$$

The optimal relaxation parameter of (linear) SOR theory is given by [YOUNG-GREGORY], [GOLUB-VanLOAN],

$$\omega_{optimal} = \frac{2}{1 + \sqrt{1 - \bar{\rho}^2}}, \tag{51}$$

where $\bar{\rho}$ denotes the spectral radius (largest magnitude of eigenvalues) of the "Jacobi iteration matrix" $D^{-1}(L + U) = \dfrac{1}{1 + 2\mu}(L + U)$. Fortunately, the eigenvalues of (50) are known and the value of $\bar{\rho}$ turns out to be

$$\bar{\rho} = \frac{2\mu}{1 + 2\mu} \cos \frac{\pi}{M + 1}. \tag{52}$$

In phase-change problems, however, the parameter $\mu = \theta \dfrac{\alpha \Delta t}{\Delta x^2}$ changes value because the diffusivity α changes from α_L to α_S, thus invalidating the theoretical conditions of SOR theory. [ELLIOTT-OCKENDON, p. 86] have suggested that ω be calculated from (51-52) by taking as α the average

$$\alpha = \frac{N^L}{M}\alpha_L + (1 - \frac{N^L}{M})\alpha_S, \tag{53}$$

where N^L is the number of liquid nodes. Clearly, the resulting SOR algorithm (45-48) is somewhat empirical, and some experimentation with various values of ω may be advisable.

4.3.F Implicit scheme with Newton iteration

Since (30) is a nonlinear system, we may try to solve it by the Newton-Raphson method which has the promise of up to quadratic convergence, whereas the nonlinear SOR can achieve linear convergence at best. The computational drawbacks are that a linear system must be formed and solved at each iteration and the quadratic rate can only be attained in a small neighborhood of the solution. Thus, the amount of work needed to gain quadratic convergence must be balanced against the efficiency of the nonlinear SOR. In addition, a Newton iteration may fail to converge in some cases.

The Newton-Raphson method involves the derivative of the $u = u(E)$ curve, (28), in the Jacobian of the system. Since the derivative experiences jumps at $E = 0$ and at $E = \rho L$, we are faced with conceptual difficulties as to the meaning and existence of this derivative. The mathematical issues are dicsussed in [KELLEY-RULLA] who give precise meaning to the Jacobian as a Gateaux derivative. They show that superlinear convergence is to be expected, to the weak solution (§4.4) of the elliptic problem whose discrete form is (30).

A practical way to overcome the conceptual difficulty is to approximate the curve by a smooth one near those two points. For example, we could use parabolas to smooth the corners of the curve using the approximation, for small $\varepsilon > 0$,

$$u_\varepsilon(E) = \begin{cases} \dfrac{E}{\rho c_S}, & E \le -\varepsilon \\ -\dfrac{(E-\varepsilon)^2}{4\varepsilon\rho c_S}, & -\varepsilon < E < \varepsilon \\ 0, & \varepsilon \le E \le \rho L - \varepsilon \\ \dfrac{(E - \rho L + \varepsilon)^2}{4\varepsilon\rho c_L}, & \rho L - \varepsilon < E < \rho L + \varepsilon \\ \dfrac{E - \rho L}{\rho c_L}, & E \ge \rho L + \varepsilon \end{cases}$$

It turns out however that the smoothing plays essentially no role numerically, so we may simply use the curve itself. The slopes of the $u = u(E)$ are

$$u'(E) = \begin{cases} 1/\rho c_S, & E \le 0, \\ 0, & 0 \le E \le \rho L, \\ 1/\rho c_L, & \rho L \le E. \end{cases} \tag{54}$$

We think of the system to be solved, (30), as the equation $\vec{F}(\vec{E}) = \vec{0}$, by setting

$$F_j(E_j^{n+1}) = E_j^{n+1} - \theta\frac{\Delta t_n}{\Delta x_j}\left[\frac{1}{R_{j+½}^{n+1}}u_{j+1}^{n+1} - \left(\frac{1}{R_{j+½}^{n+1}} + \frac{1}{R_{j-½}^{n+1}}\right)u_j^{n+1} + \frac{1}{R_{j-½}^{n+1}}u_{j-1}^{n+1}\right] - b_j^n,$$
$$j = 1, 2, \ldots, M,$$

with b_j^n, from (29), containing the old-time values. We view u_j^{n+1} as $u(E_j^{n+1})$, given by (28), and we want to find $\vec{E^{n+1}}$. The Newton algorithm applied to the system $\vec{F}(\vec{E}) = \vec{0}$ consists of the iteration ($p = 0, 1, 2, \ldots$):

Find $\Delta\vec{E}$ by solving the linear system

$$\frac{\partial\vec{F}}{\partial\vec{E}}(\vec{E^{(p)}})\,\Delta\vec{E} = -\,\vec{F}(\vec{E^{(p)}})\,; \tag{55}$$

correct $\vec{E}$ via

$$\vec{E^{(p+1)}} = \vec{E^{(p)}} + \Delta\vec{E}; \tag{56}$$

and then compute u from (28); repeat until convergence is attained (see below). Here $\frac{\partial\vec{F}}{\partial\vec{E}}(\vec{E^{(p)}})$ denotes the Jacobian matrix of $\vec{F}(\vec{E})$ with respect to $\vec{E}$. This is a tridiagonal matrix, with components

$$\begin{aligned} \frac{\partial F_j}{\partial E_{j-1}} &= -\frac{\theta\Delta t_n}{\Delta x_j}\frac{1}{R_{j-½}^{(p)}}u'(E_{j-1}^{(p)}), \\ \frac{\partial F_j}{\partial E_j} &= 1 + \frac{\theta\Delta t_n}{\Delta x_j}\left(\frac{1}{R_{j+½}^{(p)}} + \frac{1}{R_{j-½}^{(p)}}\right)u'(E_j^{(p)}), \\ \frac{\partial F_j}{\partial E_{j+1}} &= -\frac{\theta\Delta t_n}{\Delta x_j}\frac{1}{R_{j+½}^{(p)}}u'(E_{j+1}^{(p)}), \qquad j = 1, \ldots, M. \end{aligned} \tag{57}$$

Since this matrix is diagonally dominant (§**4.1.F**), the system in (55) may be solved very efficiently by the direct tridiagonal algorithm.

As it is typical with the Newton method, the values of the unknown $\vec{E}$ may appear to have converged even though $\vec{E}$ is far from the solution. It is necessary to check convergence not only of the unknowns but also of the residual. Hence, the appropriate stopping criteria are

$$\max_{1\le j\le M}|\Delta E_j| < TOLERANCE_1 \quad \text{or} \quad \max_{1\le j\le M}|u_j^{(p+1)} - u_j^{(p)}| < TOLERANCE_1\,,$$

and

$$\max_{1 \le j \le M} | E_j^{(p)} + C_j^{(p)} u_j^{(p)} - z_j^{(p)} | < TOLERANCE_2 .$$

Convergence of the residuals need not be as strict. With $TOLERANCE_1$ taken as 10^{-6}, $TOLERANCE_2 = 10^{-3}$ ensures convergence without wasting iterations.

The Newton iteration will converge if the initial guess, taken as $\vec{E}^{(0)} = \vec{E}^n$, is within the radius of convergence. This can be ensured by taking the time step, Δt_n, small enough. How small is not obvious, but typically the time step can be quite large (see next section). Note that each iteration involves forming the Jacobian and solving the linear system in (55), so it is roughly at least twice as expensive as an S.O.R. iteration. Nevertheless, the faster convergence makes the scheme considerably more efficient than S.O.R. as the experiments reported in the next section indicate.

4.3.G Performance of the schemes on the two-phase Stefan problem

To see how the schemes of **§4.3.C,E,F** perform, we test them on the two-phase Stefan Problem with constant imposed temperature at $x = 0$ by comparison with the exact solution. The exact (Neumann) solution in dimensionless variables appears in **§2.2.B**. It depends on the dimensionless quantities $\boldsymbol{St}_L$, $\boldsymbol{St}_S$, ν, and U. To retain direct physical meaning, we implement the enthalpy scheme on the original formulation, (1)**§2.2**, which can be made identical to the dimensionless formulation (representing a generic two-phase problem) by choosing

$$T_m = 0 , \quad T_L = 1 , \quad T_S = -1 , \qquad \rho = c_L = k_L = 1 . \tag{58}$$

Input values for $\boldsymbol{St}_L$, $\boldsymbol{St}_S$, ν and U then determine all the other physical variables.

We simulate melting in the slab $0 \le x \le 1$, until the melt front reaches the middle of the slab, for three different Stefan numbers: $\boldsymbol{St}_L = \boldsymbol{St}_S =: \boldsymbol{St} = 0.1$, $1.$, and 5, with $\nu = 1$, $U = 1$. In view of the normalizations in (58), we are simulating the melting of a solid, initialy at $T_S = -1$, by imposing $T_L = 1$ at $x = 0$, when all the thermophysical parameters are equal to 1 and the latent heat is $L = 1/\boldsymbol{St}$. Thus, the melting process is slow for $\boldsymbol{St} = 0.1$, moderate for $\boldsymbol{St} = 1$, and fast for $\boldsymbol{St} = 5$. The corresponding transcendental roots for the Neumann solution are $\lambda = 0.189134$, $\lambda = 0.3777598$, and $\lambda = 0.450161$, and the half-melt-times are found to be $t_{½melt} = 1.75$, 0.438 and 0.308, respectively. At the back face $x = 1$ we impose the temperature values of the Neumann solution itself, evaluated at $x = 1$ at any desired time t. This way we are solving numerically the same problem that the Neumann solution solves and we can make valid comparisons between the numerical and exact (Neumann) solutions.

On each of these three model problems, we compare the performance of the *explicit* scheme with that of the *fully implicit* ($\theta = 1$) schemes with S.O.R. and Newton iterations, for $M = 40$ and for $M = 80$ nodes, as we increase the time-step Δt of the implicit schemes in multiples of the explicit time-step ($\Delta t_{expl} = \Delta x^2 / 3\alpha$), i.e. using $\Delta t = factor \times \Delta t_{expl}$ with $factor$ = 1, 5, 10, 20, ... The implicit schemes remain stable but the accuracy degrades eventually, as expected (see §**4.3.I**). The results are reported in the tables that follow. We output three measures of error for each run, obtained by comparison with the Neumann solution:

max X *error* = maximum error in melt-front location over $0 \le t \le t_{½melt}$,
max T *error* = maximum error in Temperature at the nodes over $0 \le t \le t_{½melt}$,
max L^1*error* = maximum L^1-error in Temperature at the nodes over $0 \le t \le t_{½melt}$, (computed via trapezoidal rule integration).

To test the effect of the S.O.R. iteration, we make runs with both $\omega = 1$ (Gauss-Seidel) and with various other values of ω, including the $\omega_{optimal}$, (51), of the linear S.O.R. (value marked with * in the Tables). Note that the parameter μ does

Table 4.3.3: Performance of the Explicit and Implicit Schemes
For $St_L = St_S = 0.1$ with $M = 40$ nodes

factor	time steps	ω	max X error	max T error	max L^1*error*	# of iterations	implicit advantage	CPU
explicit								
1	9321		.007	.09	.012			30.0u
implicit								
1	9321	1.0	.008	.087	.014	43619	0.20	132.7u
		*1.07	.008	.087	.014	39236	0.20	88.0u
		1.2	.008	.087	.014	27692	0.30	60.0u
		Newton	.008	.087	.014	18666	0.50	72.8u
10	933	1.0	.007	.084	.018	18728	0.50	22.2u
		1.3	.007	.084	.018	12501	0.70	15.7u
		*1.4	.007	.084	.018	13969	0.70	22.4u
		Newton	.007	.084	.019	1886	4.94	7.4u
20	467	1.0	.007	.084	.024	16603	0.60	19.6u
		1.4	.007	.084	.024	9126	1.00	11.3u
		*1.52	.007	.084	.024	10579	0.90	13.6u
		Newton	.004	.084	.024	981	9.50	13.7u
40	235	1.0	.007	.070	.017	14845	0.63	16.1u
		1.5	.007	.070	.019	6203	1.50	7.2u
		*1.62	.007	.070	.019	6787	1.40	7.7u
		Newton	.007	.070	.019	519	18.0	7.4u
80	118	1.6	.015	.110	.031	3753	2.51	4.4u
		Newton	.004	.110	.031	301	31.1	4.2u
200	49	1.7	.037	.100	.043	1974	4.86	2.4u
		*1.77	.037	.100	.043	2319	4.1	2.7u
		Newton	.007	.100	.043	160	58.8	2.3u

not vary here since $\alpha_L = \alpha_S$ in these runs. It is observed that this $\omega_{optimal}$ is not necessarily optimal ! We keep increasing the time-step *factor* and vary the ω until one of the errors exceeds 10%.

As measures of performance, we report the number of time-steps executed, the total number of iterations performed (tolerance set to 10^{-6}, in single precision arithmetic), and also the ratio

$$\textit{implicit-advantage} = \frac{\textit{time-step factor}}{\textit{iterations per time-step}}.$$

as a way of gauging if the implicit schemes do better or worse than the explicit scheme. This compares the number of time-steps executed by the explicit scheme with the total number of iterations. Since the computation performed during an S.O.R. iteration is essentially equivalent to an explicit time-step computation, this ratio compares the computational costs. A value less than 1 indicates that the explicit scheme performed better than the implicit scheme.

Table 4.3.4: Performance of the Explicit and Implicit Schemes
For $St_L = St_S = 0.1$ with $M = 80$ nodes

factor	time steps	ω	max X error	max T error	max L^1 error	# of iterations	implicit advantage	CPU
explicit								
1	37286		.003	.049	.006			276.0u
implicit								
1	37286	1.0	.004	.049	.007	142019	0.26	565.0u
		Newton	.004	.049	.007	74715	0.50	555.5u
20	1865	1.4	.004	.048	.012	31423	1.19	99.9u
		*1.53	.004	.048	.012	35805	1.04	84.0u
		Newton	.002	.049	.017	3794	9.83	28.5u
40	934	1.5	.004	.048	.009	23365	1.60	52.8u
		*1.64	.003	.04	.009	27520	1.36	65.0u
		Newton	.004	.048	.009	2014	18.5	94.9u
100	374	1.6	.026	.073	.019	16065	2.33	37.4u
		*1.75	.026	.073	.019	18795	1.99	38.1u
		Newton	.002	.074	.019	927	40.2	42.8u
200	188	*1.81	.018	.097	.023	10619	3.54	22.4u
		Newton	.003	.097	.023	563	66.4	25.5u
400	95	1.8	.016	.103	.033	5718	6.65	11.5u
		Newton	.014	.104	.033	10342	3.6	438.8u

Table 4.3.5: Performance of the Explicit and Implicit Schemes
For $St_L = St_S = 1.$ with $M = 40$ nodes

factor	time steps	ω	max X error	max T error	max L^1 error	# of iterations	implicit advantage	CPU
explicit	2359		.009	.056	.007			9.9u
implicit								
1	2359	1.0	.014	.057	.011	9942	0.24	31.2u
		*1.07	.014	.057	.011	8710	0.27	18.7u
		1.2	.014	.057	.011	9796	0.24	20.3u
		Newton	.014	.057	.011	4692	0.50	20.8u
5	472	1.2	.033	.060	.015	5214	0.45	7.7u
		*1.28	.033	.060	.015	5408	0.44	7.7u
		Newton	.008	.060	.015	957	2.45	4.6u
10	237	1.3	.033	.060	.016	3735	0.63	4.8u
		*1.406	.330	.060	.016	4042	0.59	5.2u
		Newton	.008	.060	.016	489	4.79	2.2u
20	119	1.4	.051	.074	.019	2063	1.15	4.5u
		*1.52	.051	.074	.019	2979	0.80	3.6u
		Newton	.017	.074	.019	257	9.11	1.2u
30	80	1.0	.080	.111	.027	3595	0.67	7.1u
		Newton	.012	.113	.027	198	11.8	2.9u

Table 4.3.6: Performance of the Explicit and Implicit Schemes
For $St_L = St_S = 1.$ with $M = 80$ nodes

factor	time steps	ω	max X error	max T error	max L^1 error	# of iterations	implicit advantage	CPU
explicit	9370		.004	.035	.003			71.5u
implicit								
1	9370	1.0	.007	.036	.005	44902	0.21	158.5u
		*1.07	.007	.036	.005	40848	0.23	160.6u
		Newton	.007	.036	.005	18726	0.50	169.9u
10	939	1.0	.016	.033	.008	20476	0.46	59.0u
		*1.41	.016	.033	.008	13953	0.67	36.8u
		Newton	.004	.033	.008	1911	4.89	17.2u
40	235	*1.64	.065	.052	.014	7931	1.18	18.4u
		Newton	.009	.052	.014	546	17.14	5.3u
100	95	*1.75	.073	.079	.021	4934	1.92	11.4u
		Newton	.014	.079	.037	(a)	0.05	8522.2u
200	48	*1.81	.070	.146	.033	2964	3.24	6.8u
		Newton	.015	.146	.041	500192	0.02	2312.0u

(a) exceeded the limit of 10^5 iterations at two of the time steps

Table 4.3.7: Performance of the Explicit and Implicit Schemes For $St_L = St_S = 5.$ with $M = 40$ nodes

factor	time steps	ω	max X error	max T error	max L^1 error	# of iterations	implicit advantage	CPU
explicit	1697		.035	.025	.003			8.0u
implicit								
1	1697	1.0	.032	.032	.008	10056	0.17	19.6u
		*1.07	.032	.032	.008	8901	0.19	16.3u
		1.1	.032	.032	.008	8776	0.19	20.3u
		Newton	.013	.032	.008	3310	0.50	15.1u
10	171	1.3	.087	.061	.018	2764	0.62	4.4u
		*1.41	.087	.061	.018	2923	0.58	4.0u
		Newton	.018	.061	.018	352	4.69	1.6u
20	87	1.52	.078	.109	.025	2190	0.79	2.8u
		Newton	.017	.109	.025	204	8.14	4.2u

Table 4.3.8: Performance of the Explicit and Implicit Schemes For $St_L = St_S = 5.$ with $M = 80$ nodes

factor	time steps	ω	max X error	max T error	max L^1 error	# of iterations	implicit advantage	CPU
explicit	6669		.017	.019	.0014			63.1u
implicit								
1	6669	1.0	.016	.019	.004	33894	0.20	141.9u
		*1.07	.016	.019	.004	29535	0.23	107.6u
		Newton	.006	.020	.004	13200	0.50	137.5u
10	668	1.3	.044	.026	.009	9553	0.70	26.5u
		*1.41	.044	.026	.009	10077	0.66	25.1u
		Newton	.009	.026	.009	1358	4.85	11.5u
40	168	*1.64	.078	.066	.017	5694	1.18	14.0u
		Newton	.014	.066	.017	406	16.3	3.3u
50	135	*1.67	.062	.074	.019	5206	1.30	11.6u
		Newton	.012	.074	.019	331	19.9	2.9u
55	123	*1.68	.102	.079	.020	5002	1.35	11.2u
		Newton	.015	.079	.020	323	20.4	2.6u

The computations were performed on a SUN 4/110 workstation, and we report the CPU units (user time) given by the UNIX time command. Note that since the temperature range is normalized to $-1 \le T \le 1$, the temperature errors (5th and 6th columns) may be viewed as percent errors.

Several observations may be gleaned from the tables.

1. The implicit schemes are more avantageous for slow phase change process than for faster ones. Thus, for $St = 0.1$, large time steps may be used to good advantage, but for $St = 5$ the Δt factor can only be pushed up to about 15 (for $M = 40$) or 50 (for $M = 80$).
2. Performance is poor when Gauss-Seidel iterations ($\omega = 1$) are used in solving the nonlinear system. S.O.R ($1 < \omega < 2$) speeds up convergence considerably. Thus, for the slow-melting problem ($St = 0.1$), the S.O.R. scheme with $\Delta t = 40 \times \Delta t_{expl}$ and $\omega = 1.5$ achieves the same accuracy as the explicit scheme at lower cost.
3. The Newton iteration scheme outperforms the others, provided the time-step is not pushed up too high. At about $40 \times \Delta t_{expl}$ it achieves very high efficiencies.

4.3.H Performance on problems with unequal properties

When the thermophysical properties of the phases are not equal, the maximum error in temperature may degrade, but in all other aspects the performance is similar to that with equal properties. Sample results are shown in **Tables 4.3.9 - 12.**

Keeping the conductivities equal to 1, we examine the case of different Stefan numbers first (different specific heats). In **Table 4.3.9** we take $St_L = 0.1$ and $St_S = 5$, for which the trancendental root is $\lambda = 0.0697816$, and we run up to *time* = 30 at which 3/4ths of the slab has melted. In **Table 4.3.10** we take $St_L = 5$, $St_S = 0.1$, for which the root is $\lambda = 0.946083$, and we run up to *time* = .1 by which half the slab has melted. We observe that the SOR scheme is very efficient on both the slow and the fast problem. These were run on a SUN/Sparcstation2.

Table 4.3.9: Performance of the Explicit and Implicit Schemes
For $St_L = 0.1$, $St_S = 5$. with $M = 40$ nodes, up to time= 30

factor	time steps	ω	max X error	max T error	max L^1 error	# of iterations	implicit advantage	CPU
explicit	160700		.016	.058	.008			207.5u
implicit								
1	160700	1.0	.016	.058	.008	417942	0.38	322.2u
		Newton	.016	.058	.008	320423	0.50	2176.8u
10	16070	1.1	.017	.058	.009	104307	1.53	50.6u
		Newton	.016	.058	.009	32070	5.0	134.2u
20	8040	1.5	.022	.057	.009	95261	1.68	38.2u
		Newton	.017	.057	.009	16077	10.0	67.3u
80	2010	1.6	.038	.061	.011	40689	3.93	14.6u
		Newton	.017	.061	.011	4062	39.4	16.9u

Table 4.3.10: Performance of the Explicit and Implicit Schemes For $St_L = 5.$, $St_S = 0.1$ with $M = 40$ nodes, up to time= .1

factor	time steps	ω	max X error	max T error	max L^1 error	# of iterations	implicit advantage	CPU
explicit	29340		.008	.039	.009			31.8u
implicit								
1	29340	1.0	.007	.036	.009	95408	0.28	61.0u
		Newton	.008	.036	.009	53377	0.50	238.1u
20	1470	1.5	.006	.045	.014	23342	1.14	8.5u
		Newton	.006	.045	.014	2717	9.8	12.2u
60	491	1.6	.028	.047	.018	11425	2.33	4.1u
		Newton	.004	.047	.018	944	28.3	4.2u
100	295	1.7	.016	.076	.024	8656	3.08	3.1u
		Newton	.003	.076	.024	587	45.5	2.6u

Table 4.3.11: Performance of three methods for unequal conductivities Explicit Scheme, for $k_L = 1$, $k_S = 0.25$, up to time=3

Nodes / Method	max X error	max T error	max L^1 error	CPU
$M = 40$				
resistance method	.022	.189	.030	75.4u
averaging method	.020	.148	.021	68.8u
Kirchoff temp. method	.007	.088	.012	78.2u
$M = 80$				
resistance method	.011	.113	.015	505.4u
averaging method	.010	.089	.011	507.7u
Kirchoff temp. method	.004	.048	.006	509.2u

Next we keep $St_L = St_S = 0.1$ and take different conductivities: $k_L = 1$, $k_S = 0.25$ or 0.04 (using $\nu \equiv \sqrt{\alpha_L / \alpha_S} = 2$ or 5), for which $\lambda = 0.200574$ and 0.207326 respectively. We implement three choices for effective conductivities: the resistance method (harmonic averaging) (14a), the (arithmetic) averaging method (14b), and the Kirchoff temperature method (16)-(17). The maximum error (numerical − exact) in front location, temperature at the output points, and the (integrated) L^1-error up to time=3 (when almost 70% of the slab has melted) are shown in the tables. The three implementations of the explicit scheme are compared in **Table 4.3.11**, for the case $k_S = 0.25$. The results from the SOR scheme were very similar. We notice that the maximum error in temperature (third column) is an order of magnitude larger than the error in front location or the L^1-error. This arises at mushy nodes and it is an exageration of the error in front

Table 4.3.12: Performance of three methods for unequal conductivities
Implicit Scheme ($\omega = 1$), for $k_L = 1$ $k_S = 0.04$, up to time=3

Nodes / Method	max X error	max T error	max L^1 error	CPU
$M = 40$				
resistance method	.141	.816	.171	48.2u
averaging method	.044	.513	.054	47.3u
Kirchoff temp. method	.014	.279	.030	30.3u
$M = 80$				
resistance method	.074	.734	.087	340.9u
averaging method	.022	.301	.027	321.5u
Kirchoff temp. method	.007	.167	.015	212.1u

location, due to the steepness of the temperature profile. Indeed, k_S being only 1/4th of k_L implies that heat cannot conduct as well into the solid, resulting in a very steep temperature profile ahead of the front. Thus, the exact temperature at the node (center of the control volume) can be substantially colder than T_m, which is the value asigned there by the enthalpy scheme, whence the large temperature error. The integrated (L^1-) error averages this local extreme error over the mesh, and it is a better indicator of the accuracy of the computed solution than the maximum pointwise error in this case of unequal conductivities. The same behavior is seen in **Table 4.3.12**, where we compare the three methods on the implicit scheme for the case $k_S = 0.04\, k_L$. The superiority of the Kirchoff temperature method for handling unequal conductivities is apparent for both the explicit (**Table 4.3.11**) and especially the implicit scheme (**Table 4.3.12**).

4.3.I Implicit versus explicit schemes

The choice of whether to use an explicit or implicit numerical scheme is one of almost eternal contention. On the one hand, explicit schemes harbor inherent limitations on the time step size imposed by stability conditions. On the other hand, even for unconditionally stable schemes, such as the fully implicit scheme, the time-step must be restricted to retain accuracy. This is particularly true for phase-change problems, since there is an inherent physical time-scale present in such problems, namely the time it takes for the interface to travel through a control volume. Thus, one would expect loss in accuracy whenever the physical restriction:

$$|X'|\Delta t \le \Delta x \tag{59}$$

is violated. Of course, with the front speed $X'(t)$ being unknown (and varying with time), this relation does not tell us how to choose Δt. However, for the classical Stefan problem, we have $X'(t) = \lambda\sqrt{\alpha_L / t}$, which is largest at the very first time-step, so it suffices to restrict Δt by

$$\Delta t \leq \Delta x^2 / \lambda^2 \alpha_L = 3 \, \Delta t_{expl} / \lambda^2. \tag{60}$$

For the three Stefan numbers of .1, 1., and 5, the factor $3/\lambda^2$ has the values 84, 21, and 15, respectively. The computations reported in **§4.3.G** seem to agree with this, even though the factors seem somewhat conservative when $M = 80$ nodes are used. Note that the faster the melting process the smaller the time-step factor can be for reasonable accuracy, and the less advantageous the implicit scheme becomes over the explicit one.

The overall advantage of implicit schemes depends strongly on the cost of the iteration scheme used to solve their nonlinear updating relations. How many iterations are needed to provide an acceptably accurate result, and is the resulting computing effort less than that required by the stable explicit scheme? As the tables in **§4.3.G** indicate, the answer to this question depends strongly on the thermal conditions of the process, e.g., temperature gradients, rates of change, and the computational effort required to evaluate the thermophysical parameters. SOR speeds up the convergence and thus always outperforms Gauss-Seidel.

Which method is preferable also strongly depends on the intended use. For a research code, intended primarily for exploratory studies, it is much easier to program and debug an explicit scheme. On the other hand, for a "production" type code intended to be run hundreds of times (especially on machines of limited computing speed), the overhead associated with the development of an implicit code may be worth the possible savings on execution times.

4.3.J Use of the enthalpy method for a multi-layered slab

The enthalpy approach may be applied to any geometry or structure. An important application is to composite regions composed of different materials. The materials present might include one or more phase changing materials with different properties, or a phase change material within a container. The former situation is of particular interest in examining nuclear meltdown scenarios where hot radioactive "sludge" falls onto a metal plate. Cooled by the plate the sludge at first solidifies and then, due to its own radioactive decay, heats up again with the plate now getting hotter as well. At some point the either the solid sludge or the plate begin to melt, depending on which has the lower melt temperature, and finally the second material also melts. The second situation is of interest in Latent Heat Thermal Energy Storage where the material used for latent heat storage is in a container of a second material. The two materials might, for example, be a hydrated salt (the PCM) and a plastic. Let us describe briefly how the enthalpy approach can be extended to this case. Consider a slab $l_w \leq x \leq l$ as consisting of the wall: $-l_w \leq x \leq 0$ and the PCM: $0 \leq x \leq l$. We subdivide the wall thickness into M_w subintervals: I_{-1} , I_{-2} , $\cdots I_{-M_w}$ of lengths Δx_{-i} , $i = 1, 2, 3, \ldots, M_w$, and the PCM into M subintervals: I_1 , I_2 , $\cdots$, I_M of lengths Δx_i , $i = 1$, $\cdots$, M. The midpoint of I_j is marked as the node x_j, $j = -M_w, -M_w + 1, \cdots, -1, 1, 2, \cdots, M$. The total number of nodes (and subintervals) is $M_{TOT} = M_w + M$.

The enthalpy approach is flexible enough to allow us to treat the whole slab (wall plus PCM) as one region, but of course with different thermophysical properties for the wall, liquid and solid. This can be achieved by taking the enthalpy to depend on location as well as on temperature:

$$E(x,T) = \begin{cases} \rho_w c_w [T - T_m], & -l_w \le x \le 0 \\ \rho c_S [T - T_m], & 0 \le x \le l, \quad T < T_m \\ [0, \rho L], & 0 \le x \le l, \quad T = T_m \\ \rho L + \rho c_L [T - T_m], & 0 \le x \le l, \quad T > T_m \end{cases}$$

Then, $x = 0$ appears as an "interface" moving at speed zero, and energy conservation across it will be ensured if the flux jump is zero there (by the Stefan condition). Hence, the temperature T_0 at $x = 0$ can be determined from the equality of fluxes: $-k_w \dfrac{T_0 - T_{-1}}{\Delta x_{-1} / 2} = -k_1 \dfrac{T_1 - T_0}{\Delta x_1 / 2}$ where k_1 is k_L or k_S, according to the phase of V_1. This yields

$$T_0 = \frac{\dfrac{k_w}{\Delta x_{-1}} T_{-1} + \dfrac{k_1}{\Delta x_1} T_1}{\dfrac{k_w}{\Delta x_{-1}} + \dfrac{k_1}{\Delta x_1}} = \frac{\tilde{R}_1 T_{-1} + \tilde{R}_{-1} T_1}{\tilde{R}_{-1} + \tilde{R}_1}$$

where $\tilde{R}_{-1} = \dfrac{\Delta x_{-1}}{2 k_w}$, $\tilde{R}_1 = \dfrac{\Delta x_1}{2 k_1}$, and the flux at $x = 0$ can be expressed as $q_0 = -\dfrac{T_1 - T_{-1}}{\tilde{R}_{-1} + \tilde{R}_1}$ representing both $q_{-1+½}$ and $q_{1-½}$.

In all other aspects the enthalpy scheme remains the same as in the previous two-phase example. Namely, we update E_j from the global conservation law

$$E_j^{n+1} = E_j^n + \frac{\Delta t_n}{\Delta x_j} [q_{j-½}^{n+\theta} - q_{j+½}^{n+\theta}], \qquad j = -M_w, \cdots, -1, 1, 2, \cdots, M,$$

and then we update T_j^{n+1} from

$$T_j^{n+1} = \begin{cases} T_m + \dfrac{E_j^{n+1}}{\rho_w c_w}, & j = -M_w, \cdots, -1 \\ T_m + \dfrac{E_j^{n+1}}{\rho c_S}, & j = 1, 2, \cdots, M, \quad \text{if} \quad E_j^{n+1} < 0 \\ T_m, & j = 1, 2, \cdots M, \quad \text{if} \quad 0 \le E_j^{n+1} \le \rho L \\ T_m + \dfrac{E_j^{n+1} - \rho L}{\rho c_L}, & j = 1, 2, \cdots M, \quad \text{if} \quad E_j^{n+1} > \rho L . \end{cases}$$

The wall-PCM surface temperature can then be found from

$$T_0^{n+1} = \frac{\tilde{R}_1 T_{-1}^{n+1} + \tilde{R}_{-1} T_1^{n+1}}{\tilde{R}_1 + \tilde{R}_{-1}},$$

the wall-fluid surface temperature from (see (29), §**4.1.C**)

$$T_{surf}^{n+1} = \frac{T_{M_w}^{n+1} + h\tilde{R}_{M_w} T_\infty^{n+1}}{1 + h\tilde{R}_{M_w}}, \qquad \tilde{R}_{M_w} = \frac{\Delta x_{-M_w}}{2k_w},$$

and the back (insulated) PCM surface temperature from (see (27), §**4.1.C**)

$$T_{PCMsurf}^{n+1} = T_M^{n+1}.$$

PROBLEMS

PROBLEM 1. Verify (16) and (17).

PROBLEM 2. (a) Construct an explicit enthalpy algorithm for a 1-phase Stefan problem in a cylinder $R_{in} \le r \le R_{out}$, with imposed temperature at $r = R_{out}$ and insulated at $r = R_{in}$. (b) Determine the stability criterion.

PROBLEM 3. Derive the form of $R_{1-½}$, u_0, $R_{M+½}$, u_{M+1} appropriate for the boundary conditions: convective flux at $x=0$, zero flux at $x=l$, so that (25) be valid for $j=1,2,\ldots,M$. [Verify (26)].

PROBLEM 4. Repeat PROBLEM 3 for the boundary conditions: imposed temperature at $x=0$ and imposed flux at $x=l$.

PROBLEM 5. Construct an implicit enthalpy algorithm for a 1-phase Stefan problem in a cylinder $R_{in} \le r \le R_{out}$, with boundary conditions
(a) imposed temperature at $r=R_{out}$ and insulated at $r=R_{in}$.
(b) convective flux at $r=R_{out}$ and insulated at $r=R_{in}$.
(c) imposed flux at $r=R_{out}$ and convective flux at $r=R_{in}$.

PROBLEM 6. Determine the conditions under which a nodal volume in one dimension that contains a sharp front will nevertheless have a total enthalpy that assigns the liquid or solid phase to it. Assume that the temperature is linear in the space variable, with distinct solid and liquid gradients.

PROBLEM 7. Prepare an explicit 2-dimensional code for melting and freezing of a rectangle with imposed boundary temperature. Check the code using the contrived solution of §**2.7**.

PROBLEM 8. Use the code of PROBLEM 7 to examine the process of freezing of two ice cubes placed near to each other in warm water. Can you find a situation in which they will form a bridge between them? What if they are both at the melt temperature?

PROBLEM 9. Replace the temperature by the "Kirchoff temperature" (15) in the conservation law and in the equation of state and discretize to derive the analogs of (12), of the explicit scheme, and of (24)-(26).

PROBLEM 10. Prepare an explicit and an implicit/SOR code simulating the two-phase, one-dimensional melting problem in a slab with imposed constant temperature at $x = 0$ and imposed time-varying temperature at $x = l$. Impose the Neumann explicit solution at $x = l$ and compare the numerical solution to the explicit one. Examine the sensitivity of your results to the choices (14a-d),(16)-(17) for effective conductivity.

PROBLEM 11. Define an explicit algorithm and a Crank-Nicolson/Gauss-Seidel algorithm for simulating a charge-discharge Latent Heat Thermal Energy Storage system consisting of a rectangular iron container filled with water. The layer of water/ice is 1 ft wide and the iron container wall is half an inch thick. Across the front surface of the iron wall convective heating/cooling from an ambient transfer fluid takes place, while all other sides are insulated. During the heating period, lasting 2 hours, the fluid temperature is 200 °F , and the heat transfer coefficient is 40 BTU/ft^2-hr-°F; during the cooling period, lasting 3 hours, the fluid temperature is −100°F, and the heat transfer coefficient is 20 BTU/ft^2-hr-°F.

PROBLEM 12. Describe the steps that you would take to estimate the performance of the composite system of PROBLEM 11 before you begin to simulate the system. In particular, under what conditions could you replace the effect of the container wall by an enhanced resistance? Similarly, what considerations would enter into your choice of the number of grid points?

PROBLEM 13. Define explicit and Crank-Nicolson/Gauss-Seidel algorithms for simulating the thermal behavior of finite slabs of radioactive sludge and metal, given that initially the sludge is liquid and the metal is solid.

4.4. MATHEMATICAL FRAMEWORK OF THE ENTHALPY FORMULATION

4.4.A Introduction

Traditionally, the concept of *solution* of a partial differential equation is taken to mean a *continuous* function having *continuous* derivatives up to the order of the PDE, which satisfies the PDE at each point of an appropriate region. Thus, a **classical solution** to the Heat Equation in a region G is a function $u(x,t)$ with u, u_x, u_{xx}, and u_t continuous in G and such that $u_t(x,t) = u_{xx}(x,t)$ at each point (x,t) of the region G. The need to weaken these conditions and admit a larger class of functions as possible "solutions" arose already in the late 1800's in the context of the classical Dirichlet Problem for the Laplace Equation. In treating this problem, David Hilbert first introduced the idea of **generalized solutions** as limits (in an appropriate sense) of "classical solutions". The ideas further evolved in the 1930's and 1940's, mainly in the context of the Calculus of Variations and hyperbolic problems (shocks), leading to the concepts of **weak derivatives** and **weak solutions** within the framework of **Sobolev spaces** (see **§4.4.B**). These developments revolutionized the whole theory of Partial Differential Equations and eventually were brought to bear on moving boundary problems.

The development of a mathematical theory for Stefan-type problems had a slow start. Being highly nonlinear and non-standard PDE problems, they remained outside the mainstream theoretical PDE developments for almost fifty years after the original work of C. Neumann and Stefan.

Existence results on the Stefan Problem *for small times* were obtained by Rubinstein and others in the late 1940's and 1950's, see [RUBINSTEIN]. The results were limited and the problems were regarded as intractable within the confines of the theory of classical solutions. In fact, the well-posedness of the classical one-dimensional two-phase Stefan Problem for the heat equation was established correctly (and under reasonable assumptions on the data) only in 1972 [MEIRMANOV, 1973], also [CANNON-HENRY-KOTLOW, 1974, 1976]. More general (but still one-dimensional) moving boundary problems were studied by [MEIRMANOV, 1973], [FASANO-PRIMICERIO, 1977a,b,c], and others (see the Gatlinburg Conference Proceedings [WILSON-SOLOMON-BOGGS] for surveys of results up to mid-1977). By 1980, examples of reasonable Stefan problems lacking classical solutions were given [MEIRMANOV, 1980] thus establishing the inherent shortcomings of classical formulations of moving boundary problems and of their front-tracking numerical counterparts.

In the meantime, the weak formulation gained the ground lost by the classical one. It was S. Kamin in 1958 [KAMENOMOSTSKAYA], and Oleinik [OLEINIK] who first introduced a weak reformulation of the Stefan problem and its finite-difference discretization, while enthalpy-based numerical methods were first proposed by Dusinberre in the heat transfer literature [DUSINBERRE], and by [ROSE] and [SOLOMON, 1966] in the computing literature. In 1968, A. Friedman established the well-posedness of the weak solution in multidimensional Stefan Problems [FRIEDMAN, 1968], providing confidence in the weak formulation. A solution is interpreted in Sobolev-space sense, to be described below. Since that time the attention of theoretical studies on the Stefan problem shifted to questions of *regularity* of the weak solution and of the set of all points $(\vec{x},t)$ where $T(\vec{x},t)=T_m$, representing the interface. In 1-dimension success came early, when Cannon-Henry-Kotlow established the continuous differentiability of the "weak" interface without undue restrictions on the data, thus proving the well-posedness of the classical problem. Their direct proof came afterwards ([C-H-K,1976]). In higher dimensions however, regularity results have proved very difficult to obtain. Even the *continuity* of the weak temperature was established only in 1979 for one-phase problems (by variational inequalities methods, [CAFFARELLI-FRIEDMAN]), and in 1982 by [DiBENEDETTO, 1982] and [CAFFARELLI-EVANS, 1983] for the 2-phase problem! Clearly, regularity of the interface is a very difficult mathematical problem.

These theoretical results have established the credibility and generality of the weak formulations so that we can use them with complete confidence.

4.4.B Weak derivatives

As we have already remarked, the characteristic of the enthalpy formulation is that the interface location is not brought out explicitly. Instead, we retreat to the basic local conservation law

$$\rho\, e_t(\vec{x}, t) + div q(\vec{x}, t) = 0$$

which is to be valid **throughout** the material. The phases will only be distinguished by the equation of state relating enthalpy and temperature, (5)§**1.2** or (7)§**4.3**. But then e experiences a jump at the points where $T(\vec{x}, t) = T_m$, so we are faced with the question of the **meaning** of the derivative e_t. Similarly, $\vec{q} = -k_S \cdot \nabla T$ when $T < T_m$ and $\vec{q} = -k_L \cdot \nabla T$ when $T > T_m$, so what does $div\vec{q}$ mean ? Obviously this is precisely the reason why, since the time of Lagrange, whenever discontinuous quantities arise, the regions are split into subregions along the discontinuities, and then the classical meaning can be assigned to derivatives.

Fortunately, since Dirac introduced his "mathematically meaningless δ-function" full mathematical meaning has been given to such quantities. Already in the late 1930's, S.L. Sobolev, C.B. Morrey and others were led to concepts of "weak derivatives" in the context of the Calculus of Variations and hyperbolic problems (shocks). Then, in 1948, Laurent Schwartz published his "Theorie des Distributions" in which the Dirac δ was finally put on firm mathematical foundations as a **functional** acting on test functions. An explosion of mathematical research followed, in which the whole theory of partial differential equations was recast into the Sobolev space framework of weak derivatives. The enthalpy formulation is a direct consequence of these developments, as is the entire field of Finite Elements.

Let us briefly discuss some of these ideas, as they pertain to the enthalpy method. The classical concept of derivative is a local concept (requires smoothness pointwise) and a very stringent one, as it eventually became clear when obvious solutions to some partial differential equations could not be given meaning within its confines. The concept of weak derivative frees us from it, by replacing continuity at every point of a set by integrability over the set, a much more general and looser concept (and a global one as opposed to local).

To introduce the concept of weak derivative, we need the idea of distribution and distributional derivative of Laurent Schwartz.

Let Ω be an open bounded set in $\mathbf{R}^N$ with piecewise differentiable boundary, $\partial\Omega$. The *closure* of Ω is the union of the set and its boundary, $\bar{\Omega} = \Omega \cup \partial\Omega$. The set of all functions with continuous derivatives of orders up to m in Ω is denoted by $\mathbf{C^m}(\Omega)$, while for $1 \le p < \infty$, $\mathbf{L^p}(\Omega)$ denotes the space of functions whose p-th power is integrable on Ω, with norm

$$\|f\|_{L^p(\Omega)} = \left(\int_\Omega |f(x)|^p \, dx \right)^{1/p} .$$

In particular, $L^2(\Omega)$ is a Hilbert space with inner product $< f, g > := \int_\Omega f(x)\, g(x)\, dx$.

The **support** of a function ϕ is the closure of the set of points $\vec{x}$ on which $\phi(\vec{x}) \neq 0$ (so that ϕ vanishes identically outside its support). We denote by $C_0^\infty(\Omega)$ the set of infinitely differentiable functions with compact (closed and bounded) support in Ω . The **space** $\mathbf{D}(\Omega)$ of **test functions on** Ω consists of the $C_0^\infty(\Omega)$ functions with convergence defined as follows: $\phi_n \to \phi$ in $\mathbf{D}(\Omega)$ if the supports of all the ϕ_n's and of ϕ are contained in a common bounded subset of Ω, and every partial derivative of the ϕ_n's converges uniformly to the corresponding derivative of ϕ.

Clearly, $\mathbf{D} \subset \mathbf{C}^\infty \subset \cdots \subset \mathbf{C}^1 \subset \mathbf{C} \subset \cdots \subset \mathbf{L}^2 \subset \mathbf{L}^1$ and $\mathbf{D}$ is *dense* in any one of these spaces (meaning that every function in, say, $\mathbf{L}^2$ is the $\mathbf{L}^2$-limit of a sequence of test functions, etc).

A real-valued function defined on a function-space is called a **functional.** Thus, a functional U on $\mathbf{D}(\Omega)$ assigns a unique real number $U(\phi)$ to each $\phi \in \mathbf{D}(\Omega)$. A functional is **linear** if $U(\alpha\phi + \beta\psi) = \alpha\, U(\phi) + \beta\, U(\psi)$ for all scalars α, β and all ϕ, ψ in $\mathbf{D}(\Omega)$. A functional is **continuous** if $\lim_{n\to\infty} U(\phi_n) = U(\lim_{n\to\infty} \phi_n)$ for any convergent (in $\mathbf{D}(\Omega)$) sequence $\{\phi_n\}$. Finally, *a continuous linear functional on* $\mathbf{D}(\Omega)$ *is called a* **distribution** *on* Ω. The value of a distribution U on a test function ϕ is a number denoted by $< U , \phi >$. The space of distributions on Ω (consisting of all continuous linear functionals on $\mathbf{D}(\Omega)$) is denoted by $\mathbf{D}'(\Omega)$.

Any function u, integrable on Ω, determines a distribution U via

$$< U , \phi > \; := \int_\Omega u(\vec{x})\, \phi(\vec{x})\, d\vec{x} \, , \qquad \phi \in \mathbf{D}(\Omega). \tag{1}$$

Such a distribution is called **regular**. By identifying the function u with the distribution U it generates we may view (integrable) functions as (regular) distributions, so that $\mathbf{L}^1(\Omega) \subset \mathbf{D}'(\Omega)$.

However, there are distributions which do **not** arise from functions, e.g., the Dirac δ_a, defined by

$$< \delta_a , \phi > \; := \; \phi(a) \, , \qquad \phi \in \mathbf{D}(\Omega), \qquad a \text{ fixed in } \Omega. \tag{2}$$

Indeed, *there is no integrable function* $\delta(x)$ with the property $\int \delta(x)\, \phi(x) = \phi(a)$ for $\phi \in \mathbf{D}(\Omega)$, and therefore δ *cannot* be a *regular* distribution.

The great advantage of distributions is that they possess derivatives of all orders. The distributional derivative of $U \in \mathbf{D}'(\Omega)$ with respect to x_j is defined to be the distribution $\frac{\partial U}{\partial x_j}$ such that

$$< \frac{\partial U}{\partial x_j}, \phi > \;:=\; -< U, \frac{\partial \phi}{\partial x_j} >, \qquad \phi \in \mathbf{D}(\Omega). \tag{3}$$

EXAMPLE 1: The distributional derivative of the "unit jump" (Heaviside) function $u(x)=1$ for $x>0$, $=0$ for $x<0$, is the Dirac δ. Indeed, for any $\phi \in \mathbf{D}(-\infty,\infty)$ we have

$$<u,\phi'> \;=\; \int_{-\infty}^{\infty} u(x)\,\phi'(x)\,dx \;=\; \int_{-\infty}^{0} 0\cdot\phi' dx + \int_{0}^{\infty} 1\cdot\phi' dx$$

$$= \phi(\infty) - \phi(0) \;=\; -\phi(0) \;=\; -<\delta_0, \phi>,$$

which, by (3), means $u' \overset{\mathbf{D}'}{=} \delta_0$.

Of interest for us are *integrable distributional derivatives of integrable functions*, called **weak derivatives**.

DEFINITION 1 Let $u \in \mathbf{L}^1(\Omega)$. We say that an integrable function $v \in \mathbf{L}^1(\Omega)$ is the **weak derivative** of u on Ω with respect to x_j if

$$\int_\Omega u \frac{\partial \phi}{\partial x_j}\,dx \;=\; -\int_\Omega v\,\phi\,dx \qquad \text{for all } \phi \in \mathbf{C}_0^\infty(\Omega). \tag{4}$$

This is in fact a generalization of the classical continuous derivative concept, because if the classical derivative $\frac{\partial u}{\partial x_j}$ exists and is continuous in Ω, then, by integration by parts, $\int_\Omega \frac{\partial u}{\partial x_j}\phi\,dx = \int_\Omega [\frac{\partial(u\phi)}{\partial x_j} - u\frac{\partial \phi}{\partial x_j}]\,dx = (u\phi)\big|_{\partial\Omega} - \int_\Omega u\frac{\partial \phi}{\partial x_j}\,dx$ $= 0 - (-\int_\Omega v\,\phi\,dx)$, implying that $\int_\Omega [v - \frac{\partial u}{\partial x_j}]\,\phi\,dx = 0$ for all $\phi \in \mathbf{C}_0^\infty(\Omega)$; this in turn implies that $v(x) = \frac{\partial u}{\partial x_j}(x)$, $x \in \Omega$ (because $\mathbf{C}_0^\infty(\Omega)$ is dense in $\mathbf{L}^1(\Omega)$), so that in fact the weak derivative does coincide with the classical derivative. Hence, we denote v by $\frac{\partial u}{\partial x_j}$ again !

Note that there is (at most) one weak derivative. Indeed, suppose v_1 and v_2 are both weak derivatives of u in $\mathbf{L}^1(\Omega)$. Then (4) implies $\int_\Omega [v_1 - v_2]\,\phi\,dx = 0$ for all $\phi \in \mathbf{C}_0^\infty(\Omega)$, which implies $v_1 - v_2 = 0$ again, so they coincide (in $\mathbf{L}^1(\Omega)$).

EXAMPLE 2: The function $u = |x|$ is not differentiable in $\Omega = (-1,1)$ but it does have weak derivative in $\mathbf{L}^1(-1,1)$ the function

$$\frac{\partial u}{\partial x} \;=\; \operatorname{sgn}(x) \;=\; \begin{cases} +1 & \text{if } x > 0 \\ -1 & \text{if } x < 0 \end{cases}$$

Indeed, for $\phi \in \mathbf{C}_0^\infty(-1\,,\,1)$ we have, $\int_{-1}^{1} |x|\phi_x dx \;=\; \int_{-1}^{0} -x\phi_x dx + \int_{0}^{1} x\phi_x dx$

$$= (-\,x\phi)\Big|_{x=-1}^{x=0} - \int_{-1}^{0}(-1)\phi dx + (x\phi)\Big|_{x=0}^{x=1} - \int_{0}^{1}(+1)\phi dx = -\int_{-1}^{1} sgn(x)\,\phi(x)\,dx\,.$$

So, in fact, the weak derivative of a function with a **corner** is a **jump** function. The derivative of a jump function, however, is not a weak derivative; it is only a distribution, containing the Dirac δ.

DEFINITION 2 The **Sobolev space** $\mathbf{H}^1(\Omega)$ consists of all square-integrable functions with square-integrable weak derivatives in Ω, i.e.

$$\mathbf{H}^1(\Omega) \;=\; \{u \in \mathbf{L}^2(\Omega):\; |\nabla u| \in \mathbf{L}^2(\Omega)\}\,.$$

It is a Hilbert space with inner product

$$<u\,,v>_{\mathbf{H}^1(\Omega)} \;:=\; \int_{\Omega} [\,u\,v + \nabla u\cdot\nabla v\,]\,dx \;\equiv\; <u\,,v>_{\mathbf{L}^2(\Omega)} + <\nabla u\,,\nabla v>_{\mathbf{L}^2(\Omega)}\,.$$

Finite difference approximation of derivatives

For any function f defined in a region $G \subset \mathbf{R}^N$, the **difference quotients**

$$\delta^h_{x_j} f(x) \;=\; \frac{f(x_1\,,\cdots,x_{j-1}\,,x_j+h\,,x_{j+1}\,,\cdots,x_N) - f(x)}{h}$$

can be defined for $x \in G'$ (a compact subset of G), and sufficiently small $|h|$.

If the function f is smooth at x, then by definition of the classical derivative $\frac{\partial f}{\partial x_j}(x)$, we have

$$\delta^h_{x_j} f(x) \;\to\; \frac{\partial f}{\partial x_j}(x) \quad \text{as } h \to 0\,,$$

and, in fact,

$$\delta^h_{x_j} f(x) \;=\; \frac{1}{h}\int_{x_j}^{x_j+h} \frac{\partial f(x_1\,,\cdots,\xi\,,\cdots,x_N)}{\partial \xi}\,d\xi\,,$$

i.e. the difference quotient is the average of the derivative. It turns out that the difference quotients actually approximate the *weak derivative.*

THEOREM 1 [MIKHAILOV, p.118} Let $f \in \mathbf{L}^2(G)$ with compact support inside G.

(i) If the weak derivative $\frac{\partial f}{\partial x_j}$ exists in $\mathbf{L}^2(G)$, then for sufficiently small $|h| \neq 0$,

$$\|\,\delta^h_{x_j} f\,\|_{\mathbf{L}^2(G)} \;\le\; \|\,\frac{\partial f}{\partial x_j}\,\|_{\mathbf{L}^2(G)} \quad \text{and} \quad \|\,\delta^h_{x_j} f - \frac{\partial f}{\partial x_j}\,\|_{\mathbf{L}^2(G)} \;\to\; 0 \;\text{ as } h \to 0\,,$$

i.e. $\delta^h_{x_j} f \to \frac{\partial f}{\partial x_j}$ as $h \to 0$ in the L^2-sense.

(ii) If there exists a constant $C>0$ such that $\| \delta_{x_j}^h f \|_{L^2(G)} \leq C$ for all sufficiently small $|h| \neq 0$, then the weak derivative $\frac{\partial f}{\partial x_j}$ exists in $\mathbf{L}^2(G)$, $\| \frac{\partial f}{\partial x_j} \|_{L^2(G)} \leq C$, and $\| \delta_{x_j}^h f - \frac{\partial f}{\partial x_j} \|_{L^2(G)} \to 0$ as $h \to 0$.

4.4.C Examples of weak formulations of problems

Weak formulations are obtained essentially by replacing classical derivatives by weak derivatives. Let us look at some examples that illustrate the procedure.

EXAMPLE 3: Consider the partial differential equation (Burger's equation)

$$u_t + u\,u_x = 0, \qquad a < x < b, \quad 0 < t < t^*.$$

A classical solution $u(x,t)$ is a function with continuous derivatives u_t, u_x in $G = (a,b) \times (0,t^*)$ such that

$$u_t(x,t) + u(x,t)\,u_x(x,t) = 0 \qquad \text{for all } (x,t) \in G.$$

In order to admit discontinuities in u (shocks), we have to relax the requirements that u_t and u_x be continuous in G. Thus we say that u is a weak solution of the equation (rewritten in divergence form)

$$u_t + (\tfrac{1}{2}u^2)_x = 0 \quad \text{in } G$$

(or that u satisfies this equation in the weak or distributional sense), if $u \in \mathbf{L}^2(G)$ and

$$\int_a^b [u\phi_t + \tfrac{1}{2}u^2\,\phi_x]\,dx\,dt = 0 \qquad \text{holds for } all \ \ \phi \in \mathbf{C}_0^\infty(G).$$

The latter expression is found by multiplying the equation by ϕ and integrating by parts, thus transferring the derivatives from u to ϕ. Note that we only need to require $u \in \mathbf{L}^2(G)$ for this relation to make sense, so such a u may be very discontinuous indeed !! That this is a reasonable concept follows from the fact that *every square-integrable classical solution will also be a weak solution* (by integration by parts), so the concept of weak solution is a *strict generalization* of the concept of classical solution.

Hence, weak formulations are obtained essentially by replacing classical derivatives by weak derivatives.

EXAMPLE 4: Consider the heat conduction problem

$$T_t = \alpha\,T_{xx}, \qquad 0 < x < l, \quad 0 < t < t^* \tag{5a}$$

$$T(x,0) = T_{init}(x), \qquad 0 < x < l \tag{5b}$$

$$-k\,T_x(0,t) = h\,[T_\infty - T(0,t)], \qquad T(l,t) = T_l, \quad 0 < t < t^*, \tag{5c}$$

with ρ, c, k, h constants.

A classical solution in $G = (0\,,\,l) \times (0\,,\,t^*)$ is a function $T(x,t)$ such that: $T \in \mathbf{C}(\overline{G})$, $T_x \in \mathbf{C}(G \cup \{0\} \times (0\,,\,t^*))$, T_t, $T_{xx} \in \mathbf{C}(G)$ and satisfies the problem pointwise.

To formulate the concept of a weak solution in G, we multiply the partial differential equation by a $\mathbf{C}^\infty$ function ϕ (to be restricted further later) and integrate by parts:

$$0 = < \rho c T_t - (kT_x)_x\,,\,\phi > = \int_0^{t^*}\int_0^{l} (\,\rho c\,[\,(\phi T)_t - T\phi_t\,] - (kT_x\phi)_x + (kT_x)\phi_x\,)\,dxdt$$

$$= \int_0^{l} \rho c \phi T\big|_{t=t^*} dx - \int_0^{l} \rho c \phi T\big|_{t=0} dx - \iint_G \rho c T \phi_t dxdt - \int_0^{t^*} kT_x\phi\big|_{x=l} dt$$

$$+ \int_0^{t^*} kT_x\phi\big|_{x=0} dt + \iint_G kT_x\phi_x dxdt.$$

Now choose ϕ such that $\phi(x,t^*) = 0$ for all x and $\phi(l,t) = 0$ for all t. Then we find,

$$0 = -\int_0^{l} \rho c \phi(x,0)\,T_{init}(x)dx - \iint_G \rho c T \phi_t dxdt - \int_0^{t^*} h[\,T_\infty - T(0,t)\,]\phi(0,t)dt$$

$$+\iint_G kT_x\phi_x dxdt = -\int_0^{l} \rho c \phi(x,0)\,T_{init}(x)dx - \int_0^{t^*} h[\,T_\infty - T(0,t)\,]\phi(0,t)dt$$

$$-\iint_G \rho c T \phi_t dxdt + \iint_G (kT\phi_x)_x dxdt - \iint_G T(k\phi_x)_x dxdt$$

$$= -\int_0^{l} \rho c \phi(x,0) T_{init}(x)dx - \int_0^{t^*} h[\,T_\infty - T(0,t)\,]\phi(0,t)dt$$

$$-\iint_G T[\rho c \phi_t + (k\phi_x)_x]dxdt + \int_0^{t^*} k\phi_x T\big|_{x=l} dt - \int_0^{t^*} k\phi_x T\big|_{x=0} dt\,.$$

Let us now further restrict ϕ to satisfy the condition $\phi_x(0,t) = 0$ for all t. Then we conclude that for every such test function ϕ, the classical solution of problem (5) satisfies

$$\iint_G T\,[\,\rho c \phi_t + (k\phi_x)_x\,]dxdt + \int_0^{l} \rho c T_{init}(x)\,\phi(x,0)\,dx$$
$$= \int_0^{t^*} \Big(h[\,T(0,t) - T_\infty(t)\,]\phi(0,t) + k\phi_x(l,t)T_l(t) \Big) dt \tag{6}$$

This expression will make sense as long as $T(x,t)$ is a bounded measurable function on G, and admits boundary values on $x = 0$ which are at least square integrable in time. From the (trace) theory of Sobolev spaces ([KUFNER-JOHN-FUCIK], [TREVES]), we know that it is enough to require $T \in \mathbf{L}^2(\,(0,t^*)\,;\,\mathbf{H}^1(0,l)\,)$ which means

$$\int_0^{t^*}\left[\int_0^l \left(|T|^2 + |T_x|^2\right)dx\right]dt < \infty,$$

which simply amounts to T, $T_x \in \mathbf{L}^2(G)$. Thus we can convert (6) into

DEFINITION 3 We say T is a **weak solution** of problem (5) in G, with data $T_{init} \in \mathbf{L}^2(0,l)$, $T_l \in \mathbf{L}^2(0,t^*)$, if T and its (distributional) derivative T_x are in $\mathbf{L}^2(G)$ and the relation

$$\iint_G T\,[\rho c\phi_t + (k\phi_x)_x]dxdt + \int_0^l \rho c T_{init}(x)\,\phi(x,0)\,dx$$
$$= \int_0^{t^*} \left(h[\,T(0,t) - T_\infty(t)\,]\phi(0,t) + k\phi_x(l,t)T_l(t) \right)dt \tag{6}$$

holds for any $\mathbf{C}^\infty$ function $\phi(x,t)$ such that $\phi\big|_{t=t^*} = 0$, $\phi_x\big|_{x=0} = 0$, $\phi\big|_{x=l} = 0$. Clearly, every *classical* solution such that T, $T_x \in \mathbf{L}^2(G)$ is also such a *weak* solution and every smooth *weak* solution is *classical.*

This weak solution concept is one of several that can be formulated (see [TREVES], [FRIEDMAN, 1964]).

4.4.D Classical formulation of Stefan problems in 3-dimensions

Consider a region Ω in $\mathbf{R}^3$, with piecewise smooth boundary, occupied by a phase-change material. On part of the boundary, say $\partial_I\Omega$, the temperature is imposed, while on the rest of the boundary, $\partial_{II}\Omega$, the flux is imposed. Assuming there is a change of phase with the two phases separated by a sharp, smooth interface, the problem is to determine the temperature field and the interface location satisfying a heat equation inside the solid, a heat equation inside the liquid, interface conditions, and initial and boundary conditions. In order to state these precisely, we need to introduce some notation.

Let us denote by Ω the region occupied by the phase-change material, by G the space-time cylinder

$$G = \Omega \times (0\,,t^*)\,, \quad \text{with } t > 0,$$

and by Σ the interface (assumed to be a smooth surface) with equation

$$\Sigma:\ \Sigma(\vec{x},t) = 0\,.$$

Moreover, let

$$\Omega^+(t) = \{\,\vec{x} \in \Omega:\ \Sigma(\vec{x},t) > 0 \ \text{ and } \ u(\vec{x},t) > 0\,\} \quad (\text{liquid region at time } t),$$

$$\Omega^-(t) = \{\,\vec{x} \in \Omega:\ \Sigma(\vec{x},t) < 0 \ \text{ and } \ u(\vec{x},t) < 0\,\} \quad (\text{solid region at time } t),$$

$$\Sigma(t) = \{\vec{x} \in \Omega:\ \Sigma(\vec{x},t) = 0\,\} \quad (\text{interface at time } t), \quad \Sigma = \bigcup_{0<t<t^*} \Sigma(t),$$

$$\Omega(t) = \Omega^+(t) \cup \Omega^-(t) \cup \Sigma(t),$$

$$G^+ = \bigcup_{0<t<t*} \Omega^+(t), \qquad G^- = \bigcup_{0<t<t*} \Omega^-(t), \qquad G = G^+ \cup G^- \cup \Sigma .$$

Let us consider the Stefan Problem with thermophysical properties constant in each phase.

DEFINITION 4 The function pair $T(\vec{x},t)$, $\Sigma(\vec{x},t)$ constitutes a **classical solution** to the Stefan problem if

(i) T, $\dfrac{\partial \Sigma}{\partial t}$, $\nabla\Sigma$ are continuous in $G = \Omega \times (0, t*)$,

(ii) $\dfrac{\partial T}{\partial t}$, ∇T, $\nabla^2 T$ are continuous in $G^+ \cup G^-$,

(iii) at each point $(\vec{x}, t)$ they satisfy the following

$$T_t = \alpha_L \nabla^2 T \qquad (\vec{x}, t) \in G^+ ,$$

$$T_t = \alpha_S \nabla^2 T \qquad (\vec{x}, t) \in G^- ,$$

$$T(\vec{x}, t) = T_m , \qquad (\vec{x}, t) \in \Sigma ,$$

$$\rho L V_n = [\![-k \frac{\partial T}{\partial n}]\!]_-^+ \quad \text{on } \Sigma, \quad V_n = \text{normal component of velocity of } \Sigma ,$$

$$T(\vec{x},0) = T_0(\vec{x}) \begin{cases} > T_m & \text{for } \vec{x} \in \Omega^+(0), \\ = T_m & \text{for } \vec{x} \in \Sigma(0), \\ < T_m & \text{for } \vec{x} \in \Omega^-(0), \end{cases} \qquad \text{(initial condition)},$$

$$T(\vec{x},t) = T_I(\vec{x},t) \quad \text{on } \partial_I G := \partial_I \Omega \times (0, t*) \quad \text{(imposed temperature on part of the boundary)},$$

$$-k \frac{\partial T}{\partial \vec{n}} = Q_{II}(\vec{x},t) \quad \text{on } \partial_{II} G := \partial_{II} \Omega \times (0, t*) \quad \text{(imposed flux on part of the boundary)}.$$

We assume that the initial and boundary data are smooth and compatible for simplicity. The questions of existence and uniqueness of solution are still open for such a problem (and likely to remain open except under severe assumptions on data).

Let us emphasize that the nature of the interface was decided before the problem was even stated! In fact, the fundamental limitation of the classical formulation is precisely that we need to *assume* that the interface will be a sharp surface in order to even *state* the problem. In a cyclic freezing/melting process where several interfaces are expected to appear, we would not know a-priori the structure of the interfaces and the very formulation of the problem becomes problematic.

Now let us consider Stefan problems for the general nonlinear Heat Conduction Equation, i.e. when the heat capacity and conductivity are temperature-dependent. In this case we can apply the so-called Kirchoff transformation to simplify the dependence on conductivity and thus reduce the problem to a simpler normalized form.

Specifically, we introduce the "Kirchoff temperature" u as

$$u = K(T) := \int_{T_m}^{T} k(\tau)d\tau = \begin{cases} \int_{T_m}^{T} k_S(\tau)d\tau & \text{for } T < T_m \text{ (solid)}, \\ \int_{T_m}^{T} k_L(\tau)d\tau & \text{for } T > T_m \text{ (liquid)}. \end{cases} \tag{7}$$

Then we have $\nabla u = k(T)\nabla T = -\vec{q}$ and therefore the energy conservation law $\rho e_t + div\vec{q} = 0$ takes the simpler form

$$\rho e_t = \nabla^2 u, \tag{8}$$

Since the conductivity is positive, the Kirchoff temperature $K(T)$ is an increasing function of T, and the transformation is invertible,

$$T = K^{-1}(u). \tag{9}$$

The enthalpy (per unit volume) may then be expressed as a function of the Kirchoff temperature, $\rho e(T) = \rho e(K^{-1}(u))$, which we shall denote by $\beta(u)$,

$$\beta(u) = \begin{cases} \int_{T_m}^{T} \rho c_S(\tau)d\tau = \int_{K^{-1}(0)}^{K^{-1}(u)} \rho c_S(\tau)d\tau = \int_0^u \rho \dfrac{c_S(K^{-1}(w))}{k_S(K^{-1}(w))} dw & \text{for } u < 0, \\ [0, \rho L] & \text{for } u = 0, \\ \int_{T_m}^{T} \rho c_L(\tau)d\tau + \rho L = \int_0^u \rho \dfrac{c_L(K^{-1}(w))}{k_L(K^{-1}(w))} dw + \rho L & \text{for } u > 0. \end{cases} \tag{10}$$

$\beta(u)$ is a strictly increasing *multivalued* mapping, differentiable except at $u = 0$, where it has a finite jump of magnitude ρL,

$$\beta'(u) \geq \alpha_0 > 0 \text{ for } u \neq 0 \quad \text{and } [\![\beta(u)]\!]_{0-}^{0+} = \rho L > 0.$$

Let's also note for future reference that there exist constants α_0, $\alpha_1 > 0$ such that

$$\alpha_0(r_1 - r_2) \leq \beta(r_1) - \beta(r_2) \leq \alpha_1(r_1 - r_2) + \rho L, \quad \forall r_1, r_2 \in \mathbf{R}. \tag{11}$$

Since the enthalpy $\beta(u)$ is multivalued at $u = 0$, the notation

$$H \in \beta(u), \tag{12a}$$

is a convenient way of expressing the *equation of state*

$$\begin{aligned} H &= \beta(u) && \text{for } u \neq 0, \\ H &\in [0, \rho L] && \text{for } u = 0. \end{aligned} \tag{12b}$$

With $H(\vec{x}, t)$ the enthalpy density at $\vec{x}$ at time t, the heat conduction equation, in each phase, may be written as

$$H_t = \nabla^2 u. \tag{13}$$

The classical formulation of the Stefan Problem may now be stated as follows.

DEFINITION 5 The function triple $u(\vec{x},t)$, $H(\vec{x},t)$, $\Sigma(\vec{x},t)$ constitutes a **classical solution** to the Stefan problem if

(i) u is continuous in $\bar{G}$, $\dfrac{\partial \Sigma}{\partial t}$, $\nabla\Sigma$ are continuous in G, ∇u is continuous in $G^+ \cup \Sigma \cup \partial_{II} G$ and in $G^- \cup \Sigma \cup \partial_{II} G$,

(ii) $\dfrac{\partial u}{\partial t}$, $\nabla^2 u$ are continuous in $G^+ \cup G^-$,

(iii) $$H(\vec{x},t) = \beta(u(\vec{x},t)) \quad \text{for } u(\vec{x},t) \neq 0, \quad (\vec{x},t) \in \bar{G}, \tag{14}$$

(iv) H, u, Σ satisfy the following at each point $(\vec{x},t)$

$$H_t = \nabla^2 u \quad \text{for } (\vec{x},t) \in G - \Sigma \quad (\text{i.e. inside each phase}), \tag{15}$$

$$u(\vec{x},t) = 0 \qquad \text{for } (\vec{x},t) \in \Sigma, \tag{16a}$$

$$\rho L V_n = -[\![\frac{\partial u}{\partial n}]\!]_-^+ \quad \text{on } \Sigma, \quad V_n = \text{normal component of velocity of } \Sigma, \tag{16b}$$

$$u(\vec{x},0) = u_0(\vec{x}) \begin{cases} >0 & \text{for } \vec{x} \in \Omega^+(0), \\ =0 & \text{for } \vec{x} \in \Sigma(0), \\ <0 & \text{for } \vec{x} \in \Omega^-(0), \end{cases} \qquad (\text{initial condition}), \tag{17}$$

$$u(\vec{x},t) = u_I(\vec{x},t) =: K(T_I) \quad \text{on } \partial_I G := \partial_I \Omega \times (0,t^*), \tag{18a}$$

$$\frac{\partial u}{\partial \vec{n}} = -Q_{II}(\vec{x},t) \qquad \text{on } \partial_{II} G := \partial_{II}\Omega \times (0,t^*). \tag{18b}$$

The weak formulation of this general Stefan Problem will be presented in the following subsection.

4.4.E Weak formulation of the Stefan Problem

The classical formulation (14)-(18) demands that the conditions of the problem be satisfied *pointwise*. We wish to relax this requirement to achieve a more general and flexible formulation. The idea is to retreat to the primitive physical law of energy conservation throughout the material (irrespectively of phase), and interprete the derivatives as weak derivatives. Thus, we interprete (13) in distributional sense:

$$< H_t - \nabla^2 u \,, \phi > \; = 0 \qquad \text{for } \textit{all test functions } \phi \in C_0^\infty(G), \tag{19}$$

globally in G, but for regular distributions, in $\mathbf{L}^2(G)$ (i.e. with $<,>$ the $\mathbf{L}^2(G)$-inner product). To arrive at a precise formulation, we integrate by parts to move the derivatives over to the test function and incorporate the initial and boundary conditions, as in EXAMPLE 4.

Let $\{u, H, \Sigma\}$ be a *classical* solution (satisfying (14)-(18)) and let ϕ be a smooth (test) function on G (to be restricted further later). From (19) we have

$$0 = \langle H_t - \nabla^2 u , \phi \rangle \equiv \iint_G \Big[H_t - \nabla^2 u\Big]\phi \, dxdt$$

$$= \iint_{G^+ \cup G^-} \Big[(H\phi)_t - H\phi_t - div(\nabla u\phi) + div(u\nabla\phi) - (u\nabla^2\phi)\Big] dxdt \tag{20}$$

$$= -\iint_G \Big[H\phi_t + u\nabla^2\phi\Big] dxdt + \iint_{G^+ \cup G^-} \Big[(H\phi)_t + div(u\nabla\phi - \phi\nabla u)\Big] dxdt ,$$

whence

$$\iint_G \Big(H\phi_t + u\nabla^2\phi\Big) dxdt = \iint_{G^+ \cup G^-} [(H\phi)_t + div(u\nabla\phi - \phi\nabla u)] \, dxdt ,$$

which, by the Divergence Theorem applied in each phase,

$$= \iint_{\partial G^+ \cup \partial G^-} \Big(H\phi \, N_t + (u\nabla\phi - \phi\nabla u)\cdot\vec{N}_x\Big) dS , \tag{21}$$

with N_t and $\vec{N}_x$ the time and space components of the outward unit normal to the boundaries. Now, ∂G^+ consists of $\Omega^+(t^*)$, $\Omega^+(0)$, Σ, and $\bigcup_{0<t<t^*} \partial\Omega^+(t)$ (the lateral surface of the space-time region G^+), on which respectively we have $N_t dS =$ $(+1)dx$, $(-1)dx$, $N_t^\Sigma dS$, and 0, and $\vec{N}_x dS = 0, 0, \vec{N}_x^\Sigma dS$, and $\vec{n}_x ds$. Here, N_t^Σ and $\vec{N}_x^\Sigma$ denote the components of the unit normal to the interface Σ, pointing into the solid, and $\vec{n}_x$ denotes the spatial component of the outgoing normal to the lateral surface, i.e. the normal to $\partial\Omega(t)$. Similar relations hold for ∂G^-. Hence, the integrals in the right-hand side of (21) yield

$$\iint_{\partial G^+ \cup \partial G^-} H\phi \, N_t \, dS = \int_{\Omega^+(t^*) \cup \Omega^-(t^*)} H\phi \, dx - \int_{\Omega^+(0) \cup \Omega^-(0)} H\phi \, dx$$

$$+ \int_\Sigma H\phi \, N_t^\Sigma dS - \int_\Sigma H\phi \, N_t^\Sigma dS \tag{22}$$

$$= \int_{\Omega(t^*)} H(x,t^*)\,\phi(x,t^*)\,dx - \int_{\Omega(0)} H(x,0)\,\phi(x,0)\,dx + \int_\Sigma [\![H\phi]\!]_-^+ N_t^\Sigma dS ,$$

and

$$\iint_{\partial G^+ \cup \partial G^-} (u\nabla\phi - \phi\nabla u)\cdot\vec{N}_x \, dS = \int_0^{t^*} \int_{\partial\Omega^+(t) \cup \partial\Omega^-(t)} \Big[u\nabla\phi - \phi\nabla u\Big]\cdot\vec{n}_x ds dt$$

$$+ \int_\Sigma \Big[u\nabla\phi - \phi\nabla u\Big]\cdot\vec{N}_x^\Sigma dS - \int_\Sigma \Big[u\nabla\phi - \phi\nabla u\Big]\cdot\vec{N}_x^\Sigma dS \tag{23}$$

$$= \int_0^{t^*} \int_{\partial\Omega} \Big[u\frac{\partial\phi}{\partial\vec{n}_x} - \phi\frac{\partial u}{\partial\vec{n}_x}\Big] ds dt + \int_\Sigma [\![u\nabla\phi - \phi\nabla u]\!]_-^+ \cdot \vec{N}_x^\Sigma dS .$$

Now, $H(x,0) \in \beta(u(x,0))$ is known from the initial condition, while $u = u_I$ on $\partial_I G$ and $\frac{\partial u}{\partial \vec{n}_x} = -Q_{II}$ on $\partial_{II} G$ are known from the boundary conditions. We eliminate the unknown terms by choosing the test function ϕ such that $\phi(x,t^*)=0$ for $x \in \Omega$, $\phi=0$ on $\partial_I G$, and $\frac{\partial \phi}{\partial \vec{n}_x}=0$ on $\partial_{II} G$. We also note that, u and $\nabla\phi$ being continuous across Σ, the term $[\![u\nabla\phi]\!]_-^+$ vanishes along Σ. Then, (21) becomes

$$\iint_G \left[H\phi_t + u\nabla^2\phi \right] dxdt = -\int_{\Omega(0)} H(x,0)\,\phi(x,0)\,dx + \int\int_{\partial_I G} u_I \frac{\partial \phi}{\partial \vec{n}_x}\, dsdt$$
$$+ \int\int_{\partial_{II} G} \phi\, Q_{II}\, dsdt + \int_\Sigma \phi \left([\![H]\!]_-^+ N_t^\Sigma - [\![\nabla u]\!]_-^+ \cdot \vec{N}_x^\Sigma \right) dS\,. \tag{24}$$

By (16a), $u=0$ on Σ so by (12b) $[\![H]\!]_-^+ = \rho L$ and the integrand in the last integral becomes ϕ times $(\rho L N_t^\Sigma - [\![\nabla u]\!]_-^+ \vec{N}_x^\Sigma)$, which vanishes, thanks to the Stefan Condition (16b). Indeed, the normal velocity of the *moving* surface Σ is $V_n = \frac{d\vec{x}}{dt} \cdot \vec{n}^\Sigma$, with $\vec{n}^\Sigma = \frac{\nabla\Sigma}{|\nabla\Sigma|}$ the unit normal to the *moving* surface $\Sigma(\vec{x},t) = 0$, whence $\Sigma_t + \nabla\Sigma \cdot \frac{d\vec{x}}{dt} = 0$ implies $V_n = -\frac{\Sigma_t}{|\nabla\Sigma|}$. Hence, the Stefan Condition (16b) may be written as

$$\rho L \Sigma_t - [\![\nabla u]\!]_-^+ \cdot \nabla\Sigma = 0\,. \tag{25}$$

Noting that the time and space components of the unit normal $\vec{N}^\Sigma$ to the *fixed* in space-time surface Σ are

$$N_t = \frac{\Sigma_t}{\sqrt{\Sigma_t^2 + |\nabla\Sigma|^2}}, \quad \text{and} \quad \vec{N}_x = \frac{\nabla\Sigma}{\sqrt{\Sigma_t^2 + |\nabla\Sigma|^2}},$$

we see that (25) indeed implies the vanishing of the last integral in (24).

Summarizing, we have shown that the classical solution $\{H, u, \Sigma\}$ satisfies the relation

$$\iint_G \left[H\phi_t + u\nabla^2\phi \right] dxdt = -\int_{\Omega(0)} H(x,0)\,\phi(x,0)\,dx + \int\int_{\partial_I G} u_I \frac{\partial \phi}{\partial \vec{n}_x}\, dsdt$$
$$+ \int\int_{\partial_{II} G} \phi\, Q_{II}\, dsdt \tag{26}$$

for any test function ϕ in the set

$$\Phi = \{ \phi \in C^2(\bar{G}) : \ \phi|_{t=t^*} = 0, \ \ \phi|_{\partial_I G} = 0, \ \ \frac{\partial \phi}{\partial \vec{n}}|_{\partial_{II} G} = 0 \}.$$

Note now that (26) makes sense for H, u merely bounded (square) integrable functions in G, and that Σ does not appear at all in (26). Thus we are led to the following concept of weak solution.

DEFINITION 6 The function pair u, H is said to be a **weak solution** of the Stefan Problem (14-18) if

(i) u and H are bounded (square) integrable functions in G;

(ii) $H(\vec{x},t) \in \beta(u(\vec{x},t))$ almost everywhere in $\bar{G}$;

(iii) the relation

$$\iint_G \left[H\phi_t + u\nabla^2\phi\right]dxdt + \int_{\Omega(0)} H_0(x)\,\phi(x,0)\,dx = \iint_{\partial_I G} u_I \frac{\partial\phi}{\partial\vec{n}_x}\,dsdt + \iint_{\partial_{II} G} \phi\, Q_{II}\,dsdt, \tag{27}$$

with $H_0(\vec{x}) \in \beta(u_0(\vec{x}))$, holds for every test function

$$\phi \in \Phi = \left\{ \phi \in C^2(\bar{G}) : \phi|_{t=t^*}=0,\ \ \phi|_{\partial_I G}=0,\ \ \frac{\partial\phi}{\partial\vec{n}}|_{\partial_{II} G}=0 \right\}.$$

THEOREM 2 (a) A classical solution is also a weak solution.

(b) A sufficiently smooth weak solution is also a classical solution, i.e. if $\{H, u\}$ constitute a weak solution (according to DEFINITION 6), and if the set $\{(\vec{x},t)\in G: u(\vec{x},t)=0\}$ is a smooth surface (the graph of some $\Sigma(\vec{x},t)=0$), with u, Σ satisfying (i),(ii) of DEFINITION 5, then $\{H, u, \Sigma\}$ constitute a classical solution.

Proof: Part (a) was established above. To prove Part (b), let $\{H, u\}$ be a weak solution which is smooth and whose "isotherm" $\Sigma=\{(\vec{x},t)\in G: u(\vec{x},t)=0\}$ $\equiv \{(\vec{x},t)\in G: 0\le H(\vec{x},t)\le \rho L\}$ is a well-defined surface with equation $\Sigma(\vec{x},t)=0$. Choosing $\phi \in C_0^\infty(G^\pm)\subset\Phi$ in (27) (i.e. with support inside G^+ or inside G^-), we get

$$\iint_{G^\pm} [H\phi_t + u\nabla^2\phi]\,dxdt = 0,$$

which, after integration by parts, implies

$$-\iint_{G^\pm} \phi\,[H_t - \nabla^2 u]dxdt = 0 \qquad \text{for all } \phi\in C_0^\infty(G^\pm),$$

which, thanks to the smoothness of H, u, finally implies that

$$H_t = \nabla^2 u \quad \text{in } G^\pm.$$

Hence the PDE is satisfied pointwise inside G^+ and inside G^-.

Similarly, taking $\phi\in\Phi$ with support a neighborhood of $\Omega(0)$, we obtain the integral identity

$$\int_\Omega [H_0(x) - H(x,0)]\,\phi(x,0)\,dx = 0$$

for all such ϕ, whence we conclude that $H(x,0) = H_0(x)$, $x \in \Omega(0)$, or, equivalently, $u(x,0) = \beta^{-1}(H_0(x)) = u_0(x)$, $x \in \Omega(0)$, and so $u(x,t)$ satisfies the initial condition.

Similarly, taking $\phi \in \Phi$ with support in a neighborhood of $\partial_I G$, or in a neighborhood of $\partial_{II} G$, we recover the boundary conditions.

Finally, choosing $\phi \in \Phi$ with support in a neighborhood of the interface Σ, we recover the Stefan Condition (16b). We conclude that a smooth weak solution is also a classical solution.

This theorem justifies the assertion that a weak solution is indeed a generalized concept of solution, capable of existing even when the classical solution may not exist at all. In addition, a weak solution satisfies the equation $H_t = \nabla^2 u$ in the sense of distributions in G. Note that in this formulation there is absolutely no assumption whatsoever regarding the nature of the interface; in fact the interface is not even mentioned ! It is only recovered a posteriori as the set of points $(\vec{x}, t)$ where H lies between 0 and ρL. It may happen to be a sharp, smooth surface (in which case the weak solution is in fact a classical solution), or it may be an extended mushy zone. The great advantage of the weak formulation is that we do not need to know the answer in order to state the problem (which is the case with the classical formulation)!

Having succeeded in broadening the concept of solution (which should make existence "easier"), the question of uniqueness becomes paramount. Is this class of solutions so broad that it may possibly contain too many solutions? The answer is no, as shown by the following

THEOREM 3 (Uniqueness of Weak Solutions)

There exists at most one weak solution of the Stefan Problem.

Proof: For given data $H_0(x)$, $u_I(x,t)$, $Q_{II}(x,t)$, suppose there are two weak solutions, $\{u_1 , H_1\}$ and $\{u_2 , H_2\}$, with $H_i \in \beta(u_i)$. Subtracting the two integral identities satisfied by $\{u_1 , H_1\}$ and $\{u_2 , H_2\}$ we have, for all $\phi \in \Phi$,

$$\iint_G [(H_1 - H_2)\phi_t + (u_1 - u_2)\nabla^2\phi]\, dxdt = 0,$$

or

$$\iint_G (H_1 - H_2)[\phi_t + w\nabla^2\phi]\, dxdt = 0, \tag{28}$$

where

$$w(x,t) = \begin{cases} \dfrac{u_1(x,t) - u_2(x,t)}{H_1(x,t) - H_2(x,t)}, & u_1(x,t) \neq u_2(x,t) \\ 0 & u_1(x,t) = u_2(x,t) \end{cases}$$

is a *bounded* measurable function on G (recall that, from (11), $\beta(u_1) - \beta(u_2) \geq \alpha_0 (u_1 - u_2)$, for all $u_1 , u_2 \in \mathbf{R}$). We would like to show that

$$\iint_G (H_1 - H_2)\psi\, dxdt = 0 \quad \text{for } all \ \psi \in C_0^\infty(G),$$

which then would imply that $H_1 = H_2$ almost everywhere in G, implying that $u_1 = u_2$ almost everywhere in G. To do this we need to find $\phi \in \Phi$ such that

$$\phi_t + w\nabla^2\phi = \psi \quad \text{for any given} \quad \psi \in C_0^\infty(G). \tag{29}$$

However, the coefficient w in (29) is too wild to let us solve such an equation. So, instead, we approximate w by a sequence of smooth functions $w_m \in C^\infty(\bar{G})$ as follows. Using standard mathematical tools (mollifiers) we construct a sequence of functions $\bar{w}_m \in C^\infty(\bar{G})$, such that $0 \le \bar{w}_m(x,t) \le l_1$ = bound on w, and

$$\|\bar{w}_m - w\|_{L^2(G)} \equiv \left(\iint_G |\bar{w}_m(x,t) - w(x,t)|^2\, dxdt\right)^{1/2} \le \frac{1}{m}, \qquad m = 1, 2, \ldots$$

Take $w_m = \bar{w}_m + \dfrac{1}{m}$. Now, for each $\psi \in C_0^\infty(G)$ and each $m = 1, 2, \ldots$ consider the problem

$$\begin{cases} \dfrac{\partial \phi_m}{\partial t} + w_m \nabla^2 \phi_m = \psi & \text{in } G \quad \text{(backward heat equation)} \\ \phi_m(x, t^*) = 0, & x \in \Omega, \\ \phi_m\big|_{\partial_I G} = 0, & \dfrac{\partial \phi_m}{\partial \vec{n}}\Big|_{\partial_{II} G} = 0. \end{cases}$$

It can be shown [LADYZHENSKAYA] that each such problem has **unique** classical solution ϕ_m and that

$$|\phi_m(x,t)| \le l_2 \ \text{ on } \bar{G}, \qquad \| w_m \nabla^2 \phi_m \|_{L^2(G)} \le l_3, \qquad \left\| \frac{w}{w_m} \right\|_{L^2(G)} \le l_4, \quad \text{for all } m.$$

Clearly $\phi_m \in \Phi$, so from (28) we have

$$\iint_G (H_1 - H_2)\,[(\phi_m)_t + w\nabla^2\phi_m]\, dxdt = 0 \quad \text{for } m = 1, 2, \ldots,$$

which in turn implies

$$\iint_G (H_1 - H_2)\psi\, dxdt = \iint_G (H_1 - H_2)\,[(w_m - w)\nabla^2\phi_m]\, dxdt,$$

and the above estimates on w_m, ϕ_m can be used, [OLEINIK], to show that the right-hand side tends to zero as $m \to \infty$, establishing our goal.

Existence of weak solutions was established first by [KAMENOMOSTKAYA, 1960] (for the imposed temperature problem), by showing that the solutions of the explicit enthalpy scheme converge to the weak solution (see **§4.5**)! An analytic (non-constructive) existence proof was given by [FRIEDMAN, 1968], and several other proofs followed for more general nonlinear equations and with nonlinear boundary conditions, e.g. [DAMLAMIAN], [NIEZGODKA-PAWLOW], [ALEXIADES-CANNON]). Thus the weak solution approach succeeded in providing us with both theoretical foundations and numerical methods for the Stefan problem in its full generality.

Note however, that a weak solution u is, in general, only a bounded integrable function, and even its *continuity* has not been established in general (see **§4.4.A**).

Let us also mention that finite-element discretizations converge to a slightly different weak solution ([ELLIOTT-OCKENDON]), defined as follows.

DEFINITION 7 The pair $\{u, H\}$ is said to be a **variational solution** of the above Stefan Problem if

(i) u , H are bounded (square) integrable functions in G,

(ii) the weak gradient, ∇u, is integrable in G,

(iii) $H(\vec{x},t) \in \beta(u(\vec{x},t))$ a.e. in G

(iv) $u\big|_{\partial_I G} = u_I$,

(v) $$\iint_G [H\phi_t - \nabla u \cdot \nabla \phi]\, dxdt + \int_\Omega H_0(x)\phi(x,0)dx - \int\int_{\partial_{II}G} \phi Q_{II} dS = 0 \qquad (30)$$

holds for every test function ϕ in $\tilde{\Phi} := \{ \phi \in C^1(\bar{G}) : \phi(\vec{x},t^*) = 0 \}$.

It is easy to see that every such "variational" solution is also a weak solution. Indeed, for $\phi \in \Phi \subset \tilde{\Phi}$ we have

$$0 = \iint_G [H\phi_t - \mathrm{div}(u\nabla\phi) + u\nabla^2\phi]\, dxdt + \int_\Omega H_0(x)\phi(x,0)dx - \int\int_{\partial_{II}G} \phi Q_{II} dS =$$

$$\iint_G [H\phi_t + u\nabla^2\phi]\, dxdt + \int_\Omega H_0(x)\phi(x,0)dx - \int\int_{\partial_I G} u_I \frac{\partial\phi}{\partial\vec{n}} dS - \int\int_{\partial_{II}G} \phi Q_{II} dS,$$

which is exactly the integral identity (27), defining the weak solution. Conversely, a weak solution with integrable ∇u is also a "variational solution". In analytic proofs of existence of weak solutions (e.g. [FRIEDMAN, 1968]), the integrability of ∇u is established, and therefore the two concepts of solution are in fact the same.

Keeping the concept of weak derivatives in mind, we can state the weak formulation briefly as:

$$H_t = \nabla^2 u \quad \text{in the weak sense in } G,$$

$$H\big|_{t=0} = H_0(\vec{x}), \quad \vec{x} \in \Omega$$

$$u\big|_{\partial_I G} = u_I, \qquad \frac{\partial u}{\partial \vec{n}}\Big|_{\partial_{II}G} = -Q_{II}.$$

With this interpretation, we see that the problem appears formally identical with a plain heat conduction problem, so its natural discretization is in fact the enthalpy scheme, presented in **§4.3**. In the next section we show that the connection is not merely formal, but that in fact the numerical solution does converge to the weak solution as Δx, $\Delta t \to 0$.

PROBLEMS

PROBLEM 1. Consider the heat equation in a slab: $u_t = \alpha u_{xx}$, $a < x < b$, $t > 0$, (apart from any initial and boundary conditions).

(a) Give a precise definition of a classical solution to the equation in $G = (a, b) \times (0, t^*)$.

(b) Formulate a concept of weak solution of the equation on G and give a precise definition.

(c) Prove that every bounded classical solution is also a weak solution.

(d) Prove that every smooth (how smooth?) weak solution is also a classical solution.

PROBLEM 2. Consider the problem

$$\rho c T_t = \operatorname{div}(k \nabla T), \quad \vec{x} \in \Omega, \; t > 0, \quad \text{for } \Omega \text{ a bounded region in } \mathbf{R}^3,$$

$$T(\vec{x}, 0) = T_{init}(\vec{x}), \quad \vec{x} \in \Omega,$$

$$T(\vec{x}, t) = T_I(\vec{x}, t), \quad \vec{x} \in \partial_I \Omega \quad (\text{on part of the boundary})$$

$$-k \frac{\partial T}{\partial \vec{n}} = Q(\vec{x}, t), \quad \vec{x} \in \partial_{II} \Omega \quad (\text{on the rest of the boundary}).$$

(a) Give a precise definition of classical solution in $G = \Omega \times (0, t^*)$.

(b) Formulate a concept of weak solution in G.

(c) Prove that every bounded classical solution is also a weak solution.

(d) Prove that every smooth (how smooth?) weak solution is also a classical solution.

4.5. CONVERGENCE OF THE ENTHALPY SCHEME AND EXISTENCE OF THE WEAK SOLUTION

4.5.A Introduction

In the previous section we presented the weak formulation of the Stefan Problem and established the uniqueness of the weak solution. Here we shall prove its existence by showing that the time-explicit enthalpy scheme of **§4.3.B** in fact converges to the weak solution as $\Delta x, \Delta t \to 0$ (with $\alpha \Delta t / \Delta x^2 < 1/2$). For simplicity, we restrict ourselves to a one-dimensional Stefan Problem and follow the original proof of [KAMENOMOSTSKAYA], [ATTHEY]. This will establish both the well-posedness of the weak formulation and the validity of the enthalpy scheme as a method of approximating the (unique) weak solution of the problem.

A convergence proof for the implicit enthalpy scheme of **§4.3.D** may be found in [ELLIOTT-OCKENDON]. Abstract (non-constructive) existence proofs, in any number of dimensions, appear in e.g. [FRIEDMAN, 1968], [MEIRMANOV, 1991].

4.5.B Structure of the proof

We present the convergence proof for the case of the model phase-change problem of **§4.3.A**. Thus we consider a slab $0 \le x \le l$, initially solid at $T(x,0) = T_{init}(x) \le T_m$, whose face at $x = 0$ is heated convectively with an ambient temperature $T_\infty(t) \ge T_m$: $q(0,t) = h\,[\,T_\infty(t) - T(0,t)\,]$, and insulated at $x = l$: $q(l,t) = 0$. For simplicity we assume thermophysical properties *constant* in each phase.

For convenience in notation we employ the Kirchoff transformation (see **§4.4.D**) to introduce the "Kirchoff temperature"

$$u := \begin{cases} k_S\,[\,T - T_m\,], & \text{for } T < T_m \quad (\text{solid}), \\ 0, & \text{for } T = T_m \quad (\textit{interface}), \\ k_L\,[\,T - T_m\,], & \text{for } T > T_m \quad (\textit{liquid}). \end{cases} \tag{1}$$

and the corresponding enthalpy

$$\beta(u) = \begin{cases} \rho c_S\, u \,/\, k_S & \text{for } u < 0 \quad (\text{solid}), \\ [\,0\,,\,\rho L\,] & \text{for } u = 0 \quad (\text{interface}), \\ \rho c_L\, u \,/\, k_L + \rho L & \text{for } u > 0 \quad (\text{liquid}). \end{cases} \tag{2}$$

For our one-dimensional problem, Ω is simply the interval $(0\,,l)$, $G \equiv (0\,,l) \times (0\,,t^*)$, $G^- \equiv \{\,(x,t)\!: 0 < x < X(t),\ 0 < t < t^*\,\}$, $G^+ \equiv \{\,(x,t)\!: X(t) < x < l,\ 0 < t < t^*\,\}$, and $\Sigma(x,t) \equiv x - X(t)$. Let us also set $G_*^- \equiv \{\,(x,t)\!: 0 \le x \le X(t),\ 0 < t < t^*\,\}$, and $G_*^+ \equiv \{\,(x,t)\!: X(t) \le x \le l,\ 0 < t < t^*\,\}$. Our problem may be stated as follows (PROBLEM 1, see DEFINITION 2, **§4.4.B**).

CLASSICAL FORMULATION Find $H(x,t)$, $u(x,t)$, $X(t)$ such that

(i) u is continuous in $\bar{G}$, X' is continuous in $(0\,,t^*)$, u_x is continuous in G_*^- and in G_*^+,

(ii) u_t, u_{xx} are continuous in G^- and in G^+,

(iii) $H(x,t) = \beta(\,u(x,t)\,)$ for $u(x,t) \ne 0$, $(x,t) \in \bar{G}$,

(iv) H, u, X satisfy

$$H_t = u_{xx} \quad \text{for } (x,t) \in G,\ x \ne X(t) \quad (\text{inside each phase}) \tag{3}$$

$$u(X(t)\,,t) = 0\,, \quad 0 < t < t^*, \tag{4a}$$

$$\rho L\, X'(t) = -\,u_x(X(t)^-\,,t) + u_x(X(t)^+\,,t)\,, \quad 0 < t < t^*, \tag{4b}$$

$$u(x,0) = u_{init}(x), \quad 0 \le x \le l, \qquad X(0) = 0, \tag{5}$$

$$u_x(0,t) = \frac{h}{k_L}[u(0,t) - u_\infty(t)], \quad 0 < t < t^*, \tag{6a}$$

$$u_x(l,t) = 0, \quad 0 < t < t^*. \tag{6b}$$

The data $u_{init}(x)$ and $u_\infty(t)$ correspond to $T_{init}(x)$ and $T_\infty(t)$ respectively, via (1), and they are assumed to be bounded continuous functions.

To arrive at a weak formulation (**§4.4.E**), we multiply by a test function ϕ and integrate by parts. In this one-dimensional case, the computations concerning the interfacial terms are much simpler than those in **§4.4.E**, and those concerning the boundary conditions are similar to EXAMPLE 4 **§4.4.C**. In particular, the convective boundary condition at $x = 0$ results in an integral containing $u(0,t)$, so we need u to admit boundary values at $x = 0$ belonging to $\mathbf{L}^2(0,t^*)$. Thus we arrive (PROBLEM 2) at the following

WEAK FORMULATION Find $H(x,t)$, $u(x,t)$ such that

(i) H and u are bounded (square) integrable functions in G, $u_x \in \mathbf{L}^2(G)$,

(ii) $H(x,t) \in \beta(u(x,t))$ almost everywhere in $\bar{G}$,

(iii) the relation

$$\int_0^{t^*}\!\!\int_0^l \Big[H\phi_t + u\phi_{xx}\Big]dxdt + \int_0^l H_{init}(x)\,\phi(x,0)\,dx = \int_0^{t^*} \frac{h}{k_L}\Big[u(0,t) - u_\infty(t)\Big]\phi(0,t)\,dt, \tag{7}$$

where $H_{init}(x) \in \beta(u_{init}(x))$, holds for every test function

$$\phi \in \Phi = \{\phi \in C^2(\bar{G}) : \phi\big|_{t=t^*} = 0, \quad \phi_x(0,t) = 0, \quad \phi_x(l,t) = 0\}, \tag{8}$$

The data $u_{init}(x)$ and $u_\infty(t)$ are assumed to be bounded square-integrable functions.

The enthalpy numerical scheme for this problem was constructed in **§4.3**. Here we consider the explicit scheme (**§4.3.A,B**) with uniform $\Delta x := l/M$ and $\Delta t = t^*/N$ satisfying the CFL stability condition ((21) **§4.3.B**)

$$\mu := \frac{\alpha_{\max}\Delta t}{\Delta x^2} \le \mu_0 < \frac{1}{2}, \tag{9}$$

where $\alpha_{\max} = \max\{\frac{k_L}{\rho c_L}, \frac{k_S}{\rho c_S}\}$. Thus, for any integer $M > 0$, we choose $\Delta x := \frac{l}{M}$ and an integer $N > \frac{1}{2}\frac{\alpha_{\max} t^*}{\Delta x^2}$ (so that the CFL condition (9) holds for $\Delta t := t^*/N$), and consider the explicit enthalpy scheme for (3)-(6). Note that in order to have (10d) valid also for $j = 1$ and $j = M$ we take $x_j = (j - ½)\Delta x$, $j = 0, 1, \ldots, M, M+1$ and assign u_0^n to the fake node $x_0 = -½\Delta x$, with $u(0,t)$ the average of u_0^n, u_1^n. The scheme is (PROBLEM 3) :

initial values: $\quad u_j^0 = u_{init}(x_j), \quad j = 0, 1, \ldots, M, M+1, \quad$ (10a) *boundary condition at* $x = 0$:

$$u_0^n = \frac{\left[1 - \dfrac{h\Delta x}{2k_L}\right] u_1^n + \dfrac{h\Delta x}{k_L} u_\infty(n\Delta t)}{1 + \dfrac{h\Delta x}{2k_L}}, \qquad n = 1, \ldots, N, \tag{10b}$$

boundary condition at $x = l$: $\qquad u_{M+1}^n = u_M^n, \qquad n = 1, \ldots, N, \qquad$ (10c)

interior values:

$$H_j^{n+1} = H_j^n + \frac{\Delta t}{\Delta x^2}\left[u_{j-1}^n - 2u_j^n + u_{j+1}^n \right], \; j = 1, \ldots, M, \; n = 0, 1, \ldots, N-1, \tag{10d}$$

and for $j = 1, \ldots, M$, $n = 0, 1, \ldots, N$:

$$u_j^n = \beta^{-1}(H_j^n) \equiv \begin{cases} \alpha_S H_j^n, & \text{if } H_j^n \le 0 \quad (\text{solid}) \\ 0, & \text{if } 0 < H_j^n < \rho L \quad (\text{interface}) \\ \alpha_L\left[H_j^n - \rho L \right], & \text{if } H_j^n \ge \rho L \quad (\text{liquid}) \end{cases} \tag{11}$$

For each mesh of size M, this explicit algorithm generates the finite sequences

$$\{H_j^n\}_{j=1,\ldots,M}^{n=0,\ldots,N-1}, \qquad \{u_j^n\}_{j=1,\ldots,M}^{n=0,\ldots,N-1},$$

from which we define the piecewise-constant functions (discrete solutions)

$$\vec{u}^M(x,t) := u_j^n, \quad \vec{u}_x^M(x,t) := \frac{u_{j+1}^n - u_j^n}{\Delta x}, \quad \vec{H}^M(x,t) := H_j^n$$

$$\text{for } (j-1)\Delta x \le x < j\Delta x, \quad j = 1, \ldots, M,$$

$$n\Delta t \le t < (n+1)\Delta t, \; n = 0, \ldots, N-1, \quad \text{if } (x,t) \in \bar{G};$$

$$\vec{u}^M(x,t) := 0, \quad \vec{H}^M(x,t) := 0, \quad \text{if } (x,t) \notin \bar{G}.$$

The superscript M indicates the dependence on the mesh-size M (and the corresponding N). Letting $M \to \infty$ (implying $N \to \infty$, $\Delta x \to 0$, $\Delta t \to 0$), we generate a sequence of discrete solutions { $\vec{u}^M$, $\vec{H}^M$ } on finer and finer meshes, and we hope these tend to the actual weak solution of the continuous problem. From the uniqueness theorem of **§4.4.E**, we know that there can be at most one weak solution, but the *existence* of such a solution is yet to be established. In fact, the goal of this section is to establish the following

THEOREM (existence) If the data $u_{init}(x)$ and $u_\infty(t)$ are bounded functions and satisfy

$$u_{init}, \; (u_{init})_x \in \mathbf{L}^2(0, l) \quad \text{and} \quad u_\infty, \; (u_\infty)_t \in \mathbf{L}^2(0, t^*),$$

then the problem has unique weak solution $\{u, H\}$, and the discrete solutions of the enthalpy scheme converge to it as the mesh gets finer.

The strategy of the proof is to establish the following claims:

CLAIM 1 : The sequence of discrete solutions $\{ \vec{u}^M , \vec{H}^M \}$ posesses a subsequence which converges to some $\{ u , H \}$ (in an appropriate sense).

CLAIM 2 : Any such limiting $\{ u , H \}$ is in fact a weak solution of the continuous problem, (thus establishing existence of weak solutions).

CLAIM 3 : No matter how $M \to \infty$, the discrete solutions $\{ \vec{u}^M , \vec{H}^M \}$ converge to the (unique) weak solution of the problem (therefore the numerical scheme does in fact provide us with approximations to the actual weak solution).

4.5.C Proof of CLAIM 1

We wish to show that the sequence of discrete solutions $\{ \vec{u}^M , \vec{H}^M \}_{M=1,2,...}$, posesses a convergent subsequence. To this end we prove

LEMMA 1 The sets $\{ \vec{u}^M : M = 1,2,..., \}$ and $\{ \vec{H}^M : M = 1,2,..., \}$ are (equi)bounded in $\mathbf{L}^2(G)$, i.e. there exist constants $C_1 > 0$, $C_2 > 0$ (independent of M) such that

$$\| \vec{u}^M \|_{\mathbf{L}^2(G)} \le C_1 \text{ and } \| \vec{H}^M \|_{\mathbf{L}^2(G)} \le C_2 \text{, for } any\ M = 1,2,... \tag{13}$$

Proof : For each integer M, $\vec{u}^M$ is the piecewise-constant function defined in (12), so we have

$$\| \vec{u}^M \|^2_{\mathbf{L}^2(G)} \equiv \iint_G |\vec{u}^M(x,t)|^2 dxdt = \sum_{n=0}^{N-1} \sum_{j=1}^{M} \int_{n\Delta t}^{(n+1)\Delta t} \int_{(j-1)\Delta x}^{j\Delta x} |\vec{u}^M(x,t)|^2 dxdt$$
$$= \sum_{n=0}^{N-1} \sum_{j=1}^{M} |u_j^n|^2 \, \Delta x \Delta t , \tag{14}$$

so we want to show that this sum is bounded independently of M, N.

From (2), we see that for any $j = 1, ..., M$, $n = 0, ..., N-1$,

$$H_j^{n+1} - H_j^n = \beta(u_j^{n+1}) - \beta(u_j^n) \ge \frac{1}{\alpha_{\max}} [u_j^{n+1} - u_j^n] ,$$

so (10d) implies

$$u_j^{n+1} \le [1 - 2\mu] u_j^n + \mu [u_{j-1}^n + u_{j+1}^n] .$$

The CFL condition (9) guarantees $1 - 2\mu > 0$, so replacing u_j^n, u_{j-1}^n, u_{j+1}^n by the largest one, we get

$$u_j^{n+1} \le (1 - 2\mu + \mu + \mu) \max \{u_{j-1}^n , u_j^n , u_{j+1}^n\} \le \max_{0 \le m \le M+1} \{ u_m^n \} .$$

Now, by (10b), $u_0^n \le \max \{u_1^n , u_\infty(t_n)\}$, and since $u_\infty(t)$ is assumed bounded $|u_\infty(t_n)| \le C_\infty$, so $u_0^n \le \max \{u_1^n , C_\infty\}$. Also $u_{M+1}^n = u_M^n$ from (10c), so we have

$$u_j^{n+1} \leq \max_{0 \leq m \leq M+1} \{ u_m^n \} \leq \max_{1 \leq m \leq M} \{ u_m^n , C_\infty \}, \; j = 1, ..., M, \; n = 0, ..., N-1. \tag{15a}$$

But initially $u_j^0 = u_{init}(x_j) \leq C_0$ (since $u_{init}(x)$ is assumed bounded), so $u_j^1 \leq \max\{C_0 , C_\infty\}$, $j = 1, \ldots, M$, and applying (15a) repeatedly we get

$$|u_j^n| \leq \max\{C_0 , C_\infty\} =: C, \quad 1 \leq j \leq M, \quad 0 \leq n \leq N. \tag{15b}$$

Then (14) yields the M-independent bound

$$\| \bar{u}^M \|^2_{\mathbf{L}^2(G)} \leq C^2 (N\Delta t)(M\Delta x) \equiv C^2 t^* l =: C_1 .$$

Finally, $|H_j^n| = |\beta(u_j^n)| \leq \dfrac{1}{\alpha_{\max}} + \rho L =: C'$ implies a similar M-independent bound on $\| \bar{H}^M \|_{\mathbf{L}^2(G)}$. □

From a basic result about Hilbert spaces, which states: "Any bounded set in a Hilbert space has a weakly-convergent subsequence", we then obtain

COROLLARY 1 The set $\{(\bar{u}^M , \bar{H}^M) : M = 1, 2, ...,$ contains a weakly-convergent subsequence, i.e. there exists a subsequence $\{M_i'\}$ of integers and functions u, $H \in \mathbf{L}^2(G)$ such that, as $M_i' \to \infty$,

$$\iint_G [\bar{u}^{M_i'}(x,t) - u(x,t)] \, \psi(x,t) \, dxdt \to 0 ,$$

$$\text{and} \iint_G [\bar{H}^{M_i'}(x,t) - H(x,t)] \, \psi(x,t) \, dxdt \to 0 , \tag{16}$$

for any $\psi \in \mathbf{L}^2(G)$.

We would like to show that such a $\{u, H\}$ is a weak solution, which however requires that $u_x \in \mathbf{L}^2(G)$. To establish this, we need the following stronger a priori estimates.

LEMMA 2 There exist constants $C_3 > 0$, $C_4 > 0$, independent of M (and N), such that

$$\Delta x \Delta t \sum_{n=0}^{N-1} \sum_{j=1}^{M} \left(\frac{u_j^{n+1} - u_j^n}{\Delta t} \right)^2 \leq C_3 , \tag{17a}$$

and

$$\Delta x \Delta t \sum_{n=0}^{N-1} \sum_{j=1}^{M} \left(\frac{u_{j+1}^n - u_j^n}{\Delta x} \right)^2 \leq C_4 . \tag{17b}$$

Proof : Let $U_j^n := \dfrac{u_{j+1}^n - u_j^n}{\Delta x}$, $j = 0, ..., M$, $n = 0, ..., N$, $W_j^n := \dfrac{u_j^{n+1} - u_j^n}{\Delta t}$, $j = 1, ..., M$, $n = 0, ..., N$. Multiplying (10d) by $W_j^n \Delta x \Delta t$ we get

$$\Delta x \Delta t \, \frac{H_j^{n+1} - H_j^n}{\Delta t} \, W_j^n \; = \; \Delta x \Delta t \, \frac{u_{j-1}^n - 2u_j^n + u_{j+1}^n}{\Delta x^2} \, W_j^n \, .$$

Using the lower bound (11)§4.4 on the left-hand side, this becomes

$$\frac{\Delta x \Delta t}{\alpha_{\max}} (W_j^n)^2 \; \leq \; \Delta x \Delta t \, \frac{U_j^n - U_{j-1}^n}{\Delta x} \, W_j^n \; = \; \Delta t \, [U_j^n - U_{j-1}^n] \, W_j^n \, . \tag{18}$$

We manipulate the right-hand side as follows. Adding and subtracting $U_j^n W_{j+1}^n$ we get

$$\begin{aligned} \Delta t \, [\, U_j^n - U_{j-1}^n \,] \, W_j^n \; &= \; \Delta t \, [\, U_j^n W_{j+1}^n - U_{j-1}^n W_j^n \,] \; + \; \Delta t \, U_j^n \, [\, W_j^n - W_{j+1}^n \,] \\ &= \quad S_1 \quad + \quad S_2 \, . \end{aligned} \tag{19a}$$

In the last term, S_2, we rearrange the differences:

$$\begin{aligned} S_2 &\equiv \Delta t \, U_j^n \, [\, W_j^n - W_{j+1}^n \,] = U_j^n \, [\, u_j^{n+1} - u_j^n - u_{j+1}^{n+1} + u_{j+1}^n \,] = \Delta x \, U_j^n \, [\, U_j^n - U_j^{n+1} \,] \\ &= \frac{1}{2} \Delta x \, [\, (U_j^n)^2 - (U_j^{n+1})^2 \; + \; (\, U_j^n - U_j^{n+1} \,)^2 \,] \\ &= \quad S_3 \quad + \quad S_4 \, . \end{aligned} \tag{19b}$$

We rearrange the last term, S_4, again, and estimate it:

$$\begin{aligned} S_4 &\equiv \frac{1}{2} \Delta x \, (\, U_j^n - U_j^{n+1} \,)^2 = \frac{1}{2\Delta x} \Delta t^2 \, [\, W_j^n - W_{j+1}^n \,]^2 \leq \frac{\Delta t^2}{\Delta x} \, [\, (W_j^n)^2 + (W_{j+1}^n)^2 \,] \\ &= \Delta x \Delta t \, \frac{\mu}{\alpha_{\max}} \, [\, (W_j^n)^2 + (W_{j+1}^n)^2 \,] , \qquad (\, \mu := \alpha_{\max} \Delta t \, / \, \Delta x^2 \,) \, . \end{aligned} \tag{19c}$$

Hence, (18) yields

$$\begin{aligned} &\frac{\Delta x \Delta t}{\alpha_{\max}} (W_j^n)^2 \; \leq \; S_1 \; + \; S_3 \; + \; S_4 \\ &\leq \Delta t \, [\, U_j^n W_{j+1}^n - U_{j-1}^n W_j^n \,] + \frac{1}{2} \Delta x \, [\, (U_j^n)^2 - (U_j^{n+1})^2 \,] + \frac{\Delta x \Delta t}{\alpha_{\max}} \, \mu \, [\, (W_j^n)^2 + (W_{j+1}^n)^2 \,]. \end{aligned}$$

Summing over j, n, the differences collapse and we find

$$\begin{aligned} \frac{\Delta x \Delta t}{\alpha_{\max}} \sum_{n=0}^{N-1} \sum_{j=1}^{M} (W_j^n)^2 \; &\leq \; \Delta t \sum_{n=0}^{N-1} [\, U_M^n W_M^n - U_0^n W_1^n \,] \; + \frac{1}{2} \Delta x \sum_{j=1}^{M} [\, (U_j^0)^2 - (U_j^N)^2 \,] \\ &\quad + \frac{\Delta x \Delta t}{\alpha_{\max}} \cdot 2\mu \sum_{n=0}^{N-1} \sum_{j=0}^{M} (W_j^n)^2 \, . \end{aligned}$$

Since $U_M^n \equiv (u_{M+1}^n - u_M^n)/\Delta x = 0$, this becomes

$$\frac{\Delta x \Delta t}{\alpha_{\max}} \, [\, 1 - 2\mu \,] \sum_{n=1}^{N-1} \sum_{j=1}^{M} (W_j^n)^2 \; \leq \; -\Delta t \sum_{n=0}^{N-1} U_0^n W_1^n + \frac{1}{2} \Delta x \sum_{j=1}^{M} [\, (U_j^0)^2 - (U_j^N)^2 \,] \, ,$$

or

$$\frac{\Delta x \Delta t}{\alpha_{\max}}[1-2\mu]\sum_{n=0}^{N-1}\sum_{j=1}^{M}(W_j^n)^2+\frac{1}{2}\Delta x\sum_{j=1}^{M}(U_j^N)^2$$

$$\le \sum_{n=0}^{N-1}\left[\frac{u_0^n-u_1^n}{\Delta x}\right][u_1^{n+1}-u_1^n] \;+\; \frac{1}{2}\Delta x\sum_{j=1}^{M}\left(\frac{u_{j+1}^0-u_j^0}{\Delta x}\right)^2$$

$$=: \quad S_5 \quad + \quad S_6 . \tag{20}$$

Now, using the boundary condition (10b),

$$S_5 = \frac{h}{k_L\left(1+\dfrac{h\Delta x}{2k_L}\right)}\sum_{n=0}^{N-1}[u_\infty^n-u_1^n][u_1^{n+1}-u_1^n]$$

$$\le \frac{h}{k_L}\left\{\left|\sum_{n=0}^{N-1}u_\infty^n(u_1^{n+1}-u_1^n)\right|+\left|\sum_{n=0}^{N-1}u_1^n(u_1^{n+1}-u_1^n)\right|\right\}$$

$$= \frac{h}{k_L}\left\{\left|\sum_n[u_\infty^{n+1}u_1^{n+1}-u^{\infty^n}u_1^n-(u_\infty^{n+1}-u_\infty^n)u_1^{n+1}]\right|\right.$$

$$\left.+\left|\sum_n\frac{1}{2}[(u_1^{n+1})^2-(u_1^n)^2-(u_1^{n+1}-u_1^n)^2]\right|\right\}$$

$$= \frac{h}{k_L}\left\{\left|u_\infty^N u_1^N-u_\infty^0 u_1^0-\Delta t\sum_n u_1^{n+1}\frac{u_\infty^{n+1}-u_\infty^n}{\Delta t}\right|\right.$$

$$\left.+\frac{1}{2}\left|(u_1^N)^2-(u_1^0)^2-\sum_n(u_1^{n+1}-u_1^n)^2\right|\right\}$$

$$\le \frac{h}{k_L}\left\{|u_\infty^N u_1^N|+|u_\infty^0 u_1^0|\Delta t\sum_n|u_1^{n+1}|\left|\frac{u_\infty^{n+1}-u_\infty^n}{\Delta t}\right|+\frac{1}{2}|(u_1^N)^2-0-0|\right\}$$

$$\le \frac{h}{k_L}\left\{C_\infty C+C_\infty C+C\int_0^{t^*}|(\bar{u}_\infty^M)_t|\,dt+\frac{1}{2}C^2\right\}$$

which, by collecting the constants, and using the Cauchy-Schwarz inequality on the integral, is

$$\le C_5 + C\sqrt{t^*}\,\|(\bar{u}_\infty^M)_t\|_{L^2(0,t^*)} .$$

From THEOREM 1 **§4.4.B**, the norm in the last term is bounded by the norm of $(u_\infty)_t$, which is assumed to be in $\mathbf{L}^2(0,t^*)$, hence the last term is also bounded by a constant independent of M. We conclude that

$$S_5 \le C_6 , \qquad \text{(independently of } M\text{)}.$$

On the other hand, from the initial condition (10a), we have

$$S_6 := \frac{1}{2}\Delta x \sum_{j=1}^{M}\left(\frac{u_{j+1}^0 - u_j^0}{\Delta x}\right)^2$$

$$= \frac{1}{2}\sum_{j=1}^{M}\int_{(j-1)\Delta x}^{j\Delta x}\left(\bar{u}_x^M(x,0)\right)^2 dx = \frac{1}{2}\int_0^l \left(\bar{u}_x^M(x,0)\right)^2 dx = \frac{1}{2}\int_0^l |(\bar{u}_{init}^M)_x|^2 dx$$

$$\leq \frac{1}{2}\|(u_{init})_x\|^2_{\mathbf{L}^2(0,l)} =: C_7 .$$

For the last inequality we have used the assumption that $(u_{init})_x \in \mathbf{L}^2(0,l)$ in conjunction with THEOREM 1 **§4.4.B**.

Using the estimates for S_5 and S_6, (20) becomes

$$\frac{\Delta x \Delta t}{\alpha_{\max}}[1-2\mu]\sum_{n=0}^{N-1}\sum_{j=1}^{M}(W_j^n)^2 + \frac{1}{2}\Delta x\sum_{j=1}^{M}(U_j^N)^2 \leq C_6 + C_7 =: C_3 \tag{21}$$

Since each term on the left-hand side is nonnegative, we obtain

$$\Delta x \Delta t \sum_{n=0}^{N-1}\sum_{j=1}^{M}\left(\frac{u_j^{n+1} - u_j^n}{\Delta t}\right)^2 \leq C_3, \tag{22}$$

which establishes (17a), as well as

$$\Delta x \sum_{j=1}^{M}\left(\frac{u_{j+1}^N - u_j^N}{\Delta t}\right)^2 \leq C_3 . \tag{23}$$

Noticing that (21) holds also for any $N' \leq N$, we see that estimate (23) also holds with N replaced by $n = 0, 1, \ldots, N$, whence

$$\Delta x \Delta t \sum_{n=0}^{N-1}\sum_{j=1}^{M}\left(\frac{u_{j+1}^n - u_j^n}{\Delta x}\right)^2 \leq N\Delta t\, C_3 = t^* C_3 =: C_4$$

which establishes (17b). □

LEMMA 3 Any sequence in $\{\bar{u}^M : M = 1, 2, \ldots\}$ contains a subsequence which converges weakly to a u in $\mathbf{L}^2(G)$ whose weak derivative u_x is in $\mathbf{L}^2(G)$.

Proof. As in COROLLARY 1, the equiboundeness of $\|\bar{u}^M\|_{\mathbf{L}^2(G)}$ implies that there exists a subsequence $\{M_i''\}$ and a function u in $\mathbf{L}^2(G)$ to which $\{\bar{u}^{M_i''}\}$ converges weakly. Next we show the equiboundeness of $\|\bar{u}_x^M\|_{\mathbf{L}^2(G)}$. We have,

$$\| \bar{u}_x^M \|^2_{L^2(G)} \equiv \iint_G | \bar{u}_x^M |^2 \, dxdt = \sum_{n=0}^{N-1} \sum_{j=1}^{M} \int_{n\Delta t}^{(n+1)\Delta t} \int_{(j-1)\Delta x}^{j\Delta x} \left(\frac{u_{j+1}^n - u_j^n}{\Delta x} \right)^2 dxdt$$

$$= \sum_{n=0}^{N-1} \sum_{j=1}^{M} \left(\frac{u_{j+1}^n - u_j^n}{\Delta x} \right)^2 \Delta x \Delta t \ < \ C_3$$

by LEMMA 2. It follows, as above, that there exists a subsequence $\{M_i'''\}$ and a function w in $\mathbf{L}^2(G)$ to which $\{\bar{u}_x^{M_i'''}\}$ converges weakly. We claim that w is in fact the weak derivative of u. Indeed, for any $\psi \in \mathbf{C}_0^\infty(G)$, one can show (PROBLEM 4) that

$$\iint_G \bar{u}_x^M \, \psi \, dxdt = -\iint_G \bar{u}^M \, \psi_x \, dxdt \, .$$

Taking the limit $M \to \infty$ through the subsequence $\{M_i''' \}$ we get $\iint_G w \, \psi \, dxdt = -\iint_G u\psi_x \, dxdt$ for any $\psi \in \mathbf{C}_0^\infty(G)$. By the very definition of weak derivative for u, we see that $u_x = w \in \mathbf{L}^2(G)$. □

4.5.D Proof of CLAIM 2

We wish to show that the $\{u, H\}$ from CLAIM 1 constitutes a weak solution to our Stefan Problem, i.e. that (7) holds for any test function ϕ as in (8).

To this end, for any such test function ϕ and any integer M (and corresponding N, Δx, Δt), we define the discrete values

$$\phi_j^n := \phi(x_j, t_n), \ j = 1, ..., M; \ \ \phi_0^n := \phi_1^n, \ \ \phi_{M+1}^n := \phi_M^n, \ n = 0, ..., N, \tag{24}$$

and the piecewise constant functions (their dependences on M being understood)

$$\bar{\phi}(x,t) := \phi_j^n, \ \ \bar{\phi}_{xx}(x,t) := \frac{\phi_{j+1}^n - 2\phi_j^n + \phi_j^n}{\Delta x^2}, \ \ \bar{\phi}_t(x,t) := \frac{\phi_j^n - \phi_j^{n-1}}{\Delta t},$$
$$\text{for } (j-1)\Delta x \le x < j\Delta x, \quad n\Delta t \le t < (n+1)\Delta t \, . \tag{25}$$

LEMMA 4 As $M \to \infty$, the piecewise constant functions $\bar{\phi}$, $\bar{\phi}_{xx}$, $\bar{\phi}_t$ converge strongly in $\mathbf{L}^2(G)$ to ϕ, $\dfrac{\partial^2 \phi}{\partial x^2}$, $\dfrac{\partial \phi}{\partial t}$, respectively.

Proof. By the Mean Value Theorem, for some $(j-1)\Delta x < x_j^* < j\Delta x$, $n\Delta t < t_n^* < (n+1)\Delta t$, we have

$$\iint_G | \bar{\phi}(x,t) - \phi(x,t) |^2 dxdt = \sum_{n=0}^{N-1} \sum_{j=1}^{M} | \phi_j^n - \phi(x_j^*, t_n^*) |^2 \Delta x \Delta t \, ,$$

As $M \to \infty$ (so $N \to \infty$, $\Delta x \to 0$, $\Delta t \to 0$), the intermediate points (x_j^*, t_n^*) tend to (x_j, t_n) and therefore $|\phi_j^n - \phi(x_j^*, t_n^*)| \to 0$ since $\phi(x,t)$ is continuous in G, showing that $\|\bar{\phi} - \phi\|_{L^2(G)} \to 0$. The proof for $\bar{\phi}_{xx}$, $\bar{\phi}_t$ is similar, using Taylor expansions (PROBLEM 5). □

Now we proceed with the proof of the CLAIM. Multiplying the discrete PDE (10d) by ϕ_j^n and summing over j and over n, we get

$$\sum_{n=0}^{N-1}\sum_{j=1}^{M}\frac{H_j^{n+1}-H_j^n}{\Delta t}\phi_j^n = \sum_{n=0}^{N-1}\sum_{j=1}^{M}\frac{u_{j-1}^n-2u_j^n+u_{j+1}^n}{\Delta x^2}\phi_j^n . \tag{26}$$

We apply "summation by parts" on the left-hand side as follows. Using the identity

$$\frac{H_j^{n+1}-H_j^n}{\Delta t}\phi_j^n \equiv \frac{(H\phi)_j^{n+1}-(H\phi)_j^n}{\Delta t} - H_j^{n+1}\frac{\phi_j^{n+1}-\phi_j^n}{\Delta t},$$

(which is a discrete version of $H_t\phi = (H\phi)_t - H\phi_t$), we have

$$\sum_{j=1}^{M}\sum_{n=0}^{N-1}\frac{H_j^{n+1}-H_j^n}{\Delta t}\phi_j^n = \sum_{j=1}^{M}\frac{1}{\Delta t}\Big[(H\phi)_j^N-(H\phi)_j^0\Big]-\sum_{j=1}^{M}\sum_{n=0}^{N-1}H_j^{n+1}\frac{\phi_j^{n+1}-\phi_j^n}{\Delta t}$$

and since $\phi\in\Phi$ implies $\phi_j^N := \phi(x_j,t^*)\equiv 0$, and $H_j^0 = H_{init}(x_j)$,

$$= -\frac{1}{\Delta t\Delta x}\sum_{j=1}^{M}\int_{(j-1)\Delta x}^{j\Delta x}H_{init}(x)\bar{\phi}(x,0)dx$$

$$-\frac{1}{\Delta x\Delta t}\sum_{n=0}^{N-1}\sum_{j=1}^{M}\int_{n\Delta t}^{(n+1)\Delta t}\int_{(j-1)\Delta x}^{j\Delta x}\bar{H}^M(x,t+\Delta t)\bar{\phi}_t(x,t+\Delta t)dxdt,$$

whence

$$\sum_{n=0}^{N-1}\sum_{j=1}^{M}\frac{H_j^{n+1}-H_j^n}{\Delta t}\phi_j^n = -\frac{1}{\Delta t\Delta x}\Bigg[\int_0^l \bar{H}_{init}^M(x)\bar{\phi}(x,0)dx + \int_{\Delta t}^{t^*+\Delta t}\int_0^l \bar{H}^M(x,t)\bar{\phi}_t(x,t)dxdt\Bigg]. \tag{27}$$

Next we apply summation by parts on the right-hand side of (26) via the identity

$$\frac{u_{j-1}^n-2u_j^n+u_{j+1}^n}{\Delta x^2}\phi_j^n \equiv \frac{\frac{u_{j+1}^n-u_j^n}{\Delta x}\phi_{j+1}^n-\frac{u_j^n-u_{j-1}^n}{\Delta x}\phi_j^n}{\Delta x}-\frac{u_{j+1}^n\frac{\phi_{j+1}^n-\phi_j^n}{\Delta x}-u_j^n\frac{\phi_j^n-\phi_{j-1}^n}{\Delta x}}{\Delta x}$$
$$+u_j^n\frac{\phi_{j-1}^n-2\phi_j^n+\phi_{j+1}^n}{\Delta x^2},$$

which is the discrete analog of $u_{xx}\phi\equiv(u_x\phi)_x-(u\phi_x)_x+u\phi_{xx}$. Summing over j, certain sums collapse:

$$\frac{1}{\Delta x}\sum_{j=1}^{M}\left[\frac{u_{j+1}^n-u_j^n}{\Delta x}\phi_{j+1}^n-\frac{u_j^n-u_{j-1}^n}{\Delta x}\phi_j^n\right]\equiv\frac{1}{\Delta x}\left[\frac{u_{M+1}^n-u_M^n}{\Delta x}\phi_{M+1}^n-\frac{u_1^n-u_0^n}{\Delta x}\phi_1^n\right],$$

and using (10b,c)

$$=\frac{1}{\Delta x}\left[0+\frac{h}{k_L\left(1+\dfrac{h\Delta x}{2k_L}\right)}(u_\infty^n-u_1^n)\phi_1^n\right].$$

Similarly we find

$$\frac{1}{\Delta x}\sum_{j=1}^{M}\left[u_{j+1}^n\frac{\phi_{j+1}^n-\phi_j^n}{\Delta x}-u_j^n\frac{\phi_j^n-\phi_{j-1}^n}{\Delta x}\right]=\frac{1}{\Delta x}\left[u_{M+1}^n\frac{\phi_{M+1}^n-\phi_M^n}{\Delta x}-u_1^n\frac{\phi_1^n-\phi_0^n}{\Delta x}\right]\equiv 0$$

because of (24). Thus we obtain

$$\sum_{n=0}^{N-1}\sum_{j=1}^{M}\frac{u_{j-1}^n-2u_j^n+u_{j+1}^n}{\Delta x^2}\phi_j^n$$

$$\equiv\frac{1}{\Delta x}\sum_{n=0}^{N-1}\frac{h}{k_L\left(1+\dfrac{h\Delta x}{2k_L}\right)}(u_\infty^n-u_1^n)\phi_1^n+\sum_{n=0}^{N-1}\sum_{j=1}^{M}u_j^n\frac{\phi_{j-1}^n-2\phi_j^n+\phi_{j+1}^n}{\Delta x^2}$$

$$=\frac{1}{\Delta x\Delta t}\sum_{n=0}^{N-1}\int_{n\Delta t}^{(n+1)\Delta t}\frac{h}{k_L\left(1+\dfrac{h\Delta x}{2k_L}\right)}[\bar{u}_\infty^M(t)-\bar{u}^M(0,t)]\bar{\phi}(0,t)dt$$

$$+\frac{1}{\Delta x\Delta t}\sum_{n=0}^{N-1}\sum_{j=1}^{M}\int_{n\Delta t}^{(n+1)\Delta t}\int_{(j-1)\Delta x}^{j\Delta x}\bar{u}^M(x,t)\bar{\phi}_{xx}(x,t)dxdt$$

$$=\frac{1}{\Delta x\Delta t}\left\{\int_0^{t^*}\frac{h}{k_L\left(1+\dfrac{h\Delta x}{2k_L}\right)}[\bar{u}_\infty^M(t)-\bar{u}^M(0,t)]\bar{\phi}(0,t)dt+\int_0^{t^*}\int_0^{l}\bar{u}^M(x,t)\bar{\phi}_{xx}(x,t)dxdt\right\}.$$

From this and (27), relation (26) yields

$$\int_{\Delta t}^{t^*+\Delta t}\int_0^{l}\bar{H}^M(x,t)\,\bar{\phi}_t(x,t)dxdt+\int_0^{l}H_{init}(x)\,\bar{\phi}(x,0)dx+\int_0^{t^*}\int_0^{l}\bar{u}^M(x,t)\,\bar{\phi}_{xx}(x,t)dxdt$$

$$=\frac{1}{1+\dfrac{h\Delta x}{2k_L}}\int_0^{t^*}\frac{h}{k_L}\left(\bar{u}^M(0,t)-\bar{u}_\infty^M(t)\right)\bar{\phi}(0,t)dt\,. \tag{28}$$

Now we take $M \to \infty$ through a subsequence $\{M_i\}$ for which CLAIM 1 holds and observe the convergence of each term as follows. Consider the first term in (28). Since its integrand vanishes for $t > t^*$ (see (12)), it may be written as $\int_{\Delta t}^{t^*+\Delta t}\int_0^l \tilde{H}^{M_i}\tilde{\phi}_t dxdt = \iint_G \tilde{H}^{M_i}\tilde{\phi}_t dxdt - \int_0^{\Delta t}\int_0^l \tilde{H}^{M}\tilde{\phi}_t dxdt$ and the last integral tends to zero with Δt since its integrand is bounded. Now,

$$\iint_G \tilde{H}^{M_i}\tilde{\phi}_t dxdt \equiv \iint_G \left(\tilde{H}^{M_i} - H\right)\phi_t dxdt + \iint_G \tilde{H}^{M_i}\left(\tilde{\phi}_t - \phi_t\right)dxdt$$

$$+ \iint_G H\phi_t\, dxdt \;\to\; 0 + 0 + \iint_G H\phi_t dxdt \quad \text{as } M_i \to \infty, \tag{29}$$

thanks to COROLLARY 1 and LEMMA 4 (PROBLEM 6). Hence, the first term of (28) tends to $\iint_G H\phi_t dxdt$ as $M_i \to \infty$. Similarly, the third term of (28) tends to $\iint_G u\phi_{xx}dxdt$. The second term tends to $\int_0^l H_{init}(x)\phi(x,0)dx$, and finally in the last term of (28), the factor tends to 1 as $\Delta x \to 0$, and the integral tends to $\int_0^{t^*} \frac{h}{k_L}\left(u(0,t) - u_\infty(t)\right)\phi(0,t)dt$. We conclude that (28) tends to the relation

$$\iint_G H\phi_t dxdt + \int_0^l H_{init}(x)\phi(x,0)dx + \iint_G u\phi_{xx}dxdt$$

$$= \int_0^{t^*} \frac{h}{k_L}\left(u(0,t) - u_\infty(t)\right)\phi(0,t)dt,$$

which shows that indeed $\{u,\ H\}$ constitutes a weak solution to the Stefan Problem, establishing existence of the weak solution. □

4.5.E Proof of CLAIM 3

So far we have shown that there exists (at least one) subsequence $\{\tilde{u}^{M_i},\ \tilde{H}^{M_i}\}$ converging to some $\{u,\ H\}$ which is the weak solution of our Stefan Problem. Now we wish to establish the credibility of the numerical method itself by showing that no matter how $M \to \infty$ (i.e., no matter how the mesh is refined), as long as the stability condition holds, the discrete solutions $\{\tilde{u}^{M},\ \tilde{H}^{M}\}$ will converge to the (unique) weak solution of the problem.

Actually, this follows immediately from CLAIMs 1 and 2. Indeed, for *any* sequence of integers $\{M_i^*\}$ (with corresponding N_i^*, Δx^*, Δt^* satisfying (9)), there exists a subsequence $\{M_i^{**}\}$ for which $\{\vec{u}^{M_i^{**}}, \vec{H}^{M_i^{**}}\}$ converges to a weak solution $\{u^{**}, H^{**}\}$. By uniqueness, all such solutions coincide with *the* weak solution $\{u, H\}$. Therefore, *any* mesh refinement satisfying (9) will lead to an approximation of the weak solution.

4.5.F Note on Error Estimates

Precise error estimates are difficult to derive, especially for finite-difference type discretizations like the one we have employed. [ELLIOTT] has shown that for a piecewise-linear finite element approximation of the multidimensional enthalpy formulation the error in temperature, $\| \vec{u}^{M} - u \|_{\mathbf{L}^2}$, is proportional to $\sqrt{\Delta t}$. Similar error estimates for various finite-element methods have also been obtained by [NOCHETTO], and by [VERDI-VISINTIN].

PROBLEMS

PROBLEM 1. Show that the 1-dimensional phase-change problem of **§4.3.A** transforms to (3)-(6) under the Kirchoff transformation (1)-(2).

PROBLEM 2. From the classical formulation of the model problem described at the begining of **§4.5.B** derive the weak formulation stated in **§4.5.B**.

PROBLEM 3. Construct the explicit enthalpy scheme (10)-(11) for the model problem of **§4.5.B**.

PROBLEM 4. Let $\{\vec{u}^{M}\}$, $\{\vec{u}_x^{M}\}$ be as in LEMMA 3, and let u, w be weak limits of subsequences $\{\vec{u}_i^{M''}\}$ and $\{\vec{u}_x^{M_i''}\}$ respectively.

(a) Using summation by parts, show that for any $\psi \in \mathbf{C}_0^\infty(G)$,

$$\iint_G \vec{u}_x^{M} \psi \, dxdt = -\iint_G \vec{u}^{M} \psi_x \, dxdt \quad \text{holds.}$$

(b) Take the limit $M \to \infty$ through an appropriate subsequence and show that the result is $\iint_G w\psi \, dxdt = -\iint_G u\psi_x dxdt, \quad \psi \in \mathbf{C}_0^\infty(G)$.

(c) Justify why this means $u_x = w$ in $\mathbf{L}^2(G)$.

PROBLEM 5. For any test function $\phi \in \Phi$ as in (8), let $\vec{\phi}^{M}$, $\vec{\phi}_{xx}^{M}$, $\vec{\phi}_t^{M}$ be the piece-wise-constant functions defined in **§4.5.D**. Using Taylor expansions, show that as $M \to \infty$,

$$\| \vec{\phi}_{xx}^{M} - \frac{\partial^2 \phi}{\partial x^2} \|_{\mathbf{L}^2(G)} \to 0 \quad \text{and} \quad \| \vec{\phi}_t^{M} - \frac{\partial \phi}{\partial t} \|_{\mathbf{L}^2(G)} \to 0,$$

thus completing the proof of LEMMA 4.

PROBLEM 6. Justify (29).

CHAPTER 5

APPLYING THE TECHNIQUES OF MODELING

The techniques descriped in CHAPTERs **1-4** can be used to understand phase change processes as they arise in a variety of disciplines. One such discipline is that of Latent Heat Thermal Energy Storage (LHTES), in which the modeller aims not so much at code development but rather, at understanding what is happening thermally in the possibly complex interractions between storage materials, containers, fans and other physical components, over time. This understanding can often be acquired more readily from a "back of the envelope" analytical relation than from a large computer simulation.

This chapter is meant to bring to the reader, some examples of how the tools that we have discussed in the earlier chapters can be joined together and brought to bear on "real" thermal situations arising in LHTES. Our choice of this area is guided by the fact that this has been a focus of our own work at the Oak Ridge National Laboratory for many years, and that it gives rise to the kinds of cycling and multifront situations that really do demand the balanced use of a variety of tools to reach the understanding required to make systems "work."

Energy storage is required of every living creature or mechanical system that for periods of time, must be mobile or autonomous. The forms of energy that can be stored include chemical (e.g., batteries), mechanical (e.g., flywheel storage), and thermal (or heat). The archetype thermal storage device is, of course, the ubiquitous hot water bottle, warming our feet with the heat stored in the hot water it contains.

Thermal energy is of vital importance in many respects. The most evident is domestic heating. Where once we would warm our houses with large fires, the growth of cities, the expansion of population, and our rising aspirations now rule out the use of wood fire to keep us alive and warm under low temperature conditions. In many countries people have turned to heating methods based on the use of fossil fuels, for electric generation, as well as kerosene for room heaters. In recent years, because of declining availability and rising costs associated with both oil and coal, methods have been sought for better utilization of both, as well as other energy sources such as solar and nuclear power. Inevitably, we confront a situation where the desired energy is available, but not when (or sometimes where) we want it to be. Thus, electricity generation costs are reduced if the electricity demand remains constant throughout the day; however, heating needs peak when other needs, industrial, business, etc., also peak. Solar energy is available in certain areas, during daylight hours, while we would like to use the energy at nighttime, when no sun in available. Similarly, "cold" (just like heat) is a commodity that is produced as a result of expending energy. Creating cold during summer periods requires electricity at just the peak time that heating does during winter.

Heat can be stored in a material either as *sensible* or as *latent* heat. Sensible heat storage requires materials having high density and specific heat, since the heat storage is only via temperature differences. In addition, heat transfer requires possibly large temperature gradients and ranges, and results in large temperature swings in response to external heat input and removal. Nevertheless, sensible heat storage has been successfully used in domestic heating applications for many years. Methods include rock, pebble bed, and water storage for solar and electrical energy input, as well as ceramic brick and tile units (some of them portable) that are popular in European countries and are generally heated using off-peak electricity.

The storage of thermal energy as the latent heat of fusion of a material has several attractive features. In principle, it involves the use of heat that is stored in a material at a fixed temperature, namely the melting temperature of the material. Second, if the material has a large latent heat, then it could store the heat in less volume than that required for sensible heat storage. While these features appear at first glance to be attractive, one must quantify them before being assured of their correctness. To do this requires the simulation of the processes of heat storage (melting) and heat removal (freezing) that underlie this storage approach. Thus, the quantitative study of latent heat thermal energy storage requires that we be able to model the heat transfer and phase change processes themselves. Drawbacks lie in the variety of materials problems that might arise; these include supercooling (for a variety of salts), degradation of the latent heat, changes of density, and problems related to the chemistry and containment of the material. It is important to note that the phase change does not actually occur with the PCM at a uniform temperature equal to its melt temperature; on the contrary, a possibly large thermal driving force may be needed to produce the phase change [SOLOMON, 1981b].

Latent heat thermal energy storage is under study as the possible heat storage mechanism in several additional areas. One area of special current interest concerns the powering of an engine by intermittently available solar energy as, for example, in one of the methods proposed for the power system of a space station.

Surveys of issues arising in thermal energy storage appear in [BAYLIN], [CULP], [ELLIOT et al,1976], [GOLIBERSUCH], [GRODZKA], [HOOVER, et al], [TELKES], [TOKSOY-ILKEN], [WYMAN]. See also [ABHAT], [HALE, et al,1971], [JURINAK & ABDEL-KHALIK], [STEPLER], and many other publications by organizations such as ASHRAE, ASME, SERI, etc.

In this Chapter we apply the results of our earlier work to a number of questions arising in latent heat thermal energy storage. We will be drawn first into the "world" of phase change materials (PCMs) and their properties in §**5.1**. We will then examine many different elementary configurations that arise in various latent heat storage processes (§**5.2**). We close with an examination of LHTES for the space station where we will apply all the tools at our disposal to studying a particular process. In this application we will also encounter the problem of void formation, a key problem of this area.

5.1. LATENT HEAT STORAGE AND PHASE CHANGE MATERIALS

5.1.A Introduction

Latent Heat Thermal Energy Storage (LHTES) is concerned with methods for storing energy as the phase-change latent heat of a material (the Phase Change Material or PCM). Such methods are needed for the effective use of intermittent energy sources including solar and off-peak electricity. Various LHTES devices are currently being marketed while many questions, ranging from PCM containment to heat transfer performance of systems, remain open.

In order to see the variety of LHTES options that in principle are open to us, let us list several that come to mind:

(a) *PCM Trombe Wall*: A wall filled with PCM is constructed behind a large south-facing window of a house, **Figure 5.1.1**. The wall is heated during the day by incoming solar radiation, melting the PCM. At night the heat is withdrawn to warm the house. Because of the "greenhouse effect" the heat stays in the wall during the day. The concept of a wall with large thermal mass behind a south-facing window is due to the solar energy pioneer Felix Trombe and is accordingly named a "Trombe wall." Traditional Trombe walls rely on sensible heat storage, but because of the potential for greater heat storage per unit mass, the PCM Trombe wall is an attractive concept still awaiting successful implementation. We will examine a Trombe wall storage scenario in **§5.1.C**.

(b) *PCM shutters*: Shutters containing PCM are placed outside of window areas. During the day they are opened to the outside (so we can see through the windows). The exterior side is exposed to solar radiation and heat is absorbed melting the PCM. At night we close the shutters, slide the windows used during the day aside, and heat from the PCM radiates into the room.

(c) *PCM building blocks*: Building blocks or other building materials are impregnated with a PCM. These are then used in constructing a building, resulting in a structure with a large thermal inertia (an "adobe" effect) but without the large mass associated with adobe construction.

(d) *Active PCM storage*: Using off-peak electricity a PCM mass stored in a container of some kind is melted. The heat is then available when needed. This is simply a PCM version of the popular ceramic heating devices widely available in Europe.

(e) *Solar power backups*: Any power system dependent on solar energy requires a backup power or energy storage system during periods of eclipse. This is true for the envisaged solar-based power system in the space station (see **§5.3**) and for any terrestrial solar power system. During periods of solar energy availability, some energy would be used for melting the PCM; the stored latent heat is available during darkness periods.

(f) *Air conditioning backup*: Air conditioning in summer normally imposes a huge electricity load during certain times of day. The units used are suited to size of structure and numbers of people involved. A PCM-based approach is to place a PCM storage device "in the loop" of the air conditioning, with the air conditioning unit running 24 hours per day. The unit would be cooling and dehumidifying air during daylight hours, but when not needed for this purpose (at night, when the house is cool), it would freeze the PCM. The "cold" stored in the PCM can be used to supplement the air conditioning unit during the day. The air conditioning unit required would be smaller than the unit used without PCM, permitting us to take advantage of off-peak electricity and to reduce the load during peak times.

The examination of any of these options requires an iterative mixture of experiment and simulation. Experiment is vital, for the key problems facing us include containment of the PCM, choice of the PCM, and properties of the PCM such as supercooling and degrading of latent heat. Simulation can be used to help us in designing the system, estimating system performance and building our understanding of what are the key factors in this performance.

5.1.B A simple heat storage example

In this section we focus on some aspects of the use of a particular PCM, Calcium-Chloride Hexahydrate ($CaCl_2 \cdot 6H_2O$). In S.I. units, its thermophysical properties are well approximated by the values shown in **Table 5.1.1**. Calcium Chloride Hexahydrate is a "hydrated salt," whose phase change consists of breaking the bonds linking salt and water molecules with the latter then dissolving the salt and forming the liquid phase.

Normally Calcium Chloride Hexahydrate will not freeze from its liquid state when its temperature is lowered to 27 °C. Instead it may supercool to a temperature far below this value while it remains liquid, in which case its latent heat cannot be extracted and the material will not play its storage role (**§2.4**,

Table 5.1.1: Thermophysical Properties of Calcium Chloride Hexahydrate

$T_m = 27\ °C$	$L = 190.6\ kJ/kg$
$\rho_S = 1700\ kg/m^3$ (solid)	$\rho_L = 1520\ kg/m^3$ (liquid)
$c_L = 1.42\ kJ/kg\ K$	$c_S = 2.21\ kJ/kg\ K$
$k_L = 0.53 \times 10^{-3}\ kJ/m\ s\ K$	$k_S = 1.08 \times 10^{-3}\ kJ/m\ s\ K$
$\alpha_L = 2.46 \times 10^{-7}\ m^2/s$	$\alpha_S = 2.87 \times 10^{-7}\ m^2/s$

Figure 2.4.3). This problem may be overcome by placing some nucleation agent in contact with the PCM to induce freezing at the melt temperature; the nucleation agent may be a solid crystal of a material with a high melt temperature.

The freezing front of Calcium Chloride Hexahydrate is dendritic but normally of so small a width that we can regard it as a sharp front in melting and freezing.[1]

As noted above one of the key areas of potential LHTES use is domestic heating and cooling. Using solar energy, off-peak electricity or low-powered devices for cooling that would work all the time, we wish to supplement or replace the standard heating and cooling methods for the home. The comfortable temperature range for people, depending on the relative humidity, is about 60-80 °F (15-26 °C). The melt/freeze temperature of Calcium Chloride Hexahydrate is $T_m = 27$ °C $= 80.6$ °F, making it a good candidate material for heat storage, since in principle it should store or release heat at this temperature.

Using the simple analytical relations of CHAPTER **3** we can estimate the performance of our PCM under certain idealized latent heat storage scenarios. The freeze/melt time relation based on the quasistationary approximation for an imposed surface temperature is ((4d) of §**3.5**)

$$t_{melt} = \frac{l^2 \rho L}{2k\,\Delta T\,(1+\omega)}, \tag{1}$$

with l the "radius" of the body, ω a geometrical factor equal to 0 for a slab, 1 for a cylinder and 2 for a sphere (§**3.5.B**), and t_{melt} the time of complete phase change. For freezing ρ, k are solid phase values, while for melting they are liquid values; $\Delta T = T_L - T_m$ for melting and $T_m - T_S$ for freezing, arising from surface temperatures T_L, T_S, respectively.

As a PCM with solar input and heat release to a (cold) room, we might have T_L as high as 100 °C and T_S as low as 0 °C. Thus the extreme liquid and solid Stefan numbers are, respectively,

$$St_L = \frac{c_L\,\Delta T_L}{L} = .54 \qquad \text{and} \qquad St_S = \frac{c_S\,\Delta T_S}{L} = .31 \tag{2}$$

attesting to the relevance of the quasistationary approximation (§**3.1**) and of (1).

Consider now a PCM slab 2 $cm = 2\times10^{-2}\ m$ thick; let it be insulated at one end while at the other it is to have an imposed temperature (T_L for melting, T_S for freezing). If the PCM is initially solid at its melt temperature while the surface temperature is 100 °C then its melt time will be

$$t_{melt} = \frac{l^2 \rho_L L}{2k_L\,\Delta T_L} \approx 25 \text{ minutes.}$$

The PCM slab is now storing an amount of heat roughly equal to

---

1. We note however, that under certain conditions extremely long crystals may form, extending from the cooling surface into the body of liquid. In such a case one could treat the PCM mass as a perfect mixture of solid and liquid (mushy region) at temperature T_m and obtain in this way the case opposite to that of a sharp front.

$$\rho_L L l = 5794.24 \; kJ/m^2 .$$

The discharge time for a fully melted slab assuming an imposed surface temperature of $T_S = 10$ °C will be

$$t_{freeze} = \frac{l^2 \rho_S L}{2k_S \Delta T_S} \approx 59 \text{ minutes.}$$

This example contains some interesting points concerning LHTES. On the one hand the charge process is very rapid, a reflection of the high imposed charge temperature T_L. In many cases the charging mode cannot be easily controlled: the sun is available for a certain period of time, and for simple LHTES systems the charge temperature cannot be easily reduced. We could of course make our wall thicker. But what of the discharge process? To discharge the heat stored during the 25 minute period required an hour, using an extremely low T_S. If T_S were any higher the discharge time would grow even longer. A key role of mathematical modeling of LHTES is to determine reasonable tradeoffs between charge and discharge parameters so that we succeed in storing all available heat that we wish to store, and *totally* discharge the heat during a discharge process whose duration is in accordance with our needs. This may be possible or not, according to the properties of the PCM and our ability to move heat quickly in and out of the storage material.

5.1.C Trombe wall

To place our simple example in a more realistic setting let us expand on our earlier remarks about the Trombe wall (§**5.1.A**). We are all familiar with the "greenhouse" behavior of glazing, wherein solar radiation enters a car or a room and is "trapped" there, heating the car or room. Wherever a glazed structure such as a hothouse, car, or room is exposed to the direct rays of the sun (direct gain), this effect is felt. Consider a wall in a room just behind a large south-facing glass window with a small air space between (**Figure 5.1.1**). The solar flux passing through the window will heat the wall, but if the glass is reasonably resistant to heat flow (for instance, doubly glazed) the heat will be unable to leak back out of the window and the wall temperature on the window side may become high before the sun sets. At night time we can warm the room by, say, opening vents at the floor and ceiling and having the room air flow over the wall, either by forced or natural convection. Depending on its thickness and material some heat may also be exchanged between the room-side face of the Trombe wall and the room within, but this may be negligible and can be ignored in our discussion.

Generally Trombe walls are built of stone or similar material, and store heat as sensible heat. For several years latent heat Trombe walls have been of interest, since in principle the wall could be built with smaller mass and size using a PCM. Suppose that a Trombe wall could be constructed of Calcium Chloride Hexahydrate behind a perfectly insulated window. Using our simple melting time relations of CHAPTER **3**, we wish to to examine its possible performance subject to

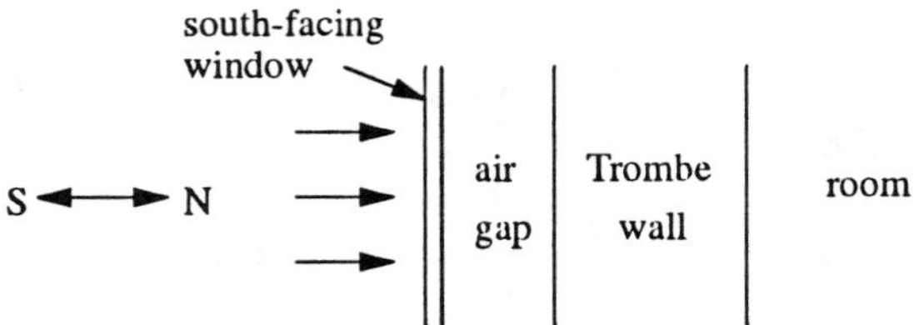

Figure 5.1.1. Cross section of a Trombe wall arrangement.

representative solar inputs. Let us examine this point.

On a particular sunny day fluxes on a vertical south facing wall in Alabama, measured on an hourly basis, had the values shown in **Table 5.1.2**. Their average value over the 11-hour period from 6 a.m. to 5 p.m. is $.33\ kJ/m^2\,s$. On the other hand, on cloudy days the heat flux may fall essentially to zero.

It is reasonable to assume that the temperature of the Trombe wall varies only as the depth into the wall; hence the heat transfer in the wall is one-dimensional and can be modeled via slab geometry. Let the x-axis be perpendicular to the face of the wall, with $x = 0$ the window-side face, and $x = l$ the room side face, for l the thickness of the wall. The heat stored during the daytime is assumed to be stored only as latent heat of melting of the material at its normal melt temperature T_m. Afterwards, it is to be released due only to cooling from the window-side face $x = 0$, arising from convective heat transfer to a room at an assumed temperature of 60 °F = 15.5 °C.

In mathematical terms, the input of heat to the wall as solar flux is modeled as

$$-k\,T_x = q_{in}\ ; \tag{3}$$

on the other hand, heat extraction by convection to a room at temperature T_S is modeled as

$$-k\,T_x = h\,[\,T_S - T(0,t)\,]\,, \tag{4}$$

where h is the heat transfer coefficient from the window-side of the Trombe wall to the flowing air stream.

Table 5.1.2: Hourly Fluxes

time	$q_{in}(kJ/m^2\,s)$	time	$q_{in}(kJ/m^2\,s)$
6 a.m.	0	Noon	.56
7 a.m.	.06	1 p.m.	.53
8 a.m.	.21	2 p.m.	.45
9 a.m.	.35	3 p.m.	.32
10 a.m.	.47	4 p.m.	.17
11 a.m.	.54	5 p.m.	0

We wish to use quasistationary melt-time relations for slabs subject to (3) and (4) to estimate the necessary thickness of the PCM wall and the manner and time frame for release of the stored energy to the room. Consider first the melting, or "charging" period, when impinging solar flux inputs heat to the wall as latent heat. From (28c) §**3.1.C**, the quasistationary approximation to the melt-time for depth l is

$$t_{melt} = \frac{\rho L}{q} l . \tag{5}$$

From **Table 5.1.2** we know that solar flux is at most available over an 11-hour time period and that its average value is $.33\ kJ/m^2\,s$ ($57\ BTU/ft^2\,hr$). If over an 11-hour period a constant flux $q_{in} = .33\ kJ/m^2\,s$ is input to a PCM slab of width l in its solid state at T_m, then we find

$$l = \frac{q_{in}}{\rho_L L} t_{melt} \approx .045\ m . \tag{6}$$

(aproximately 1.8 inches). This is an estimate of the melt-front depth obtained on a sunny day from incident beam radiation on the PCM wall.

There are two ways for discharging the heat from the Trombe wall. One is to open vents at the floor and ceiling of the wall to the room, to allow room air (at $60\,°F = 15.6\,°C$) to flow across the wall by natural convection. The second is to pump room air across the wall via forced convection. We will now find expressions for the heat transfer coefficient h in both cases.

Free- or buoyancy-driven convection of air is determined by the product of the *Prandtl* and *Grashof* numbers for air; here the Prandtl number **Pr** is the ratio of the kinematic viscosity ν of air to its thermal diffusivity,

$$\mathbf{Pr} = \nu/\alpha = c\mu/k , \tag{7}$$

where $\mu = \nu\rho$ is the viscosity (lb/ft hr), and k, c, ρ, α refer to air. For air over the temperature range under consideration we may take $\mathbf{Pr} = .7$. The Grashof number **Gr** is

$$\mathbf{Gr} = y_o^3\,\rho g \beta\,\Delta T/\mu^2 \tag{8}$$

and measures the relative effect of buoyant to viscous forces. Here y_o is the wall height (we will take $y_o = 8$ ft), ρ is the air density, g is the gravitational constant, β is the coefficient of volumetric expansion, $\Delta T = T_m - T_S$ is the wall-to-stream temperature drop, and again, μ is the air viscosity. Since **Gr** is dimensionless, we may substitute the parameter values of (8) in any system of units. In British units the values are: μ = viscosity $= .04\ lb/ft\ hr$, ρ = air density $= .08\ lb/ft^3$, β = volume expansion coefficient $= .002/°F$, c = air specific heat $= .24\ BTU/lb\,°F$, k = air conductivity $= .014\ BTU/ft\ hr\,°F$. The temperature drop of our system is $\Delta T_S = 80°F - 60°F = 20°F$, so the Grashof number is found to be $\mathbf{Gr} = 4.25 \times 10^{11}$, and the product $\mathbf{Gr}\,\mathbf{Pr} = 2.975 \times 10^{11}$. Then from [MCADAMS, p.173], we find

$$h = .19\,\Delta T_S^{1/3} \approx .5\ BTU/ft^2\,hr\,°F = .0029\ kJ/m^2\,s\,°C .$$

With this free-convection driven heat transfer, we ask how long will it take for the charged Trombe wall to completely discharge to the room. The quasistationary approximation to the *freezing* time for depth l is given by (see (35d) **§3.1.D**)

$$t_f = \frac{l^2 \rho_S L}{2k_S \, \Delta T_S} \left(1 + \frac{2k_S}{h\,l} \right). \tag{9}$$

Substitution of the thermophysical properties of the PCM and the heat transfer coefficient for natural convection yields $t_f = 134.6\ hr$. This is clearly an inordinately long time period since it spans more than 5 days for 1 day of storage !

For a sensible-heat Trombe wall the above result would not be particularly disturbing. If a few sunny days occurred in succession, then the sensible heat storage material would merely get hotter and hotter. However if this happens to a PCM wall then it may be incapable of discharging its latent heat, since to do so its temperature must fall below its melting point. Hence it will become merely a sensible heat wall with little thermal mass. Thus a PCM wall **must** discharge its latent heat in a reasonable time span. Since a free conduction discharge mode would require too long a time we must turn instead to a forced convection system.

Forced convection of air with velocity v across a vertical wall approximately obeys the relation

$$h = 1 + 0.2\,\mathrm{v}\,, \tag{10}$$

for h in $BTU/\,ft^2\,hr\,{}^\circ F$ and $\mathrm{v} < 16\ ft/s$. Solving (9) for h in terms of t_f yields

$$h = \frac{2k_S}{l} \left[\frac{2k_S t_f \, \Delta T_S}{l^2 \rho_S L} - 1 \right]^{-1}. \tag{11}$$

Positivity of h demands that $t_f > \dfrac{l^2 \rho_S L}{2k_S \, \Delta T_S} \approx 8$ hours. If we allow air speed below about 10 mph (15 ft/s) then the total discharge of the Trombe wall may be carried out in about 24-30 hours. Since sunny days may indeed be interspersed with days of little solar input, discharge periods of 1-2 days are reasonable (**Table 5.1.3**).

Clearly, $CaCl_2 \cdot 6H_2O$ has some defects as a PCM in this application because discharge in 24 hours from an 11-hour charging period is far from ideal. Its most serious defect is that its melting point T_m is too low. Thus a higher T_m would not alter the charging performance, but would possibly enhance the heat transfer without any need to have inordinately large air velocities or heat transfer coefficients.

Calculations such as the above may be used to provide order-of-magnitude predictions of system performance. Once this is done, candidate PCM's can be selected and precise, time-dependent process simulations can be carried out using the numerical methodology of CHAPTER **4**. This is the subject of the next section.

Table 5.1.3: Estimates of Discharge Times and Velocities

t_f (h)	h (BTU/ ft^2 h °F)	v (ft/ s)
24	4.0	10.1
26	3.5	8.7
28	3.2	7.5
30	2.9	6.5
32	2.7	5.7
34	2.5	5.0
36	2.3	4.4
38	2.1	3.9
40	2.0	3.4

PROBLEMS

PROBLEM 1. Suppose that it is possible to produce a cylindrical "Trombe wall" that will absorb the hourly fluxes as described in **Table 5.1.2**. The cylinder is to be 1.5 meters in height. Assume that it is to discharge radially to an exterior surface maintained at a temperature of 60 °F, with a heat transfer coefficient that is so high that it may be considered infinite. Estimate a suitable radius for the cylinder on the basis of criteria that you feel are reasonable.

PROBLEM 2. Find the radius of a sphere that can perform the task of the cylinder of PROBLEM 1.

PROBLEM 3. Discuss the possible deficiencies and drawbacks of the estimates that we have made in this section.

PROBLEM 4. All other properties being unchanged, determine an optimal melt temperature for a PCM that you would use to replace the Calcium Chloride Hexahydrate of this section.

PROBLEM 5. Derive relation (5) using the quasistationary approximation of **§3.1**.

PROBLEM 6. Derive relation (9) using the quasistationary approximation of § **3.1**.

PROBLEM 7. What is the effect of using (4a) of **§3.5** in place of (1) for our Trombe wall calculations? In particular, would one observe different qualitative behavior of the wall, or is the quasistationary approximation satisfactory?

5.2. NUMERICAL SIMULATION OF A LATENT HEAT TROMBE WALL

In §5.1 we examined the performance of a Trombe wall using a simple "back-of-the-envelope" analysis based on melting time relations of the quasistationary approximation. The accuracy of these relations is generally in doubt except in the simple "benchmark" cases from which they arise. Thus in the actual Trombe wall performance, inputs and demands on the wall are time-dependent and possibly strongly varying. We now turn to the analysis and computer simulation of the Trombe wall.

The role of a simulation code depends on the information that we seek. We have already seen that the discharge time for the PCM Calcium Chloride Hexahydrate is too long, a major difficulty, for we can then expect the wall to overheat and cease serving as a LHTES device. A simulation code helps us explore the performance of walls with other PCM's or even sensible storage materials. Similarly, in our back-of-the-envelope calculations in the previous section we took into account convective cooling from only the heated face. A simulation code makes it possible to examine alternative procedures such as pumping air over the front and rear faces instead of insulating the rear.

Beginning at 6 AM of the first day the Trombe wall is to exchange heat with a direct solar flux (via time-varying flux input) and a convective loss to the room possibly from front and back faces (**Figure 5.2.1**). On sunny days the solar flux input to the wall as direct gain q_{sun} is given by **Table 5.1.2** (**§5.1**). This starts at 6 AM and continues until 5 PM. When the sun sets, heat is given up to the adjacent room, whose temperature is $T_{ambient}$. Note that $T_{ambient}$ may be input as a function of time t. The film coefficients from wall to air are denoted by h_a for the face on the side of the window and h_b on the room side. On cloudy days, the heat flux is a (small) constant q_{cloud}. As time proceeds we can expect to observe multiple phase change fronts that cannot be predicted a priori. The most natural simulation approach is based on the use of enthalpy to characterize the state of the material. For simplicity we will assume that the densities of solid and liquid PCM phases are the same constant ρ. Denoting the Trombe wall thickness by l, our process is modeled as follows:

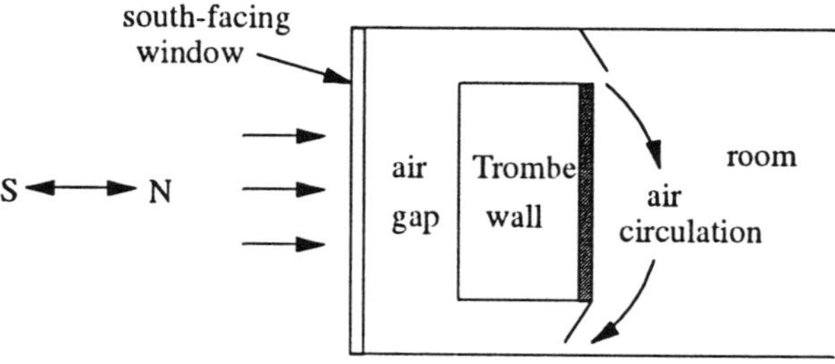

Figure 5.2.1. Trombe wall setup.

$$\rho e_t + q_x = 0\,, \qquad 0 < x < l\,, \qquad t > 0\,. \tag{1}$$

$$q_{in}(0,t) = h_a\,[\,T(0,t) - T_{ambient}\,] + q_{sun}\,, \tag{2}$$

where h_a is a time dependent heat transfer coefficient which, when set equal to zero, corresponds to not opening a vent (for natural convection) or not turning on an air pump (for forced convection). On the rear (room) side we have

$$q_{in}(l,t) = h_b\,[\,T_{ambient} - T(l,t)\,]\,, \tag{3}$$

where h_b is a time dependent heat transfer coefficient as above, and q_{sun} is the solar flux directly impinging on the Trombe wall, given by **Table 5.1.2** for a sunny day, or a small constant for a cloudy day. In both of the above $T_{ambient}$ is the ambient, or in our case, the room temperature. The relation between enthalpy and temperature is, as earlier,

$$T = \begin{cases} T_m + e\,/\,c_S, & e \le 0 \\ T_m, & 0 < e < L \\ T_m + (e - L)\,/\,c_L, & e \ge L\,. \end{cases} \tag{4}$$

The enthalpy scheme developed in **§4.3** is applied to this process in an obvious manner. Only the boundary condition at $x = l$ needs to be changed to a convective one. Because we do not expect extreme temperature gradients or rapid changes with time, uniform spatial and temporal grids Δx, Δt may be chosen. While we are free to use either an explicit or an implicit scheme, it is preferable to initially prepare an explicit code. It is much easier to program and debug, so it is more reliable. Moreover, relatively course grids will be sufficient for most cases (e.g. 8 or 16 nodal intervals), so even very long "runs" are easily done on a personal computer with little cost.

With the code written, one must be concerned with questions of validation and input/output. We test whether convergence is taking place as the mesh size tends to zero. It is good to attempt to determine the rate of convergence with decreasing mesh size, using the results of runs for successive halvings of the spatial mesh size. A second necessary step in validation is observing correct qualitative behavior of the model, especially global energy balance. Similarly, one would not want to see, e.g., violations of the maximum principle.

The program input consists of material properties, wall thickness, flux values, film coefficients, and data determining mesh sizes Δt, Δx and output times. The code operates by inputing the number M of slab subdivisions from which Δx is found as $\Delta x = l\,/\,M$. The time step Δt must be restricted to satisfy the CFL condition (39) of **§4.1**,

$$\Delta t < \frac{1}{2}\,\frac{\Delta x^2}{\alpha_{max}}\,. \tag{5}$$

Note that we use the implicit update for control volumes adjacent to wall surfaces, as described at the end of **§4.1.E**); otherwise the 1/2 of (5) should be changed to 1/3, see (42) **§4.1.E**.

The number of spatial subdivisions depends on the conditions of the problem and the material used. For solar fluxes, moderate Biot numbers $h\,l/k$ (roughly speaking below 10), and transfer-fluid temperatures not rapidly changing in time (in our case, the constant room temperature), a few uniformly spaced nodes yield sufficient accuracy to address basic questions such as those discussed below. Thus for each material a preliminary run with, say, 8 subdivisions is sufficient to provide rough energy distribution and surface temperature estimates to test the code, and then 16 or 32 nodes may be used for the studies. We note that the choice of the number of subdivisions is at times a point of controversy. A number of building simulation software packages exist with latent heat storage potential, in which very few nodal points are used and yet accuracy of the code predictions is asserted. This is a potentially dangerous step to take, because in reality, at any time when sharp gradients or rapid changes occur we do not know a priori that a small number of mesh points will suffice to resolve them. The only way to know is to execute runs with finer meshes and see if the results are close!

The output depends on the requirements of the application at hand. Since the process is one of energy storage, we need to obtain the total energy stored at a given time, and how much of it is sensible and latent heat. Similarly, we wish to know the temperature distribution as it varies in time. Since the phase change process is the important element in PCM storage, we output the state (solid, liquid, or mushy) of each nodal interval at various times. Clearly, other data such as flux values at the surface during discharge may be of interest.

Thermal storage considerations ordinarily lead to the following questions:

(i) Is there an advantage to using a phase change material as against sensible heat storage?

(ii) How long does the storage medium remain charged with available energy?

(iii) Is the phase change capability of the PCM well used?

(iv) Does the system "overheat"? That is, does the temperature grow to unreasonably high values during the process?

Let us use our simulation code to address some of these questions. Consider the case where the sunny-day heat input is given as in **Table 5.1.2, §5.1**, the cloudy-day value being $q_{cloudy} = .04\ kJ/m^2\,s$. We assume the room temperature to be $15\,°C$. Recall that for $CaCl_2 \cdot 6H_2O$ the rough estimate of the extent of melting, assuming the material initially solid at its melt temperature, is $.045\ m$. Hence, in the computer runs for the sunny day - cloudy day scenario, we chose $l = .05\ m$.

Regarding question (i), we chose two materials to which the code will be applied. These are $CaCl_2 \cdot 6H_2O$, whose properties are found in **Table 5.1.1, §5.1**, and concrete, whose properties are:

$$\rho = 2400\ kg/m^3, \quad c = .837\ kJ/kg\,°C, \quad k = 1.73\times10^{-3}\ kJ/m\,s\,°C$$

Question: *How does* $CaCl_2 \cdot 6H_2O$ *compare to concrete ?*

To explore possible differences we make a computer run with the wall insulated at $x = l$ and having the solar flux of **Table 5.1.2** from 6 AM to 5 PM, followed by convective cooling to the ambient air of the room. The film coefficient is chosen as $h_a = .0029$ in S.I. units which, as we have seen, corresponds to natural convection and is too low to discharge the PCM. However, as we see in **Figure 5.2.2**, this is sufficient to totally discharge a concrete wall of the same width as that of the PCM. Recall that the scenario being followed is that of a sunny day followed by a cloudy day. Why does the concrete lose its heat so quickly? Because its surface temperature reaches extremely high values and very much higher than that of the PCM. This is most striking in **Figure 5.2.3** in which we see the face temperature on the room-side of the wall as a function of time. The concrete face reaches values in excess of $150\,°C$ with a rapid swing occurring in a few hours. By comparison, the PCM barely edges above $50\,°C$ before dropping back to an unwavering surface temperature close to T_m for the remainder of the simulation.

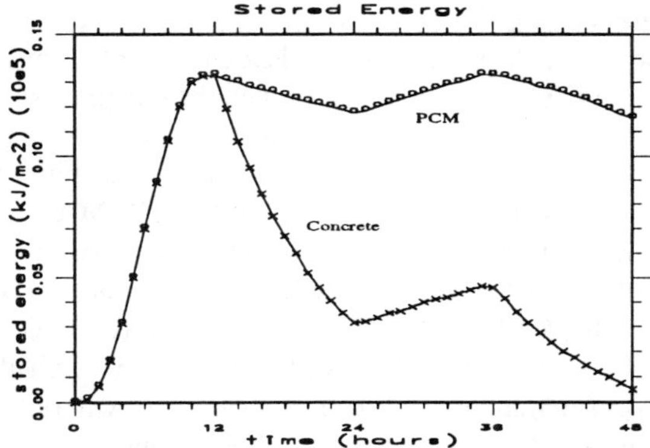

Figure 5.2.2. Stored energy in the PCM and a concrete wall.

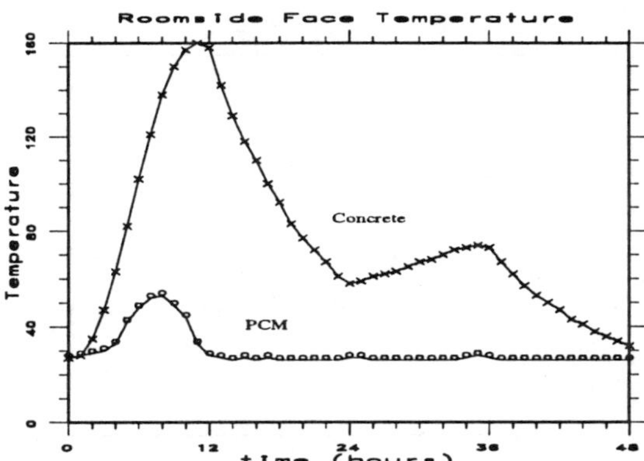

Figure 5.2.3. Face temperature of room side for Trombe wall and for concrete wall.

As we saw in §5.1, the choice of $CaCl_2 \cdot 6H_2O$ as PCM presents difficulties in discharging. Our rough estimates indicated that while charging of the Trombe wall could take place within the time of solar availability, the discharge process was far too slow if we relied on natural convection as the means for discharge. In an attempt to discharge using air with reasonable speeds, we have included in the code the possibility for either free or forced convection over both faces. We consider the case of using convection over both faces with a film coefficient of .003 in S.I. units. The discharge is to take place between the hours of 2 PM and 8 AM. Then as can be see in **Figure 5.2.4**, the system energy returns essentially to its original state after 48 hours. In this case the PCM will not overheat. Moreover, this indicates a strong degree of *controllability* through the development of an appropriate strategy for discharge.

A useful source of information is always the phase picture of the material as a function of time. This is seen for the case of convection at both walls in **Figure 5.2.5**. Note the presence of a number of mushy nodes and several regions of various phases at most times. In general, one wishes to maintain mushy nodes, i.e. phase change fronts, in the wall at all times for then the temperature throughout the wall will generally be close to the melt temperature.

Another approach to improving performance of LHTES units is through the choice of the PCM. For example, using an alternative PCM with a higher melt temperature would tend to lower the degree of overheating of the PCM, (but not nearly as well as through controlling the heat loss at the surfaces). Heat removal is proportional to the film coefficient and to the temperature gradient available to drive it. Since the film coefficient for forced convection is limited by fan sizes, comfort, and other considerations, the flow from wall to air will only be effectively large when the temperature of the wall face is high enough. We conclude that a PCM with a higher melt temperature will be a more effective agent for moving heat to the air, than one with a lower T_m.

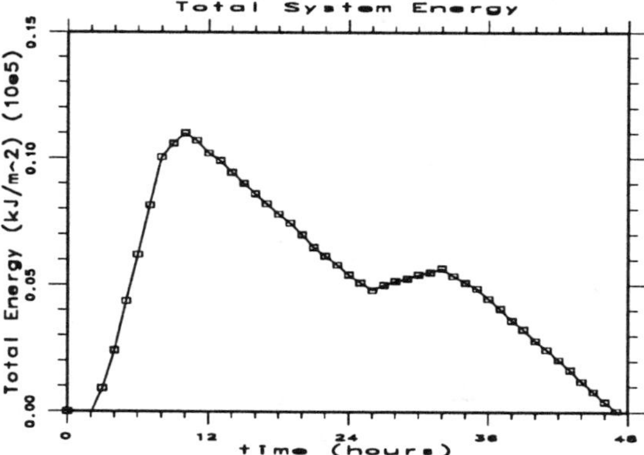

Figure 5.2.4. Total energy in the PCM wall under convective loss from both faces.

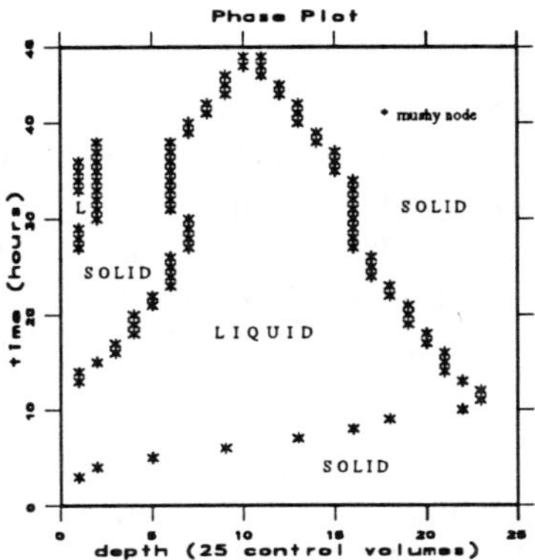

Figure 5.2.5. The phase of each control volume at one-hour intervals.

It is interesting to compare the PCM simulation results with the the simple melt-time estimates of the last section. Thus, for example, relation (6) of **§5.1.B** can be thought of as being of the form

$$X(t) = \frac{\bar{q}_0(t)\, t}{\rho_L L} \tag{6}$$

with $\bar{q}_0(t)$ the average flux from the onset of melting to time t, and $X(t)$ the location of the front at time t. This does not take into account that at the beginning of the simulation some heat loss to the (colder) room takes place. For $CaCl_2 \cdot 6H_2O$, **Table 5.2.1** compares the midpoint values of the mushy nodes with the estimates given by (6) for 16 nodal intervals, during the initial charging period. The relative error of up to about 20% is sufficiently small for many applications.

Table 5.2.1. Comparison of Computed and Predicted (by (6)) Front Locations

t (*hr*)	X (mushy node)	X (from(6))
8	--	7.08×10^{-4}
9	0	3.31×10^{-3}
10	3.125×10^{-3}	7.08×10^{-3}
11	9.375×10^{-3}	1.27×10^{-2}
12	1.5625×10^{-2}	1.89×10^{-2}
13	2.1875×10^{-2}	2.55×10^{-2}
14	2.8125×10^{-2}	3.22×10^{-2}
15	3.125×10^{-2}	3.68×10^{-2}
16	3.4375×10^{-2}	4.14×10^{-2}
17	3.75×10^{-2}	4.37×10^{-2}

PROBLEMS

PROBLEM 1. Prepare a simulation code for the Trombe wall, using the PCM and wall characteristics of this section. Examine the stability condition of your code by choosing time increments Δt that are both less than and more than the limitations imposed by your stability condition. Examine the "convergence" of the code by running it for various mesh sizes. In particular, use "extrapolation to zero" based on successive halvings of mesh size, to improve your results.

PROBLEM 2. A Trombe wall is to be constructed using two phase change materials, *A* and *B*. Both materials have the conductivity, specific heat, density and latent heat of Calcium Chloride Hexahydrate, only differing in melt temperature. Prepare an enthalpy based algorithm for simulating the thermal performance of this wall, subject to the input flux of **Table 5.1.2 (§5.1).** Use the quasistationary approximation to "size" the wall thickness, and suggest melt temperatures and layer thicknesses that will optimize thermal behavior.

PROBLEM 3. Describe the modification needed to apply an enthalpy based approach to simulating the system of PROBLEM 2 if we wish to take into account the contact resistance due to imperfect thermal contact between the layers.

5.3. THE DEVELOPMENT OF A SIMULATION CODE FOR A LHTES SYSTEM IN A SPACE STATION

5.3.A Introduction

Within this decade a manned space station is to be orbited around the earth. To meet its power requirements alternative technologies, including solar energy for direct heat-to-power conversion are being considered. As on earth, however, the sun's energy will be available to the station only for certain periods of time, between which it will be in darkness; during these periods of darkness energy will be required that will have to be stored during the sunlit periods. LHTES represents a means to this end.

Thermal energy storage in space gives rise to unique questions and constraints not normally present in earth-based systems. These include the effect of weightlessness, creating the possibility that the PCM might "float" away from its container walls insulating the PCM from external heat input and extraction. Similarly, effects that are ignorable on earth, such as convection arising from surface tension, become dominant in space. Constraints are well represented by the necessity that the storage system, and hence, the space station itself, not overheat. Unlike earth, space does not offer readily available heat sinks for the disposal of excess heat. Hence the system must be "fine tuned" to maintain a heat balance in time.

The design of a storage system for an engine in the space station is comprised of three steps: *analytical modeling*, in order to see the roles played by the various system parameters and to evaluate feasible configurations for the system; *numerical simulation*, in order to obtain information about how the system may perform under various conditions and parameter values; and *experimentation*, in order to make sure that something important has not been left out or that something unimportant has not been included to an exaggerated extent. In this section we examine the the first two parts of the design process, in order to determine possibly feasible sizes for a LHTES system with desired performance characteristics and to predict its performance in time. The system that we look at is idealized, with many simplications made, a significant one being that of not incorporating a "void" or gaseous bubble formed in the PCM as a result of the PCM density change accompanying melting and freezing.

Because of the specialized nature of the material in this section, we list the nomenclature to be used in **Table 5.3.1**.

5.3.B The system of interest

We begin by describing the system of interest: its design, energy demands, the latent heat storage unit, phase change materials (PCM's) and PCM module.

As the space station orbits around the earth, solar heat is collected by a large solar collector. This heat is to be used to power an engine. During periods of sunlight, solar energy is to be received and stored to be able to meet engine needs during periods of darkness. The power system may be regarded as an idealized Brayton cycle engine [WICHNER et al] consisting of four parts (**Figure 5.3.1**): solar collector, engine, radiator-compressor and thermal storage device. The heat exchange fluid is a gas. During periods of sunlight, assumed to be of duration t_c, the gas is heated to a temperature T_{sun} and driven through the storage device,

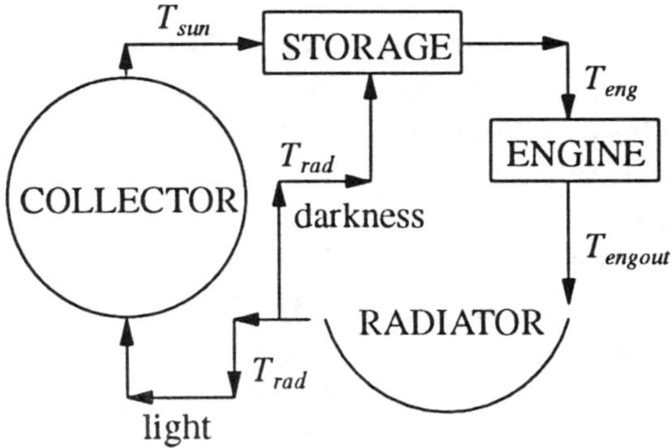

Figure 5.3.1. Power system configuration

Table 5.3.1. Nomenclature

c_g	specific heat of heat exchange fluid	BTU/lb °F
c_L	PCM specific heat in liquid phase	BTU/lb °F
c_S	PCM specific heat in solid phase	BTU/lb °F
h	heat transfer coefficient	BTU/ft^2 hr °F
k_g	conductivity of heat exchange fluid	BTU/ft hr °F
k_L	liquid PCM conductivity	BTU/ft hr °F
k_S	solid PCM conductivity	BTU/ft hr °F
L	PCM latent heat	BTU/lb
N	number of PCM jackets	
P_0	engine power output requirement	MW
r	radial variable	ft
t	time	hr
t_c	charging period (sunlight)	hr
t_d	discharge period (darkness)	hr
t_c	charging period (sunlight)	hr
T	temperature	°F
T_{eng}	minimum engine input temperature	°F
T_{rad}	radiator outlet temperature	°F
T_{sun}	solar collector outlet temperature	°F
v_0	heat exchange fluid velocity	ft/hr
α_g	heat exchange fluid diffusivity	ft^2/hr
α_L	liquid PCM diffusivity	ft^2/hr
α_S	solid PCM diffusivity	ft^2/hr
η_0	engine efficiency	
ρ_g	heat exchange fluid density	lb/ft^3
ρ_p	PCM density	lb/ft^3

reaching the engine at a temperature above its operating temperature T_{eng}. Leaving the engine at a temperature T_{engout} it is further cooled to T_{rad} by the radiator. Clearly,

$$T_{\text{rad}} < T_{\text{engout}} < T_{\text{eng}} < T_{\text{sun}} .$$

During periods of darkness, the gas emerging from the radiator at temperature T_{rad} is pumped directly through the thermal storage device that now heats the gas to a temperature above T_{eng}. The darkness period is of duration t_d, during which the PCM is discharging.

The engine is required to produce P_0 (MW) of power; at an efficiency of η_0 (%) this requires a power input of $3.413 \times 10^6 \times P_0 / \eta_0$ (BTU/hr). Hence, the engine requires a gas volumetric flow rate $\dot{V}_g$ equal to

$$\dot{V}_g = \frac{3.413 \times 10^6 \times P_0}{\eta_0 \, \rho_g \, c_g \, (T_{\text{eng}} - T_{\text{engout}})} \quad (\text{ft}^3/\text{hr}) , \tag{1}$$

For the gas flow rate $\dot{V}_g$, the thermal storage unit must succeed in heating the gas from T_{rad} to T_{eng}. How can such a storage unit be designed? First, for a darkness period of length t_d, the storage unit must provide to the gas energy at least equal to

$$S_0 = c_g \, \rho_g \, \dot{V}_g \, t_d \, (T_{\text{eng}} - T_{\text{rad}}) \qquad (\text{BTU}) \tag{2a}$$

or

$$S_0 = \frac{t_d \, (T_{\text{eng}} - T_{\text{rad}}) \times 3.413 \times 10^6 \times P_0}{\eta_0 \, (T_{\text{eng}} - T_{\text{engout}})} \qquad (\text{BTU}). \tag{2b}$$

Thermal energy storage can be accomplished via sensible or latent heat approaches. The latter may reduce the required volume and temperature swings. Thus, let V_S, V_L and ΔT_S, ΔT_L be the required volumes and temperature swings for sensible and latent heat storage, resulting in the storage of the same amount of energy. Assume that the densities ρ and specific heats c are the same for both cases. Then, equating the amounts produced via the sensible and latent heat approaches, we find

$$\rho \cdot c \, V_S \, \Delta T_S \;=\; \rho \, (L \, V_L + \, c \, V_L \, \Delta T_L) \, .$$

If $V_S = V_L$, then $\Delta T_L = \Delta T_S - (L/c)$, whence the temperature swing is possibly substantially reduced. For $\Delta T_S = \Delta T_L$, $V_L = \dfrac{St}{1 + St}$, where $St := c \, \Delta T / L$ is the Stefan number.

Some possible latent heat thermal storage geometries are shown schematically in **Figure 5.3.2**. In **Figure 5.3.2(a)**, the PCM is placed in a cylindrical jacket fitted to a pipe, permitting transfer fluid to be pumped across or along the outer cylinder boundary, and/or through the interior pipe. In operation we can, of course, alter flow directions of the exchange fluid as well. The PCM storage module may consist of arrays of such PCM jackets. The system of **Figure 5.3.2(b)** consists of PCM "plates" across which the transfer fluid is to be pumped. Other configurations include PCM-filled cylinders, possibly finned to enhance the heat exchange, and pelletized PCM, with the pellets placed in a container through which the exchange fluid is pumped. Of course, a myriad of other possibilities exist, including, e.g., tapering of the jacket of **Figure 5.3.2(a)**. Similarly, modules using different PCM's can be linked together, if a possible advantage is seen.

The PCM storage system considered here is that of the PCM jacket of **Figure 5.3.2(a)**. To enhance heat transfer we include the possibility of many jackets, all having length l_0, inner radius R_{in} and outer radius R_{out}. The number of jackets is denoted by N. We are interested in addressing the questions of how to size the module, what configurations are "feasible" in the sense of providing gas at temperature T_{eng} to the engine during periods of darkness, and how to find a feasible system of least total mass. We assume that the gas flow is always in one direction, rather than being reversed when switching between dark and sunlit periods.

Regarding candidate materials, PCM's with high melting temperatures (over 1000 °F) provide distinct advantages in efficiency for the space station power system. For a variety of reasons, the material favored for use as PCM is Lithium

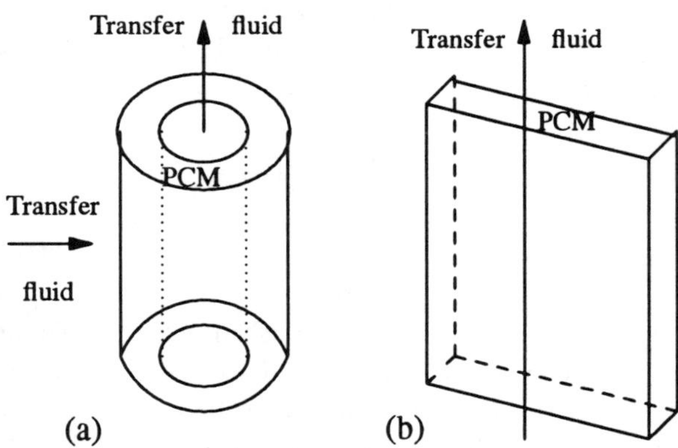

Figure 5.3.2. PCM unit configurations

Table 5.3.2. Properties of LiF and HeXe

		LiF	HeXe
density	lb/ft^3	$\rho_S = \rho_L = 112$	$\rho_g = 0.117$
specific heat	BTU/lb °F	$c_S = 0.59,\ c_L = 0.39$	$c_g = 1.25$
conductivity	BTU/ft hr °F	$k_S = 2.89,\ k_L = 0.98$	$k_g = 0.12$
melt temperature	°F	$T_m = 1560$	
latent heat	BTU/lb	$L = 450$	

Fluoride. For the illustrative purpose of this section, we will base our work on this PCM. The heat transfer fluid will be assumed to consist of a gaseous mixture of Xenon and Helium. For the sake of simplicity, we will assume the thermophysical parameters to be *constants*, listed in **Table 5.3.2.** We note that the assumption that $\rho_S = \rho_L$ for the PCM is a substantial oversimplification. Let us briefly discuss the effects of density change in this setting. Density change in a PCM may be manifested in natural convection in the melt and bulk movement of the solid. We may expect that for thin PCM jackets in low- to zero- gravity environments, natural convection will play at most a minor role. However, Lithium Fluoride is known to undergo a large density increase (about 25%) under freezing, which may lead to several possible outcomes. If the cohesive forces within the material are weaker than the wetting forces binding the material to the container walls, then internal voids (bubbles) will form; these in turn will reduce the effective solid conductivity, thus bringing closer the diffusivities of solid and liquid. On the other hand, if the internal cohesive forces are larger than the surface wetting forces, then voids will form at the surface, reducing the heat transfer to the PCM. Indeed, in a gravity-free environment, this may result in almost a complete separation of the material from the container walls, for all purposes insulating it from the transfer fluid. The effects of such voids were already seen in **§2.4**. We will not take voids into account in what follows, although work in this area is ongoing and difficult [WICHNER et al], [DRAKE, 1991].

We note also that temperature dependence of conductivity and specific heat may result in changes in the qualitative behavior of the temperature field [WICHNER et al]. While the temperature of the gas at the outlet of the solar collector (T_{sun}) may well be high, all materials have temperatures T_{max} beyond which they should not be exposed for fear of explosion, fire, corrosion, etc. For LiF, this temperature is T_{max} = 1700 °F, at which chemical reactions with potential container materials may occur.

As noted earlier, the thermal storage module must be sufficiently charged during the sunlit period t_c, to heat the gas during the darkness period t_d from the inlet temperature T_{rad} to above the temperature T_{eng} required by the engine. The configuration consists of N identical jacket assemblies (**Figure 5.3.2(a)**), each of inner and outer radii R_{in}, R_{out}, respectively. Flow velocities during the charge and discharge periods will be denoted by v_c and v_d respectively. We will not demand that they be the same. Recall that engine performance during discharge requires a minimum volumetric flow rate of $\dot{V}_g$ given in (1). Thus, for a storage unit of N jackets and a discharge gas velocity of v_d, we have the operational constraint

$$N \pi R_{in}^2 \text{v}_d \geq \dot{V}_g . \qquad (3)$$

The two overriding demands of the system are, of course, that the gas outlet temperature from the storage module not be below T_{eng}, and that at no place in the PCM does the temperature ever exceed T_{max}, which for LiF is 1700 °F.

For illustrative purposes, we assume the following system temperatures (in degrees Fahrenheit):

$$T_{eng} = 1250, \quad T_{engout} = 890, \quad T_{sun} = 1880, \quad T_{rad} = 530, \quad T_{max} = 1700, \qquad (4)$$

and assume charge and discharge periods of length t_c = 1 hr, and t_d = 38 min = 0.63 hr, respectively. Assume that the engine is required to produce 1 megawatt of power, which, estimating an overall efficiency of 15%, requires a 6.67 megawatt input. Using the values for the specific heat and density of the HeXe gas mixture, we compute from (1) a minimum throughput rate of gas into the engine equal to 430,000 cubic feet per hour.

For gasses such as HeXe at such flow rates, the flow can be assumed to be turbulent with a heat transfer coefficient [McADAMS],

$$h = 0.023 \, (2 R_{in})^{-0.2} \, k_g \, \text{v}^{0.8} , \qquad (5)$$

where v is the gas velocity. We assume that the jacket container wall is thin and has no effect on the heat transfer process. It is interesting to note that for the system temperatures given above, (2b) implies a minimum storage capacity of $S_0 = 29 \times 10^6$ BTU.

We wish to simulate the heat transfer and phase change process in a PCM jacket and hence the temperature history $T_{out}(t)$ of the gas leaving the storage module, for given input gas temperature history $T_{in}(t)$. The case of immediate interest is when $T_{in}(t)$ alternates between the values T_{sun} and T_{rad} for time periods t_c, and t_d, respectively. We will assume *axial symmetry* and ignore density changes of the PCM.

PHYSICAL PROBLEM For given T_{sun}, T_{rad}, PCM properties, and t_c, t_d, we wish to determine the length l_0 and radii R_{in}, R_{out} of the jacket, and the charge and discharge velocities v_c, v_d, so that the system has *minimal* total mass:

$$M_{\text{total}} = \pi (R_{\text{out}}^2 - R_{\text{in}}^2)\, l_0\, \rho\, N\,, \tag{6}$$

and so that the following constraints are satisfied:

(3), $T_{\text{out}}(t) \geq T_{\text{eng}}$, and PCM temperature always below T_{max}.

We now derive both simple analytical relations and numerical methods for modeling the time dependent behavior of the storage system.

5.3.C Rough sizing of the storage system via the quasistationary approximation

A first step in modeling phase change processes is to apply the quasistationary approximation to our system. For a narrow range of melting and solidification processes we can easily obtain simple analytical approximations to the key variables of the process. Let us consider one such case.

SIMPLIFIED PROBLEM A hollow PCM cylinder of inner radius R_0 is initially solid at its melt temperature T_m. At the surface $r = R_0$, heat is input via convective heat transfer with a heat transfer coefficient h from a transfer fluid at ambient temperature T_{in}. Thus a phase change front $R(t)$ separating liquid PCM $(r < R(t))$ from solid $(r > R(t))$ appears. In the liquid region, a radially symmetric, time dependent temperature function $T(r,t)$ satisfies the heat equation

$$T_t = \frac{\alpha}{r}(r\,T_r)_r\,, \qquad R_0 < r < R(t)\,, \quad t > 0\,, \tag{7}$$

while in the solid $T \equiv T_m$. Moreover, the interface conditions at the front $r = R(t)$ and the boundary condition at the inner surface are

$$T(R(t),t) = T_m\,, \tag{8a}$$

$$\rho\, L\, R'(t) = -k\, T_r(R(t),t)\,, \tag{8b}$$

$$k\, T_r(R_0,t) = h\,[\,T(R_0,t) - T_{\text{in}}\,]\,. \tag{9}$$

We now apply the quasistationary approach (§ **3.1**) to find easily computable, analytical approximations to $T(r,t)$ and $R(t)$. We replace the heat equation (7) by the steady state equation

$$(r\,T_r)_r = 0\,, \tag{7a}$$

while retaining the remaining conditions (8),(9). From (7a) and (8a) we represent $T(r,t)$ in the form

$$T(r,t) = T_m + a(t)\ln\frac{r}{R(t)}\,. \tag{10}$$

From (9) we have

$$a(t) = -\frac{h\,(T_{\text{in}} - T_m)}{k\,/\,R_0 + h\ln(R(t)\,/\,R_0)}\,. \tag{11}$$

Hence (8b) implies $k\,h\,(T_{\text{in}} - T_m)\,/\,\rho\,L \;=\; R\,R'\,[\,k\,/\,R_0 + h\ln(R\,/\,R_0)\,]$, and integration yields (PROBLEM 1)

$$\frac{k\,h}{\rho\,L}\,(T_{\text{in}} - T_m)\,t \;=\; \left\{\frac{k}{R_0} - \frac{h}{2}\right\}\frac{R^2 - R_0^2}{2} + \frac{h}{2}\,R^2\ln(\frac{R}{R_0}). \tag{12}$$

Let us consider the melting (charging) process with $T_{\text{in}} = T_{\text{sun}},\ t = t_c,\ h = h_c,\ k = k_L$; then the radius of the melt front $R_c = R(t_c)$ at $t = t_c$ obeys the relation

$$h_c\left\{\frac{k_L}{\rho\,L}\,(T_{\text{sun}} - T_m)\,t_c + \frac{R_c^2 - R_0^2}{4} - \frac{R_c^2}{2}\ln\left(\frac{R_c}{R_0}\right)\right\} \;=\; \frac{k_L}{2\,R_0}\,(R_c^2 - R_0^2)\,. \tag{13a}$$

Similarly, if the freezing process with $T_{\text{in}} = T_{\text{rad}},\ t = t_d,\ h = h_d,\ k = k_S,\ R = R_d$ is considered, then

$$h_d\left\{\frac{k_S}{\rho\,L}\,(T_m - T_{\text{rad}})\,t_d + \frac{R_d^2 - R_0^2}{4} - \frac{R_d^2}{2}\ln\left(\frac{R_d}{R_0}\right)\right\} \;=\; \frac{k_S}{2\,R_0}\,(R_d^2 - R_0^2)\,. \tag{13b}$$

Note that both (13a) and (13b) are of the form

$$h\,A \;=\; B\,, \tag{14}$$

with

$$A \;=\; \frac{k\,\Delta T}{\rho\,L}\,t + \frac{R^2 - R_0^2}{4} - \frac{R^2}{2}\ln(\frac{R}{R_0}\,, \qquad B \;=\; \frac{k}{R_0}\,\frac{R^2 - R_0^2}{2}\,. \tag{15}$$

Thus, replacing h in terms of the flow velocity v via (5), we find from (14) the relation

$$\text{v} \;=\; \left\{\frac{(2\,R_0)^{0.2}\;B}{0.023\;k_g\;A}\right\}^{1.25}\,. \tag{16}$$

To have a charge-discharge cycle that will neither overheat nor undercool, it is necessary that $R_c = R_d$. Using (16), this yields a constraining relation between the charge and discharge gas velocities v_c and v_d (PROBLEM 2). For given v_d the number of jackets N needed to satisfy the engine power requirement is found from (3).

The surface temperature approximation is given by

$$T(R_0, t) \;=\; T_m + a(t)\,\ln\,(R_0\,/R(t))\,,$$

with $a(t)$ as in (11) above. The minimum and maximum surface temperatures $T_{\text{min}}^{\text{surf}}$ and $T_{\text{max}}^{\text{surf}}$ are thus found by setting $R(t) = R_d$ and $R(t) = R_c$. The outlet gas temperature may be approximated by (PROBLEM 3)

$$T_g(l_0) \;=\; T_{\text{in}}\,e^{-\gamma\,l_0} \;+\; \bar{T}\,(1 - e^{-\gamma\,l_0})\,, \qquad \gamma = \frac{2\,h}{\text{v}\,R_0\,\rho_g\,c_g}\,, \tag{17}$$

where $\bar{T}$ is the "average" or representative surface temperature. The demand that during discharge $T_g(l_0) \;\geq\; T_{\text{eng}}$ becomes

$$l_0 \;\geq\; \frac{1}{\gamma}\,\ln\left[\frac{\bar{T} - T_{\text{rad}}}{\bar{T} - T_{\text{eng}}}\right]. \tag{18}$$

A pessimistic estimate for l_0 is thus found by setting $\bar{T} = T_{\min}^{\text{surf}}$.

As noted earlier, our system must obey a number of constraints. Let us recall and add to them:

(i) maximum PCM temperature $< T_{\max}$

(ii) minimum gas temperature $> T_{\text{eng}}$

(iii) $\dot{V}_g \leq N \mathrm{v}_d \pi R_{\text{in}}^2$

In addition, there are several reasonable conditions that should be met. These include, for example,

(iv) $0.05 \leq R_{\text{in}} \leq 0.3$ ft (vi) $\mathrm{v}_c, \mathrm{v}_d < 125,000$ft/hr

(v) $h_c, h_d < 25$BTU/ft^{2}hr°F (vii) $N < 250$.

To examine feasible configurations, a simple program can be prepared that performs the following steps:

Step 1. Given a pair $R_0 = R_{\text{in}}$, as in (iv), and $R_{\text{out}} = R_c = R_d$, find v_c, v_d from (16),(15).

Step 2. Find h_c, h_d from (13) and test via (v), (vi).

Step 3. Find l_0 from (18).

Step 4. Find N from (iii) and test via (vii).

The results of such calculations for the LiF - HeXe system with the data in (4) are shown in **Table 5.3.3**, indicating orders of magnitude of feasible configurations.

5.3.D Numerical simulation

The most obvious approach to numerical simulation of the heat storage system performance can be based on the enthaply method of CHAPTER **4**, applied to mesh volumes in cylindrical (r, z) geometry (we assume axial symmetry for simplicity). The only new feature for our system is that the heat exchange gas flowing down the tube has to be taken into account. We assume the gas temperature to be uniform in any cross-section of the pipe, so that $T_g = T_g(z, t)$.

We assume that all faces and sides of the PCM jacket are insulated, except the inner cylindrical boundary at $r = R_{\text{in}}$ where heat is exchanged via forced

Table 5.3.3. Feasible Configurations from the Quasistationary Approximation

R_{in}	R_{out}	v_c	v_d	l_0	Mass	N
0.252	0.302	27147	8653	49.3	121127	250
0.260	0.312	29100	9095	52.2	121773	223
0.270	0.324	31725	9663	56.0	123154	195
0.280	0.336	34583	10247	59.9	124388	171
0.290	0.348	37703	10847	64.1	126033	151

convection with the heat transfer coefficient given by (5) with $\mathrm{v} = \mathrm{v}_c$ or v_d for the charge or discharge modes, respectively. Positions in the PCM are described by the radial and axial variables (r, z) with $R_{\text{in}} \le r \le R_{\text{out}}$, $0 \le z \le l_0$, the inlet face being $z = 0$. The mesh subdivision to be used is defined as in CHAPTER **4**. In particular a typical mesh volume V_{ij} assigned to the point (r_i, z_j) is the region $r_i - ½\Delta r \le r \le r_i + ½\Delta r$, $z_j - ½\Delta z \le z \le z_j + ½\Delta z$, with volume $2\pi r_i \Delta r \Delta z$.

The heat balances for control volumes along the gas pipe take the form

$$\rho_g c_g \mathrm{v}\, \pi R_{\text{in}}^2 \{T_g(z+\Delta z, t) - T_g(z,t)\} = 2\pi\, R_{\text{in}} \Delta z\, h\{T^{surf}(z,t) - T_g(z,t)\}\,.$$

where $T^{surf}(z,t) := T(R_{\text{in}}, z, t)$ is the temperature of the PCM abutting on the heat exchange surface. In this relation we are ignoring heat conduction in the gas, a reasonable assumption because of the high speed flows involved. The system of heat balances for each nodal volume along the gas pipe constitutes a system of linear equations at the same time step which can be solved by, say, Gauss Elimination, given the value of the PCM surface temperature $T^{surf}(z,t)$ (found from the PCM enthalpy scheme).

5.3.E Some simulation results

For illustrative purposes we describe the results of applying the two codes described in §**5.3.C, D** to the LiF-HeXe system. The system temperatures are those in (4).

Firstly, we can use the quasistationary approximation as the starting point in the search for "feasible" configurations of the LiF-HeXe system. We input all possible pairs $(R_{\text{in}}, R_{\text{out}})$ for $0.05 \le R_{\text{in}} \le 0.3$ and $R_{\text{out}} > R_{\text{in}}$, over increments of 0.001. For each pair we find v_c, v_d, N and l_0, then comparing the results with the constraints (i),(v),(vi),(vii) (§**5.3.C**). We find that feasible configurations have:

$$0.252 \le R_{\text{in}} \le 0.292\,, \quad \text{and} \quad R_{\text{out}} = 1.2\, R_{\text{in}}\,,$$

$$27{,}000 \le \mathrm{v}_c \le 38360\,, \quad 8650 \le \mathrm{v}_d \le 11{,}000\,, \quad \text{with} \quad \mathrm{v}_c/\mathrm{v}_d \approx 3.4\,.$$

Similarly, the total mass of the PCM in the system is approximately 120,000 lb, while the number of jackets varies from 250 (for the smallest R_{in}) to 147 (for the largest R_{in}).

The values derived via the approximation code can now be used to aid in identifying truly feasible systems via the enthalpy code. With a small number of spatial nodes the enthalpy code runs rapidly on a computer of even modest computing power. For given inner and outer radii and jacket length we simulate the system performance over a range of v_c, v_d values. The code halts when either the outlet gas temperature falls below $T_{\text{eng}} = 1250$ °F (undercooling), or the PCM temperature exceeds $T_{\max} = 1700$ °F at some location (overheating). If no undercooling or overheating is predicted within 10 simulated hours, we accept the pair $(\mathrm{v}_c, \mathrm{v}_d)$ is feasible for the given values of R_{in}, R_{out} and l_0.

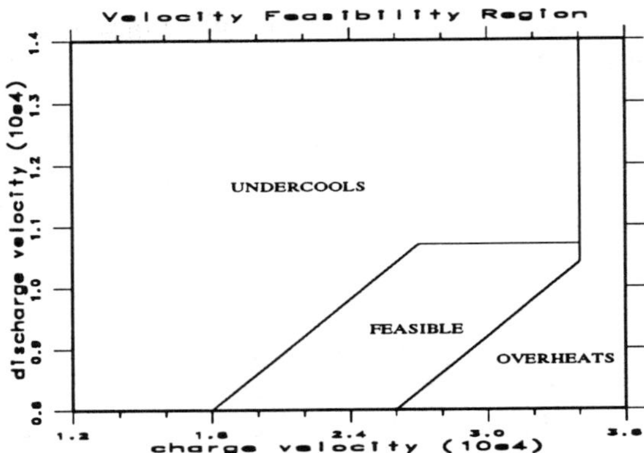

Figure 5.3.3. Velocities feasibility region for $R_{in} = .26$, $R_{out} = .312$, $l_0 = 52$ ft

In **Figure 5.3.3** we see the result of one such "feasible run" for $R_{\text{in}} = 0.26$, $R_{\text{out}} = 0.312$ and $l_0 = 52$ ft. The feasible domain consists of pairs (v_c, v_d) bounded by $\text{v}_c = 35,000$ ft/hr and $\text{v}_d = 11,000$ ft/hr, denoted in the figure by "W." The utility of such an output is that it gives us an intuitive "feeling" about zones of feasibility in which we may attempt to seek charge and discharge velocities for testing in longer, more precise computer runs with a finer mesh and degree of accuracy.

Basic questions one may ask about the system under study are

(a) can it be designed to work over long periods of time without undercooling or overheating ?

(b) can it be "optimized" over all systems that do work, aiming, for example, at identifying a configuration of least mass and smallest sensitivity to variability of conditions ?

With (a) in mind, several long-term simulations (1000 simulated hours) were executed with the enthalpy code. It was found that after a "relaxation time" of about 10 hours the system behaved periodically. Thus, in **Figures 5.3.4, 5**, we see the outlet gas temperature and the total energy (relative to solid at T_m) stored in the system as functions of time over the first 20 simulated hours. Note the sharp drops and rises of temperature accompanying the switches between charging and discharging modes. The total latent heat storage capacity of this system is 67.8×10^6 BTU (zero energy corresponds to solid at T_m). The fact that the total energy is, after 10 hours, above 40×10^6 BTU indicates that the system efficiency might possibly be improved, since apparently it is always maintaining some material in liquid phase.

The optimization issues have not been studied in detail. Our preliminary studies show that, under the simplifying assumptions we have made, the design modeled would require about 120,000 lb of PCM, and about 100 jackets of length roughly equal to 50 ft each. Moreover, the inner jacket radii would need to be

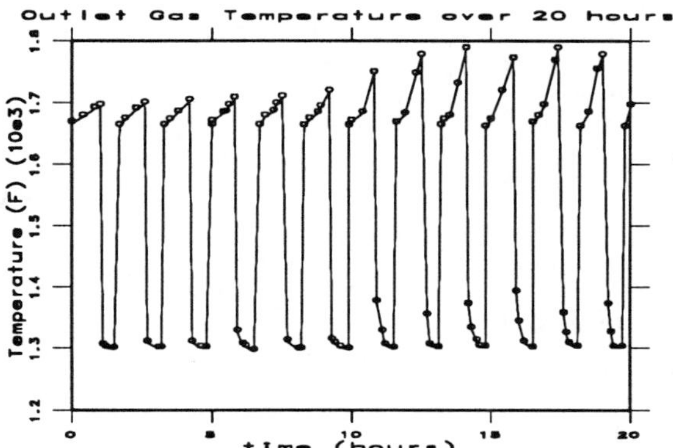

Figure 5.3.4. Outlet gas temperature during the first 20 hours of a 1000 hour simulation

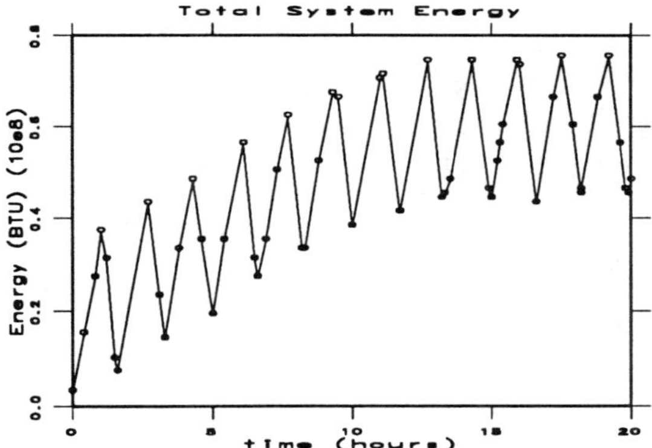

Figure 5.3.5. Stored energy during the first 20 hours of a 1000 hour simulation

about 0.25 ft, the outer in the ratio of 1.2 to the inner. Is this the "best" system of this design? Indeed, what is a suitable measure of optimality of the system? Should it be least weight, or also include some measure of comparison between the outlet temperature from the storage unit and T_{eng} (which it must exceed) and T_{max} (which it must not exceed)? Similarly, should it not include the extent of PCM utilization for latent heat storage?

The ability of the enthalpy method to handle complex phase changes, with intermixed zones of solid and liquid, coalescing domains, etc, is crucial here. **Figures 5.3.6,7** show the evolution of the solid, liquid and mushy regions, over the time interval $15.8 \le t \le 17.0$ hours, at 12 minute intervals, obtained in the 1000 hour simulation just described. For $15.8 < t \le 16.2$ the material is being discharged; thereafter, it is being charged. Note that certain portions of the PCM

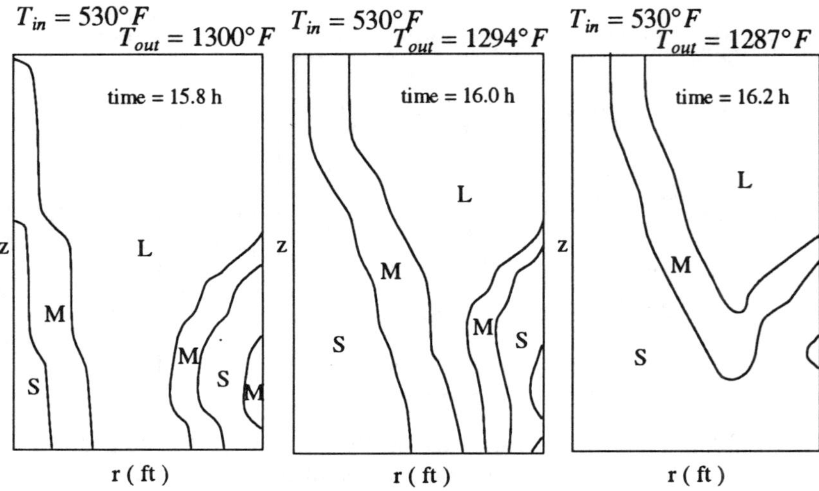

Figure 5.3.6. Distribution of phases during a discharge cycle at times: 15.8 h, 16.0 h, 16.2 h, $T_{in} = 530\,°F$

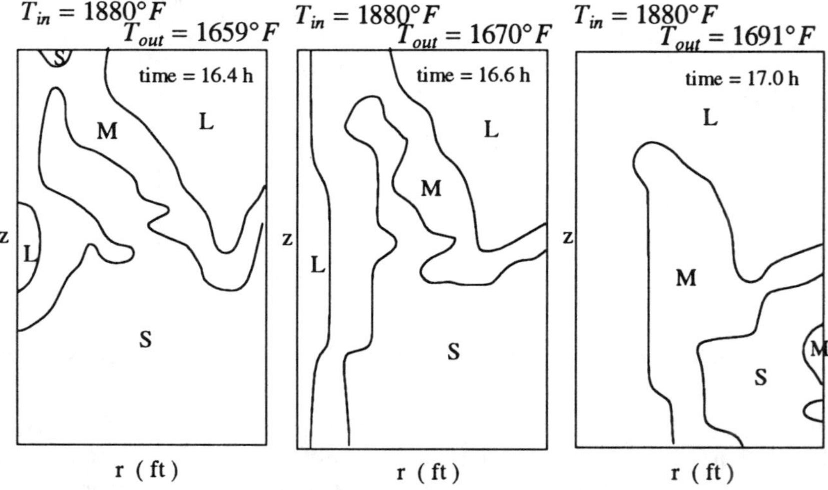

Figure 5.3.7. Distribution of phases during a charge cycle at times: 16.4 h, 16.6 h, 17.0 h, $T_{in} = 1880\,°F$

remain liquid, and others solid for all time, suggesting the possible utility of either tapering or flow reversal. We would stress again, in light of these results, that the enthalpy-based approach is the only one available for a process having this degree of geometrical complexity.

One advantage of a simulation code such as that described above is that it enables us to easily "experiment" with alternative materials and strategies.

Indeed, once the factors deemed necessary for credibility of the model have been incorporated, we can apply it to every member of the small family of high temperature PCM's to see which "works best." Let us illustrate this capability with two examples.

The first example arises from the observed "spread" of the total energy from its base value as seen in **Figure 5.3.5**. It is natural to ask what would be the effect of doubling the latent heat of the PCM, while keeping all its other thermophysical parameters the same. Using our analytical approximations a set of approximately feasible configurations can be generated, providing us with "ballpark" system sizes to which the simulation code may be applied; this is used to determine a parameter set whose system neither overheated nor undercooled for a 1000 hour run. The resulting total energy of the system varied between 30×10^6 and 60×10^6 BTU, for $R_{\text{inner}} = 0.29$, $R_{\text{out}} = 0.32$, $l_0 = 60$, total mass $= 62974$ lb, $N = 163$, $v_c = 28,000$, $v_d = 9000$. By comparison, **Figure 5.3.5** shows a total energy spread between approximately 45×10^6 and 78×10^6 BTU. Thus, we are lead to guess that a more efficient and substantially smaller system can be obtained by choosing a PCM of higher latent heat.

The second example arises from the observed sharp rises and drops of the outlet gas temperature due to the switches of the inlet temperature. One possible aid in reducing the abruptness of these changes might be by enhancing the PCM thermal conductivities. Using the simulation code we thus sought ranges of feasible configurations for the case of a PCM whose liquid and solid conductivities are twice those chosen for LiF, but whose remaining properties are unaltered. We selected our configurations with the same jacket length (75 ft). With doubled conductivities, the case run was with $R_{\text{in}} = 0.22$, $R_{\text{out}} = 0.29$, $v_c = 26,000$, $v_d = 11,000$. For unaltered LiF we chose $R_{\text{in}} = 0.26$, $R_{\text{out}} = 0.312$, $v_c = 25,000$, $v_d = 10,000$. The resulting outlet gas temperatures exhibited reduced sharpness in their changes when modes were switched.

In closing, let us take note of some of the limitations of the treatment we have presented and of possible extensions.

Recirculation. Through both charge and discharge cycles, some of the transfer fluid emerging from the engine at temperature T_{engout} could be pumped through the storage unit and/or the engine. This would mean mixing 860 °F and 530 °F gas, thus reducing the requirements on the storage unit.

Effect of container walls. The presence and heat conduction along the container has not been taken into account. This is rather simple to do in the enthalpy scheme, as described in **§4.3.J**. The walls can be considered as another "PCM" which however is not changing phase, and be discretized along with the real PCM, to compute energies and temperatures in it.

Effects of density change. Of great importance, especially for LiF with its very large density change, is the volume contraction/expansion of the PCM and the formation of voids. Under gravity, buoyancy will set up natural convection, whereas under microgravity we expect *Marangoni* flow due to surface tension at the void-liquid interface. Convection will determine the void movement. Its location adjacent to a wall may cause overheating, thermally stressing the wall and reducing its life expectancy, in addition to inhibiting heat transfer to the PCM. The presence of voids also gives rise to the possibility of radiative heat transfer across them, futher complicating matters.

Temperature dependence of thermophysical properties. The thermophysical properties of both PCM and transfer fluid may change significantly over the large temperature ranges of the power system. The incorporation of this variability, in the equations for the PCM and in the relations for the heat transfer coefficient, is straightforward in the enthalpy scheme. While one may suspect low sensitivity of results to the inclusion of this variability, this is not obvious.

Such enhancements have been incorporated in a large detailed simulation code (NORVEX), currently under development at ORNL and NASA/Lewis.

PROBLEMS

PROBLEM 1. Find the quasistationary solution to (7)-(9), and verify (10)-(12).

PROBLEM 2. Find the ratio v_c / v_d from (16) and evaluate it for $R_0 = 0.26$, $R_c = R_d = 0.312$.

PROBLEM 3. Derive a differential equation for the temperature of a rapidly flowing gas in a tube as a function of the axial direction z; ignore heat transfer by conduction, and assume heat gain and loss only through the tube walls; furthermore assume that heat transfer depends only on the distance z (see (30)§**3.2.D**). Solve this equation to derive (17) and verify (18).

BIBLIOGRAPHY

ABHAT, A., "Low temperature LHTES: Heat storage materials," *Solar Energy,* 30(4) (1983).

ABRAMOWITZ, M. AND I. STEGUN, "Handbook of Mathematical Functions," Dover Publications, New York (1965).

ADAMSON, A., "Physical Chemistry of Surfaces," 4th edition, John Wiley & Sons, New York (1982).

ALBRECHT, J., L. COLLATZ, AND K.-H. HOFFMANN, EDITORS, "Numerical Treatment of Free Boundary Problems," Birkhauser Verlag, Basel (1982).

ALEXIADES, V., "Rapid freezing of dilute alloys," *IMA J. of Applied Math.,* 30, pp. 67-79 (1983).

ALEXIADES, V., "Void formation in solidification" in *Differential Equations,* ed. C.M. Dafermos, G. Ladas, and G. Papanicolaou, Lecture Notes in Pure and Applied Math. vol.118, Marcel Dekker (1989a).

ALEXIADES, V., "Effect of a void on cyclic melting and freezing of an encapsulated PCM" in *Multiphase Flow, Heat and Mass Transfer,* ed. R.K.Shah, ASME HTD-vol.109, pp. 17-20, Amer. Soc. of Mechanical Engineers, New York (1989b).

ALEXIADES, V., "Shrinkage effects on solidification" in *Recent Trends in Welding Science and Technology,* ed. S.A. David and J.M. Vitek, pp. 171-174, ASM International, Materials Park, Ohio (1990).

ALEXIADES, V., "Casting of HgCdTe, Part II: conduction - diffusion model and its numerical simulation," ORNL/TM-11753, Oak Ridge National Laboratory (1991).

ALEXIADES, V. AND J. CANNON, "Free boundary problems in solidification of alloys," *SIAM J. of Math. Analysis,* 11, pp. 254-264 (1980).

ALEXIADES, V. AND J.B. DRAKE, "A weak formulation for phase change problems with bulk movement due to unequal densities" in *Free Boundary Problems: Theory and Applications,* ed. J. Chadam, H. Rasmussen, Pitman Lecture Notes, Longman, New York (1992).

ALEXIADES, V., J.B. DRAKE, G.A. GEIST, G.E. GILES, A.D. SOLOMON, AND R.F. WOOD, "Mathematical modeling of laser induced ultrarapid melting and solidification," ORNL-6129, Oak Ridge National Laboratory (1985a).

ALEXIADES, V., G. A. GEIST, AND A. D. SOLOMON, "Numerical simulation of a HgCdTe solidification process," ORNL-6127, Oak Ridge National Laboratory (1985b).

ALEXIADES, V. AND A. D. SOLOMON, "Critical radius for nucleation," ORNL/TM-9931, Oak Ridge National Laboratory (1986).

ALEXIADES, V. AND A. D. SOLOMON, "The formation of a solid nucleus in supercooled liquid, Part II," *J. of Non-Equilibrium Thermodynamics,* 14 (2), pp. 99-109 (1989).

ALEXIADES, V., A. D. SOLOMON, AND D. G. WILSON, "An Observation of the Total Energy of a System with Phase Changes," ORNL/CSD-72, Oak Ridge National Laboratory (1981a).

ALEXIADES, V., A. D. SOLOMON, AND D. G. WILSON, "Modeling of Phase Change Problems with Time-Varying Critical Temperature," ORNL/CSD/TM-145, Oak Ridge National Laboratory (1981b).

ALEXIADES, V., A. D. SOLOMON, AND D.G. WILSON, "The formation of a solid nucleus in supercooled liquid, Part I," *J. of Non-Equilibrium Thermodynamics,* 13, pp. 281-300 (1988).

ALEXIADES, V. AND R.P. WICHNER, "Calculation of the specific ans volumetric enthalpies of Ge-Si alloys," ORNL/TM-10548, Oak Ridge National Laboratory (1988).

ALEXIADES, V., D.G. WILSON, AND A.D. SOLOMON, "Macroscopic global modeling of binary alloy solidification processes," *Quart. of Applied Math.,* 43 (2), pp. 143-158 (1985).

ALLEN, M.B., I. HERRERA, AND G.F. PINDER, "Numerical Modeling in Science and Engineering," John Wiley & Sons, New York (1988).

ARIS, R., "Vectors, Tensors and the Basic Equations of Fluid Mechanics," Prentice-Hall (1962).

ATTHEY, D., "A finite difference scheme for melting problems," *J. of the Inst. of Math. and its Applications,* 13, pp. 353-366 (1974).

AZIZ, A. AND T.Y. NA, "Perturbation Methods in Heat Transfer," Hemisphere Pub. Co., Washington (1984).

BAYLIN, F., "Low temperature TES: A state-of-the-art survey," SERI/RR-54-164, Solar Energy Research Institute (1979).

BECKETT, P.M., "On the use of series solutions applied to solidification problems," *Mech. Research Communications,* 8, pp. 169-174 (1981).

BELL, G.E., "On the performance of the enthalpy method," *Int. J. of Heat & Mass Transfer,* 25(4), pp. 587-589 (1982).

BENNETT, H., "Industrial Waxes" in *Natural and synthetic waxes,* 1, Chemical Publishing Company, New York (1963).

BENNON, W.D. AND F.P. INCROPERA, "A continuum model for momentum, heat and species transpot in binary solid-liquid phase change systems-I. Model formulation, II. Application to solidification in a rectangular cavity," *Int. J. of Heat & Mass Transfer,* 30(10), pp. 2161-2170, 2171-2187 (1987).

BERMUDEZ, A. AND C. SAGUEZ, "Mathematical formulation and numerical solution of an alloy solidification problem" in *Free Boundary Problems: Theory and Applications,* ed. A. Fasano and M. Primicerio, I,II, pp. 237-247, Pitman, London (1983).

BIRD, R., W. STEWART, AND E. LIGHTFOOT, "Transport Phenomena," John Wiley & Sons, New York (1960).

BOLEY, B., "Time dependent solidification of binary mixtures," *Int. J. of Heat & Mass Transfer,* 21, pp. 821-824 (1978).

BOSSAVIT, A., A. DAMLAMIAN, AND M. FREMOND, EDITORS, "Free Boundary Problems: Applications and Theory," III,IV, Pitman, London (1985).

CAFFARELLI, L.A. AND L.C. EVANS, "Continuity of the temperature for the two phase Stefan problem," *Archive for Rat. Mechanics and Analysis,* 81, pp. 199-220 (1983).

CAFFARELLI, L.A. AND A. FRIEDMAN, "Continuity of the temperature in the Stefan Problem," *Indiana Univ. Math. J.,* 28(1), pp. 53-70 (1979).

CAGINALP, G., "Stefan and Hele-Shaw type models as asymptotic limits of the phase field equations," *Physical Review A,* 39, pp. 5887-5896 (1989).

CAGINALP, G. AND J. JONES, "A derivation of phase-field model with fluid properties" in *IMA Workshop on Evolution of Phase Boundaries,* IMA Volumes on Math. and its Applications, Springer-Verlag, New York (1991).

CALLEN, H B., "Thermodynamics," John Wiley & Sons, New York (1960).

CANNON, J.R., "The One-Dimensional Heat Equation," Addison-Wesley, Menlo Park (1984).

CANNON, J.R., D. HENRY, AND KOTLOW, "Continuous differentiability of the free boundary for weak solutions of the Stefan problem," *Bulletin of Amer. Math. Soc.,* 80(1), pp. 45-48 (1974).

CANNON, J.R., D. HENRY, AND KOTLOW, "Classical Solution of 1-d, 2-phase Stefan Problem," *Annali di Mat. Pura et Appl., IV,* CVII, pp. 311-341 (1976).

CARNAHAN, B., H. LUTHER, AND J. WILKES, "Applied Numerical Methods," John Wiley & Sons, New York (1969).

CARSLAW, H. AND J. JAEGER, "Conduction of Heat in Solids," 2nd edition, Clarendon Press, Oxford (1959).

CHADAM, J. AND P. ORTOLEVA, "On growth and form" in *Proc. of Conference on Math. Biology,* Southern Illinois Univ., Carbondale (1980).

CHADAM, J. AND H. RASMUSSEN, EDITORS, "Free Boundary Problems: Theory and Applications," Longman, New York (1992).

CHALMERS, B., "Principles of Solidification," John Wiley & Sons, New York (1964).

CHAMBRE, P., "On the dynamics of phase growth," *Quart. J. of Mechanics and Applied Math.,* 9, pp. 224-233 (1956).

CHAPMAN, A., "Heat Transfer," MacMillan, New York (1960).

"Chemical Engineering Guide to Heat Transfer," Hemisphere Pub. Corp. ; McGraw-Hill Chemical Engineering (1986).

CHESTER, C.R., "Techniques in Partial Differential Equations," McGraw-Hill, New York (1971).

CHO, S.H., "An exact solution of the coupled phase change problem in a porous medium," *Int. J. of Heat & Mass Transfer,* 18, pp. 1139-1142 (1975).

CHO, S.H. AND J.E. SUNDERLAND, "Heat conduction problems with melting or freezing," *J. of Heat Tranfer of the ASME,* 91C, pp. 421-426 (1969).

CHORIN, A.J. AND J.E. MARSDEN, "A Mathematical Introduction to Fluid Mechanics," Springer-Verlag, New York (1979).

CHRISTIAN, J., "The Theory of Transformations in Metals and Alloys," Pergamon, Oxford (1965).

CLYNE, T.W., "Numerical modeling of directional solidification of metallic alloys," *Metal Science,* 16, pp. 441-450 (1982).

CODDINGTON, E. AND N. LEVINSON, "Theory of Ordinary Differential Equations," McGraw Hill, New York (1955).

COOPER, T. AND G. TREZEK, "Analytical prediction of a temperature field emanating from a cryogenic surgical cannula," *Cryobiology,* 7, pp. 79-93 (1970).

COOPER, T. AND G. TREZEK, "Rate of lesion growth around spherical and cylindrical cryoprobes," *Cryobiology,* 7, pp. 183-190 (1971).

CORIELL, S.R. AND R.F. SEKERKA, "Oscillatory morphological instabilities due to non-equilibrium segregation," *J. of Crystal Growth,* 61, pp. 499-508 (1983).

COURANT, R., K. FRIEDRICHS, AND H. LEWY, "Uber die partiellen differenzengleichungen der mathematischen Physik," *Mathematische Annalen,* 100, pp. 32-74 (1928).

COURANT, R. AND D. HILBERT, "Methods of Mathematical Physics, Volume II," Interscience, New York (1962).

CRANK, J., "The Mathematics of Diffusion," Clarendon Press, Oxford (1975).

CRANK, J., "How to deal with moving boundaries in thermal problems" in *Numerical Methods in Heat Transfer,* ed. R.W. Lewis, K. Morgan, and O.C. Zienkiewicz, pp. 177-200, John Wiley & Sons, New York (1981).

CRANK, J., "Free and Moving Boundary Problems," Clarendon Press, Oxford (1984).

CROWLEY, A.B. AND J.R. OCKENDON, "On the numerical solution of an alloy solidification problem," *Int. J. of Heat & Mass Transfer,* 22, pp. 941-947 (1979).

CRYER, C.W., "A bibliography of free boundary problems," MRC # 1793, Math. Research Center, Univ. of Wisconsin (1977).

CULP, A.W., "Principles of Energy Conservation," McGraw-Hill, New York (1979).

DANILYUK, I.I., "On the Stefan problem," *Russian Math. Surveys,* 40, pp. 157-223 (1985).

DEAL, R. AND A. D. SOLOMON, "The simulation of four pure conduction paraffin-wax freezing experiments," ORNL/CSD-74, Oak Ridge National Laboratory (1981).

DEAL, R., R. IRBY, E. KESHOCK, AND A. D. SOLOMON, "On the solidification of N-Octadecane paraffin wax," ORNL/CSD-92, Oak Ridge National Laboratory (1982).

DELHAYE, J.M., "Jump conditions and entropy sources in two-phase systems. Local instant formulation," *Int. J. Multiphase Flow,* 1, pp. 395-409 (1974).

DIBENEDETTO, E., "Regularity properties of the solution of an n-dimensional two-phase Stefan problem," *Bollettino Unione Mat. Italiana* (1982).

DILLEY, J.F. AND N. LIOR, "The evaluation of simple analytical solutions for the prediction of freeze-up time, freezing, and melting" in *Heat Transfer 1986,* ed. Tien, Carey, Ferrell, 4, Hemisphere Publ. Co., New York (1986).

DOUGLAS, J. AND T. GALLIE, "On the numerical integration of a parabolic differential equation subject to a moving boundary condition," *Duke Math. J.*, 22, pp. 557-571 (1955).

DRAKE, J.B., G.A. GEIST, M.D. MORRIS, A.D. SOLOMON, J. MARTIN, AND J. TOMLINSON, "Design of PCM enhanced passive solar structures," ORNL-6281, Oak Ridge National Laboratory (1987).

DRAKE, J.B. AND ET AL, "PCMSOL-Simulation code for a passive solar structure incorporating phase change materials," ORNL-6359, Oak Ridge National Laboratory (1987).

DRAKE, J.B., "Convection in the melt," Ph.D. thesis, University of Tennessee, Knoxville (1991).

DUCHATEAU, P. AND D.W. ZACHMANN, "Partial Differential Equations," Schaum's Outline Series, McGraw-Hill, New York (1986).

DUDA, J.L. AND J.S. VRENTAS, "Heat or mass transfer-controlled dissolution of an isolated sphere," *Int. J. of Heat & Mass Transfer,* 14, pp. 395-408 (1971).

DUFFIE, J. AND W. BECKMAN, "Solar Energy Thermal Processes," Wiley-Interscience, New York (1974).

DUSINBERRE, G.M., "Numerical methods for transient heat flow," *Transactions of ASME,* 67, pp. 703-712 (1945).

DUVAUT, G., "Resolution d'un probleme de Stefan," *C.R.A.S.,* 276, pp. 1461-1463 (1973).

ECKERT, E.R.G. AND R.M. DRAKE, "Analysis of Heat and Mass Transfer," McGraw-Hill, Tokyo (1972).

ELLIOT, D.E., ET AL, "High temperature TES" in *Thermal Energy Storage,* ed. E. Kovach, NATO Science Committee Conf. (1976).

ELLIOTT, C.M., "Error analysis of the enthalpy method for the Stefan Problem," *IMA J. of Numerical Analysis,* 7, pp. 61-71 (1987).

ELLIOTT, C.M. AND J.R. OCKENDON, "Weak and Variational Methods for Moving Boundary Problems," Pitman Press, Boston (1982).

FASANO, A. AND M. PRIMICERIO, "General free boundary problems for the heat equation, I,II," *J. of Math. Analysis and Appl.,* 56, pp. 209-231 (1977a,b).

FASANO, A. AND M. PRIMICERIO, "General free boundary problems for the heat equation, III," *J. of Math. Analysis and Appl.,* 59, pp. 1-14 (1977b).

FASANO, A. AND M. PRIMICERIO, EDITORS, "Free Boundary Problems: Theory and Applications," I,II, Pitman, London (1983).

FENNEMA, O., W. POURIE, AND E. MARTHA, "Low Temperature Preservation of Foods and Living Matter," Marcel Dekker, New York (1973).

FINTEL, M., "Handbook of Concrete Engineering," Van Nostrand, New York (1974).

FIX, G.J., "Numerical methods for alloy solidification problems" in *Moving Boundary Problems,* ed. D.G. Wilson, A.D. Solomon, and P.T. Boggs, pp. 109-128, Academic Press (1978).

FIXLER, S., "Satellite thermal control using phase-change materials," *J. of Spacecraft*, 3, pp. 1362-1368 (1966).

FLEMINGS, M.C., "Solidification Processing," McGraw-Hill, New York (1974).

FORSYTHE, G. AND W. WASOW, "Finite Difference Methods for Partial Differential Equations," John Wiley & Sons, New York (1960).

FRIEDMAN, A., "Partial Differential Equations of Parabolic Type," Prentice-Hall, Englewood Cliffs (1964).

FRIEDMAN, A., "The Stefan problem in several space variables," *Transactions of the Amer. Math. Soc.*, 133, pp. 51-87 (1968).

FRIEDMAN, A., "Variational Principles and Free Boundary Problems," John Wiley & Sons, New York (1982).

FRIEDMAN, A. AND B. HU, "The Stefan problem for a hyperbolic heat equation," IMA Preprint #348, Univ. of Minn. (Oct. 1987).

GEIST, G. A., C. A. SERBIN, A. D. SOLOMON, AND J. TOMLINSON, "PCMSOL1-A computer simulation code for a direct gain passive solar structure, Part I: Theory and basic relations; Part II: Users guide," ORNL/CSD-127, Oak Ridge National Laboratory (1984).

GOLIBERSUCH, D.C., ET AL, "Thermal energy storage for utility applications," Report 75CRD256, General Electric (1976).

GOLUB, G.H. AND C.F. VAN LOAN, "Matrix Computations," Johns Hopkins Univ. Press, Baltimore (1983).

GOODMAN, T., "Integral methods for nonlinear heat transfer," *Advances in Heat Transfer*, I, pp. 51-122 (1964).

GRILL, A., K. SORIMACHI, AND J. BRIMACOMBE, "Heat flow, gap formation and breakouts in the continuous casting of steel slabs," *Metallurgical Transactions B*, 7B, pp. 177-189 (1976).

GRODZKA, P.G., ET AL, "Space thermal control by freezing and melting," Report N69-40104, Lockheed Missiles & Space Co. (1969).

GURTIN, M.E., "Towards a nonequilibrium thermodynamics of two-phase materials," *Archive for Rat. Mechanics and Anal.*, 100 (3), pp. 275-312 (1988).

HALE, D., M. HOOVER, AND M. O'NEILL, "Phase Change Materials Handbook," NASA CR-61363, Lockheed Missiles & Space Co. (1971).

HALE, N. AND R. VISKANTA, "Photographic observation of the solid-liquid interface motion during melting of a solid heated from an isothermal vertical wall," *Letters in Heat & Mass Transfer*, 5, pp. 329-337 (1978).

HERINGTON, E.F.G., "Zone Melting of Organic Compounds," John Wiley & Sons, New York (1963).

HILL, J., "One-dimensional Stefan Problems: An Introduction," Longman Scientific & Technical Publ., Burnt Mill (1987).

HILL, J. AND J. DEWYNNE, "Heat Conduction," Blackwell Scientific Publications, Oxford (1987).

HOOVER, M.J., ET AL, "Space thermal control development, Final Report," Report N72-14947, Lockheed Missiles & Space Co. (1971).

HUANG, C.L AND Y.P. SHIH, "A perturbation method for for spherical and cylindrical solidification," *Chemical Engineering Science,* 30, pp. 897-906 (1975).

HULTGREN, R., R. ORR, D. ANDERSON, AND K. KELLEY, "Selected values of thermodynamic properties of metals and alloys," John Wiley & Sons, New York (1963).

HYMAN, J.M., "Numerical methods for tracking interfaces," *Physica D,* 12D, pp. 396-407 (1984).

ISAACSON, E. AND H.B. KELLER, "Analysis of Numerical Methods," John Wiley, New York (1967).

JIJI, L. AND S. WEINBAUM, "A nonlinear singular perturbation theory for non-similar melting or freezing problems" in *Proceedings of the Fifth Int. Heat Transfer Conf., 1974,* Vol. I, Japan Soc. of Mech. Engineers (1974).

JURINAK, J.J. AND S.I. ABDEL-KHALIK, "Properties optimization for phase-change energy storage in air-based solar heating systems," *Solar Energy,* 21, pp. 377-383 (1978).

KALLUNGAL, J. AND A. BARDUHN, "Growth rate of an ice crystal in subcooled pure water," *AICHE Journal,* 23, pp. 294-303 (1977).

KAMENOMOSTSKAYA, S.L., "On the Stefan problem (in Russian)," *Mat. Sbornik,* 53(95), pp. 489-514 (1961).

KELLEY, C.T. AND J. RULLA, "Solution of the time discretized Stefan problem by Newton's method," *Nonlinear Analysis, Theory Methods and Applications,* 14, pp. 851-872 (1990).

KNIGHT, C., "The Freezing of Supercooled Liquids," D. Van Nostrand, Princeton (1976).

KOBAYASHI, R., "Modeling and numerical simulations of dendritic crystal growth," Ryukoku Univ. Report (1991).

KOBAYASHI, R., "Simulations of three-dimensional dendrites," Ryukoku Univ. Report (1992).

KUFNER, A., O. JOHN, AND S. FUCIK, "Function Spaces," Noordhoff, Leyden (1977).

KURZ, W. AND D. FISHER, "Fundamentals of Solidification," Trans. Tech. Publications, Zurich (1984).

LACEY, A.A. AND M. SHILLOR, "The existence and stability of regions with superheating in the classical two-phase one-dim. Stefan Problem with heat sources," *IMA J. of Applied Math.,* 30, pp. 218-230 (1983).

LACEY, A.A. AND A.B. TAYLER, "A mushy region in a Stefan problem," *IMA J. of Applied Math.,* 30, pp. 303-313 (1983).

LANDAU, H., "Heat conduction in a melting solid," *Quart. of Applied Math.,* 8, pp. 81-94 (1950).

LANDAU, L.D. AND E.M. LIFSHITZ, "Statistical Physics, Part 1," 3rd edition, Pergamon, Oxford (1980).

LANGER, J.S., "Instabilities and pattern formation in crystal growth," *Review of Modern Physics,* 52, pp. 1-28 (1980).

LAPIDUS, L. AND G.F. PINDER, "Numerical Solution of Partial Differential Equations in Science and Engineering," John Wiley, New York (1982).

LEVI, C. AND R. MEHRABIAN, "Heat flow during rapid solidification of undercooled metal droplets," *Metallurgical Transactions A,* 13A, pp. 221-234 (1982).

LI, DENING, "A 3-dimensional hyperbolic Stefan problem with discontinuous temperature," *Quart. of Applied Math.,* 49, pp. 577-589 (1991).

LIN, S., "An exact solution of the sublimation problem in a porous medium, Parts I, II," *J. of Heat Transfer of the ASME,* 103, 104, pp. 165-168, 808-811 (1981, 1982).

LONDON, A.L. AND R.A. SEBAN, "Rate of ice formation," *Transactions of ASME,* 65, pp. 771-778 (1943).

LUNARDINI, V.J., "Heat Transfer in Cold Climates," Van Nostrand, New york (1981).

LUPIS, C H.P., "Chemical Thermodynamics of Materials," North-Holland, New York (1983).

MAGENES, E., "Free Boundary Problems," Inst. National Alta Mat., F. Severn, Rome (1980).

MAGENES, E., "Linear schemes to approximate parabolic Stefan-type problems," N. 590, Inst. Analisi Numerica, C.N.R., Pavia (1987).

MCFADDEN, G.B. AND S.R. CORIELL, "Nonplanar interface morphologies during unidirectional solidification of a binary alloy," *Physica D,* 12 D, pp. 253-261 (1984).

MEGERLIN, F., "Geometrically unidimensional heat conduction in melting and solidification," *Forschung im Ingenieurwesen,* 34, pp. 40-46 (1968).

MEIRMANOV, A.M., "Multiphase Stefan problems for quasilinear parabolic equations," *Dinamika Sploshnoi Sredy (in Russian),* 13, Novosibirsk (1973).

MEIRMANOV, A.M., "On classical solution of Stefan multidimensional problem for quasilinear parabolic equations," *Math. Sbornik (in Russian),* 112 (154) (1980).

MEIRMANOV, A.M., "The Stefan Problem," Walter de Gruyter, Berlin-New YorK (1992).

MEYER, G.H., "An application of the method of lines to multidimensional free boundary problems," *J. of the Inst. of Math. and its Applications,* 20, pp. 317-329 (1977).

MEYER, G.H., "The numerical solution of multidimensional Stefan problems - A survey" in *Moving Boundary Problems,* ed. D.G. Wilson, A.D. Solomon, and P. Boggs, pp. 73-89, Academic Press, New York (1978).

MEYER, G.H., "A numerical method for the solidification of a binary alloy," *Int. J. of Heat & Mass Transfer,* 24, pp. 778-781 (1981).

MIKHAILOV, M., "Exact solution for freezing of humid porous half-space," *Int. J. of Heat & Mass Transfer,* 19, pp. 651-655 (1976).

MINGLE, J., "Stefan problem sensitivity and uncertainty," *Numerical Heat Transfer,* 2, pp. 387-393 (1979).

MINKOWYCZ, W.J., E.M. SPARROW, G.E. SCHNEIDER, AND R.H. PLETCHER, EDITORS, "Handbook of Numerical Heat Tansfer," John Wiley & Sons, New York (1988).

MORRIS, M. AND V. ALEXIADES, "Sensitivity analysis of a numerically simulated HgCdTe solidification process by statistical methods," ORNL-6210, Oak Ridge National Laboratory (1986).

MORSE, P. AND H. FESHBACH, "Methods of Theoretical Physics, Part I," McGraw Hill, New York (1953).

MULLINS, W.W. AND R.F. SEKERKA, "Morphological stability of a particle growing by diffusion or heat flow," *J. of Applied Physics,* 34, pp. 323-329 (1963).

MULLINS, W.W. AND R.F. SEKERKA, "Stability of a planar interface during solidification of a dilute binary alloy," *J. of Applied Physics,* 35, pp. 444-451 (1964).

MYSHKIS, A.D., V.G. BABSKII, N.D. KOPACHEVSKII, L.A. SLOBOZHANIN, AND A.D. TYUPTSOV, "Low-Gravity Fluid Mechanics," Springer-Verlag, Berlin (1987, Russian original 1976).

NARANG, H.N. AND J.B. DRAKE, "Parallel solution of a 2-D phase change problem on a hypercube," ORNL-6504, Oak Ridge National Laboratory (1989).

NASH, G.E. AND M.E. GLICKSMAN, "Capillarity-limited steady-state dendritic growth – I,II," *Acta Metallurgica,* 22, pp. 1283-1299 (1974).

NIEZGODKA, M. AND I. PAWLOW, "A generalized Stefan problem in several space variables," *Appl. Math. & Optimization,* 9, pp. 193-224 (1983).

NOCHETTO, R.H., "Error estimates for two-phase Stefan problems in several space variables, I: linear boundary conditions," *Calcolo,* 22 (1985).

OCKENDON, J. AND W. HODGKINS, EDITORS, "Moving Boundary Problems in Heat Flow and Diffusion," Clarendon Press, Oxford (1975).

ODEN, J.T. AND N. KIKUCHI, "Finite element methods for certain free boundary-value problems in Mechanics" in *Moving Boundary Problems,* ed. D. G. Wilson, A. D. Solomon, and P. Boggs, pp. 147-164, Academic Press, New York (1978).

OHANESSIAN, P. AND W. CHARTERS, "Thermal simulations of a passive solar house using a Trombe-Michel wall structure," *Solar Energy,* 20, pp. 275-281 (1978).

OLEINIK, O.A., "A method of solution of the general Stefan problem," *Soviet Math. Doklady,* 1, pp. 1350-1354 (1960).

ORTEGA, J. AND R. VOIGT, "Solution of partial differential equations of vector and parallel computers," *SIAM Review,* 27, pp. 149-240 (1985).

OZISIK, M.N., "Heat Conduction," John Wiley & Sons, New York (1980).

PARKER, R., "An Introduction to Chemical Metallurgy," 2nd edition, Pergamon, Oxford (1978).

PATANKAR, S.V., "Numerical Heat Transfer and Fluid Flow," Hemisphere, Washington (1980).

PEDROSO, R. AND G. DOMOTO, "Perturbation solutions for spherical solidification of saturated liquids," *J. of Heat Transfer of the ASME,* 95C, pp. 42-46 (1973a).

PEDROSO, R. AND G. DOMOTO, "Inward spherical solidification - solution by the method of strained coorrdinates," *Int. J. of Heat & Mass Transfer,* 16, pp. 1037-1043 (1973).

PFANN, W.G., "Zone Melting," 2nd edition, John Wiley & Sons, New York (1966).

PINSKY, M.A., "Introduction to Partial Differential Equations with Applications," McGraw-Hill, New York (1984).

PIPPARD, A., "Classical Thermodynamics," Cambridge Univ. Press, Cambridge (1957).

PITTS, D.R. AND L.E. SISSOM, "Heat Transfer," Schaum's Outline Series, McGraw-Hill, New York (1977).

PORTER, D. AND K. EASTERLING, "Phase Transformations in Metals and Alloys," Van Nostrand, New York (1981).

PRESS, W.H., B.P. FLANNERY, S.A. TEUKOLSKY, AND W.T. VETTERLING, "Numerical Recipes," Cambridge Univ. Press, Cambridge (1986).

PROTTER, M. AND H. WEINBERGER, "Maximum Principles in Differential Equations," Prentice Hall, Englewood Cliffs (1967).

RAND, R., A. RINFRET, AND H. VON LEDEN, "Cryosurgery," Charles L. Thomas Publisher, New York (1968).

REYNOLDS, W., "Thermodynamics," McGraw Hill, New York (1965).

RICHTMYER, R. AND K. MORTON, "Difference Methods for Initial-Value Problems," 2nd edition, Interscience, New York (1967).

ROSE, M., "A method for calculating solutions of parabolic equations with a free boundary," *Mathematics of Computation,* 14, pp. 249-256 (1960).

ROSENBERGER, F., "Fundamentals of Crystal Growth, Vol. I," Springer-Verlag, Berlin (1979).

RUBINSTEIN, L., "The Stefan Problem," Transl. of Math. Monographs, vol.27, Amer. Math. Society, Providence (1971).

SCHAEFFER, R.J. AND M.E. GLICKSMAN, "Fully time-dependent theory for the growth of spherical crystal nuclei," *J. of Crystal Growth,* 3-4, pp. 71-81 (1969).

SACKETT, G., "Numerical solution of a parabolic free boundary problem arising in statistical decision theory," *Mathematics of Computation,* 25, pp. 425-433 (1971).

SEENIRAJ, R.V. AND T.K. BOSE, "One-dimensional phase-change problems with radiation-convection," *J. of Heat Transfer of the ASME,* 104, pp. 811-813 (1982).

SEGEL, L.A., "Mathematics Applied To Continuum Mechanics," Dover, New York (1987).

SEKERKA, R.F., "Morphological instabilities during phase transformations" in *Phase Transformations and Material Instabilities in Solids,* ed. M.E. Gurtin, Academic Press, Orlando (1984a).

SEKERKA, R.F., "Morphological and hydrodynamic instabilities during phase transformations," *Physica D,* 12 D, pp. 212-214 (1984b).

SEKHAR, J.A., S. KOU, AND R. MEHRABIAN, "Heat flow model for surface melting and solidification of an alloy," *Metallurgical Transactions A,* 14A (1983).

SEWELL, G., "The Numerical Solution of Ordinary and Partial Differential Equations," Academic Press, Boston (1988).

SHAMSUNDAR, N. AND E. M. SPARROW, "Storage of thermal energy by solid liquid phase change temperature drop and heat flux," *J. of Heat Transfer of the ASME,* pp. 541-543 (1974).

SHOWALTER, R.E. AND N.J. WALKINGTON, *Quart. of Applied Math.,* 45, pp. 769-782 (1987).

SIMPSON, H., G. BEGGS, J. NAKAMURA, AND A. BAXTER, "The C-axis growth rate of ice crystals," *Desalination,* 18, pp. 219-230 (1976).

SMITH, G., "Numerical Solution of Partial Differential Equations," Oxford University Press, London (1965).

SOLOMON, A.D., "Some remarks on the Stefan problem," *Mathematics of Computation,* 20, pp. 347-360 (1966).

SOLOMON, A.D., "The applicability and extendability of Megerlin's method for solving paraboplic free boundary problems" in *Moving Boundary Problems,* ed. D. G. Wilson, A. D. Solomon, and P. Boggs, pp. 187-202, Academic Press, New York (1978).

SOLOMON, A.D., "A relation between surface temperature and time for a phase change process with a convective boundary condition," *Letters in Heat & Mass Transfer,* 6, pp. 189-197 (1979a).

SOLOMON, A.D., "Melt time and heat flux for a simple PCM body," *Solar Energy,* 22, pp. 251-257 (1979b).

SOLOMON, A.D., "Mathematical modeling of phase change processes for latent heat thermal energy storage," ORNL/CSD-39, Oak Ridge National Laboratory (1979c).

SOLOMON, A.D., "On the melting time of a simple body with a convection boundary condition," *Letters in Heat & Mass Transfer,* 7, pp. 183-188 (1980a).

SOLOMON, A.D., "An expression for the melting time of a rectangular body," *Letters in Heat & Mass Transfer,* 7, pp. 379-384 (1980b).

SOLOMON, A.D., "An easily computable solution to a two phase Stefan problem," *Solar Energy,* 24, pp. 525-528 (1980c).

SOLOMON, A.D., "The simulation of a PCM storage subsystem for air conditioning assist," ORNL/CSD-77, Oak Ridge National Laboratory (1981a).

SOLOMON, A.D., "A note on the Stefan number in slab melting and solidification," *Letters in Heat & Mass Transfer,* 8, pp. 229-235 (1981b).

SOLOMON, A.D., "Some approximations of use in predicting the behavior of a PCM cylinder array," *Letters in Heat & Mass Transfer,* 8, pp. 237-246 (1981c).

SOLOMON, A.D., "On the limitations of analytical approximations for phase change problems with large Biot numbers," *Letters in Heat & Mass Transfer,* 8, pp. 475-482 (1981d).

SOLOMON, A.D., "On mathematical modeling of cryosurgical processes," ORNL/CSD-89, Oak Ridge National Laboratory (1981e).

SOLOMON, A. D., "Some aspects of the computer simulation of conduction heat transfer and phase change processes" in *Computing methods in applied sciences and engineering V*, ed. R. Glowinski and J. Lions, pp. 457-470, North-Holland (1982).

SOLOMON, A.D., V. ALEXIADES, AND D.G. WILSON, "The Stefan problem with a convective boundary condition," *Quart. of Applied Math.*, 40 (2), pp. 203-217 (1982).

SOLOMON, A.D., V. ALEXIADES, AND D.G. WILSON, "A numerical simulation of a binary alloy solidification process," *SIAM J. of Scientific and Statistical Computing*, 6, pp. 911-922 (1985).

SOLOMON, A.D., V. ALEXIADES, AND D.G. WILSON, "The initial velocity of the emerging free boundary in a two-phase Stefan problem with imposed flux," *SIAM J. of Math. Analysis*, 18, pp. 1438-1452 (1987).

SOLOMON, A.D., V. ALEXIADES, D.G. WILSON, AND J.B. DRAKE, "On the formulation of hyperbolic Stefan problems," *Quart. of Applied Math.*, 43, pp. 295-304 (1985).

SOLOMON, A.D., V. ALEXIADES, D.G. WILSON, G.A. GEIST, AND J. KERPER, "An enthalpy method for the solidification of a supercooled liquid," ORNL-6217, Oak Ridge National Laboratory (1986).

SOLOMON, A.D., M.D. MORRIS, J. MARTIN, AND M. OLSZEWSKI, "The development of a simulation code for a latent heat thermal energy storage system in a space station," ORNL-6213, Oak Ridge National Laboratory (1986).

SOLOMON, A.D. AND C.A. SERBIN, "TES-A program for simulating phase change processes," ORNL/CSD-51, Oak Ridge National Laboratory (1980).

SOLOMON, A.D., D.G. WILSON, AND V. ALEXIADES, "A mushy zone model with an exact solution," *Letters in Heat & Mass Transfer*, 9 (2), pp. 319-324 (1982).

SOLOMON, A.D., D.G. WILSON, AND V. ALEXIADES, "Explicit solutions to phase change problems," *Quart. of Applied Math.*, 41, pp. 237-243 (1983).

SOLOMON, A.D., D.G. WILSON, AND V. ALEXIADES, "The quasi-stationary approximation for the Stefan problem with a convective boundary condition," *Int. J. of Mathematics and Math. Sciences*, 7, pp. 549-563 (1984).

SOLOMON, A.D., D.G. WILSON, AND V. ALEXIADES, "An approximate analysis of the formation of a buoyant solid sphere in a supercooled melt," ORNL-6212, Oak Ridge National Laboratory (1986a).

SOLOMON, A.D., D.G. WILSON, AND V. ALEXIADES, "An approximate solution to the problem of solidification of a sphere of supercooled fluid," ORNL-6292, Oak Ridge National Laboratory (1986b).

STEPLER, R., "Solar salts-New chemical systems store the sun's heat," *Popular Science* (March 1980).

STRAUSS, W.A., "Partial Differential Equations, An Introduction," John Wiley, New York (1992).

TAO, L., "On solidification of a binary alloy," *Quart. of Applied Math.*, 33, pp. 211-225 (1980).

TARZIA, D.A., "A bibliography on moving-free boundary problems for the heat-diffusion equation," Instituto Matematico "Ulisse Dini", Firenze (1988).

TAYLER, A.B., "Mathematical models in applied mechanics," Clarendon Press (1986).

TELKES, M., "Storage of solar heating/cooling," *ASHRAE Transactions*, 80(2) (1974).

THOMAS, L. AND J. WESTWATER, "Microscopic study of solid-liquid interfaces during melting and freezing Heat Transfer-Houston," *Checmial Engineering Profress Symposium Series*, 41, pp. 155-164 (1963).

TIEN, R.H. AND G.E. GEIGER, "A heat transfer analysis of the solidification of a binary eutectic system," *J. of Heat Transfer of the ASME*, 89C, pp. 230-234 (1967).

TIEN, R.H., CAREY, AND FERRELL, "Heat Transfer 1986," Vol.4, Hemisphere Publ. Co., New York (1986).

TILLER, W.A., "The Science of Crystalization: Microscopic Interfacial Phenomena," Cambridge Univ. Press, Cambridge (1991).

TOKSOY, M. AND B.Z. ILKEN, "Phase change heat transfer in cylindridal domain: Modeling and its importance in Thermal Energy Storage" in *NATO ASI on Energy Storage Systems: Fundamentals and Applications*, Cesme, Turkey (1988).

TOULOUKIAN, Y., "Thermophysical Properties of Matter," Plenum, New York (1970).

TREVES, F., "Basic Linear Partial Differential Equations," Academic Press, New York (1975).

TRIVEDI, R., "Theory of dendritic growth during the directional solidification of binary alloys," *J. of Crystal Growth*, 49, pp. 219-232 (1980).

TRUESDELL, C. AND R. TOUPIN, "The Classical Field Theories" in *Encyclopedia of Physics*, ed. S. Flugge, 3/1, pp. 226-790, Springer-Verlag, New York (1960).

TURLAND, B.D. AND R.S. PECKOVER, "The stability of planar melting fronts in 2-phase thermal Stefan problems," *J. of the Inst. of Math. & Applications*, 25, pp. 1-15 (1980).

TURNBULL, D., "Phase Changes" in *Solid State Physics*, ed. Seitz, Turnbull, 3, Academic Press, New York (1956).

TURNBULL, D., "The undercooling of liquids," *Scientific American*, 212, pp. 38-46 (1965).

VARGA, R., "Matrix Iterative Analysis," Prentice-Hall, Englewood Cliffs (1962).

VARGAFTIK, N., "Tables on the Thermophysical Properties of Liquids and Gases," 2nd edition, John Wiley & Sons, New York (1975).

VERDI, C. AND A. VISINTIN, "Error estimates for a semi-explicit numerical scheme for Stefan-type problems," *Numerische Mathematik*, 52, pp. 165-185 (1988).

WALDRAM, J.R., "The Theory of Thermodynamics," Cambridge (1985).

WEAST, R., "Handbook of Chemistry and Physics," 48th edition, Chemical Rubber Company, Cleveland (1967).

WEINER, J., "Transient heat conduction in multiphase media," *British J. of Applied Physics,* 6, pp. 361-363 (1955).

WHITAKER, S., "Elementary Heat Transfer Analysis," Pergamon Press, New York (1976).

WHITE, R.E., "The binary alloy problem: existence, uniqueness, and numerical approximation," *SIAM J. of Numerical Analysis,* 22(2), pp. 205-244 (1985).

WICHNER, R., A. D. SOLOMON, J. B. DRAKE, AND P. WILLIAMS, "Thermal analysis of heat storage canisters for a solar dynamic space power system" in *Proceedings of the ASME Solar Energy Division Conf., Golden, Colorado* (April 1988).

WILLIAMS, M.A. AND D.G. WILSON, "Vectorized difference schemes for a 3-dimensional enthalpy formulation for phase change problems," ORNL/TM-10034, Oak Ridge National Laboratory (1986).

WILSON, D. G., "Existence and uniqueness for similarity solutions of one dimensional multi-phase Stefan problems," *SIAM J. of Applied Math.,* 35, pp. 135-147 (1978).

WILSON, D.G., "Lagrangian coordinates for moving boundary problems," *SIAM J. of Applied Math.,* 42, pp. 1195-1201 (1982).

WILSON, D. G. AND A. D. SOLOMON, "A Stefan-type problem with void formation and its explicit solution," *IMA J. of Applied Math.,* 37, pp. 67-76 (1986).

WILSON, D. G., A. D. SOLOMON, AND V. ALEXIADES, "A shortcoming of the explicit solution for the binary alloy solidification problem," *Letters in Heat & Mass Transfer,* 9 (5), pp. 421-428 (1982).

WILSON, D. G., A. D. SOLOMON, AND V. ALEXIADES, "A model of binary alloy solidification," *Int. J. for Numerical Methods in Engineering,* 20, pp. 1067-1084 (1984).

WILSON, D. G., A. D. SOLOMON, AND P. BOGGS, EDITORS, "Moving Boundary Problems," Academic Press, New York (1978).

WILSON, D. G., A. D. SOLOMON, AND J.S. TRENT, "A bibliography on moving boundary problems with key word index," ORNL/CSD-44, Oak Ridge National Laboratory (1979).

WOLLKIND, D.J. AND R.D. NOTESTINE, "A non-linear stability analysis of the solidification of a pure substance," *IMA J. of Applied Math.,* 27, pp. 85-104 (1981).

WOLLKIND, D.J. AND L.A. SEGEL, "A non-linear stability analysis of the freezing of a dilute binary alloy," *Phil. Trans. Royal Soc. London,* A268, pp. 351-370 (1970).

WOOD, R.F., G.A. GEIST, A.D. SOLOMON, D.H. LOWNDES, AND G.E. JELLISON, "Modeling of undercooling, nucleation and multiple phase front formation in pulsed laser melted amorphous Silicon" in *Materials Research Society Symposium Proceedings: Energy Beam-solid interactions and Transient Thermal Processing,* 35, pp. 159-168, Materials Research Society (1985).

WOODRUFF, D.P., "The Solid-Liquid Interface," Cambridge Univ. Press, London (1973).

WOODS, L., "The Thermodynamics of Fluid Systems," Oxford University Press, Oxford (1975).

WYMAN, C., AT AL, "A review of collector and energy storage technology for intermediate temperature applications," *Solar Energy,* 24, pp. 517-540 (1980).

YOUNG, D.M. AND R.T. GREGORY, "A Survey of Numerical Mathematics," Addison-Wesley, Reading, Mass. (1973).

YUEN, W.W., "Application of the heat-balance integral to melting problems with initial sub-cooling," *Int. J. of Heat & Mass Transfer,* 23, pp. 1157-1160 (1980).

ZACHMANOGLOU, E.C. AND D.W. THOE, "Introduction to Partial Differential Equations with Applications," Dover, New York (1976,1986).

INDEX